About the Author

George Tenet was the Director of Central Intelligence from 1997 to 2004. He holds a BSFS from the Georgetown University School of Foreign Service and an MIA from the School of International Affairs at Columbia University. He was appointed to the faculty of Georgetown University in 2004 and lives outside Washington, D.C., with his wife, author Stephanie Glakas-Tenet, and their son.

At the CENTER *of the* STORM

At the CENTER *of the* STORM

The CIA During America's Time of Crisis

George Tenet

with Bill Harlow

HARPER PERENNIAL

NEW YORK • LONDON • TORONTO • SYDNEY • NEW DELHI • AUCKLAND

HARPER PERENNIAL

All statements of fact, opinion, or analysis expressed are those of the author and do not reflect the official positions or views of the CIA or any other U.S. government agency. Nothing in the contents should be construed as asserting or implying U.S. government authentication of information or Agency endorsement of the author's views. The material has been reviewed by the CIA to prevent the disclosure of classified information.

A hardcover edition of this book was published in 2007 by HarperCollins Publishers.

HarperCollins books may be purchased for educational, business, or sales promotional use. For information please write: Special Markets Department, HarperCollins Publishers, 10 East 53rd Street, New York, NY 10022.

FIRST HARPER PERENNIAL EDITION PUBLISHED 2008.

Designed by Leah Carlson-Stanisic

The Library of Congress has catalogued the hardcover edition as follows:
Tenet, George.
At the center of the storm : my years at the CIA / George Tenet with Bill Harlow.—1st ed.
xxii, 549 p., [16] p. of plates : ill. ; 24 cm.
Includes index.
ISBN: 978-0-06-114778-4
ISBN-10: 0-06-114778-8
1. Tenet, George, 1953–. 2. United States, Central Intelligence Agency—Officials and employees—Biography. 3. Intelligence officers—United States—Biography. 4. Intelligence service—United States. 4. United States—Foreign relations—1993–2001. 5. United States—Foreign relations—2001–. I. Harlow, Bill. II. Title.
JK468.I6 T42 2007
327.12730092B 22 2007280315

ISBN 978-0-06-114779-1 (pbk.)

08 09 10 11 12 DIX/RRD 10 9 8 7 6 5 4 3 2 1

For Stephanie and John Michael, my wife and son,
who accepted and shared the privilege and challenge of public service,
and reminded me each day that the sacrifices were worth it.
Their love and understanding are my greatest reward.

CONTENTS

PRINCIPAL CHARACTERS

➢ CENTRAL INTELLIGENCE AGENCY

Charles "Charlie" Allen, assistant director of central intelligence (ADCI) for collection (1998–2005).

Cofer Black, chief of CIA's Counterterrorist Center.

Helge Boes, member of the clandestine service; killed on duty in Afghanistan (2003).

Ben Bonk, deputy chief, CIA's Counterterrorist Center.

John O. Brennan, my chief of staff (2000–2001).

Lt. Gen. John "Soup" Campbell, associate director of central intelligence (ACDI) for military support (2001–2004).

Henry "Hank" Crumpton, chief of the Counterterrorist Center, Special Operations Division.

John M. Deutch, Director of Central Intelligence (DCI) (1995–1996).

Tyler Drumheller, chief of Directorate of Operations, European Division.

Alan Foley, senior CIA official.

Robert "Bob" Grenier, former Agency Islamabad senior officer, "mission manager" for Iraq.

Wilma Hall, special assistant to the Director of Central Intelligence at the Old Executive Office Building (OEOB).

Dottie Hanson, my personal assistant (1995–2004).

Bill Harlow, Agency spokesman (1997–2004).

Stephen R. "Steve" Kappes, senior officer in the clandestine service.

Richard Kerr, former senior analyst.

A. B. "Buzzy" Krongard, executive director (2001–2004).

Mark Mansfield, deputy Agency spokesman.

John E. McLaughlin, deputy director of central intelligence (DDCI) (2000–2004).

Jami Miscik, deputy director of intelligence (DDI) (2002–2005).

Michael J. "Mike" Morell, my executive assistant, presidential briefer.

John H. Moseman, my chief of staff (2001–2004).

Stanley Moskowitz, former CIA senior officer in Tel Aviv.

Rolf Mowatt-Larssen, head of the Counterterrorist Center's Weapons of Mass Destruction (WMD) branch.

Geoff O'Connell, former CIA senior officer in Tel Aviv.

Daniel "Doc" O'Connor, chief of my security detail.

James "Jim" Pavitt, deputy director of operations and head of the clandestine service.

Rob Richer, senior officer in the clandestine service.

Michael Scheuer, chief of "Alec Station," the Bin Ladin unit.

Gary Schroen, former senior officer in Islamabad, Pakistan, head of the Northern Afghanistan Liaison Team (NALT).

Johnny Micheal Spann, clandestine service officer; killed on duty in Afghanistan (2001).

Bob Walpole, national intelligence officer for strategic programs.

➢ WHITE HOUSE

Samuel "Sandy" Berger, national security advisor (1997–2001).

Robert Blackwill, U.S. ambassador to India (2001–2003), National Security Council deputy for Iraq (2003–2004).

George W. Bush, president of the United States (2001–).

Andy Card, White House chief of staff (2001–2006).

Richard B. "Dick" Cheney, vice president (2001–).

Richard "Dick" Clarke, National Security Council Counterterrorism official (1990–2003).

Bill Clinton, president of the United States (1993–2001).

Al Gore, vice president (1993–2001).

Stephen "Steve" Hadley, deputy national security advisor (2001–2005).

John Hannah, a member of the vice president's staff.

Zalmay Khalilizad, senior director for Gulf and Southwest Asia, National Security Council (2001–2003).

Anthony "Tony" Lake, national security advisor (1993–1997), Director of Central Intelligence nominee 1997.

I. Lewis "Scooter" Libby, chief of staff to the vice president (2001–2005).

John Podesta, White House chief of staff (1998–2001).

Condoleezza Rice, national security advisor (2001–2005).

➢ STATE DEPARTMENT

Madeleine Albright, secretary of state (1997–2001).

Richard Armitage, deputy secretary of state (2001–2005).

John Bolton, undersecretary of state for arms control (2001–2005).

William Burns, assistant secretary of state for Near Eastern affairs (2001–2005).

Marc Grossman, undersecretary of state for political affairs (2001–2005).

Martin Indyk, U.S. ambassador to Israel (1995–1997, 2000–2001).

Colin L. Powell, secretary of state (2001–2005).

Dennis Ross, Middle East envoy.

➢ FEDERAL BUREAU OF INVESTIGATION

Louis Freeh, FBI director (1993–2001).

Robert Mueller, FBI director (2001–).

➢ PENTAGON

Stephen Cambone, undersecretary of defense for intelligence (2003–).

William Cohen, secretary of defense (1997–2001).

Douglas Feith, undersecretary of defense for policy (2001–2005).

Vice Adm. Lowell "Jake" Jacoby, head of the Defense Intelligence Agency (2002–2005).

Gen. Richard "Dick" Myers, chairman of the Joint Chiefs of Staff (2001–2005).

Donald Rumsfeld, secretary of defense (2001–2006).

Paul Wolfowitz, deputy secretary of defense (2001–2005).

➢ WASHINGTON, D.C.

John Ashcroft, attorney general (2001–2005).

David Boren, U.S. senator from Oklahoma (1979–1994), president of the University of Oklahoma (1994–), my old boss and mentor.

George H. W. Bush, Director of Central Intelligence (1976–1977), president of the United States (1989–1993).

Norm Dicks, U.S. congressman, Washington 6th (1977–).

Richard Durbin, U.S. senator from Illinois (1997–).

Dianne Feinstein, U.S. senator from California (1992–).

Patrick Fitzgerald, special counsel in the Valerie Plame investigation (2003–2007).

Newt Gingrich, Speaker of the House of Representatives (1995–1999).

Porter Goss, U.S. congressman, Florida 13th (1989–1993), Florida 14th (1993–2004), chairman of the House Permanent Select Committee on Intelligence (1997–2004).

Robert Graham, U.S. senator from Florida (1987–2005), chairman of the Senate Select Committee on Intelligence.

Richard Haver, civilian intelligence professional.

James Hoagland, *Washington Post* columnist.

Edward Kennedy, U.S. senator from Massachusetts (1962–).

Nicholas Kristof, *New York Times* columnist.

Michael Ledeen, prominent neoconservative, scholar at the American Enterprise Institute.

Carl Levin, U.S. senator from Michigan (1979–).

Robert Novak, syndicated columnist.

Richard Perle, prominent neoconservative, chairman of the Defense Policy Board Advisory Committee (2001–2003).

Walter Pincus, veteran intelligence reporter for the *Washington Post*.

Pat Roberts, U.S. senator from Kansas (1997–), chairman of the Senate Select Committee on Intelligence.

Jay Rockefeller, U.S. senator from West Virginia (1985–), the ranking Democrat on the Senate Select Committee on Intelligence.

Janet Reno, attorney general (1993–2001).

Brent Scowcroft, national security advisor (1974–1977, 1989–1993), chairman of the president's Foreign Intelligence Advisory Board (2001–2005).

Richard C. Shelby, U.S. senator from Alabama (1987–), member of the Senate Select Committee on Intelligence (1995–2003).

Joseph Wilson, U.S. ambassador to Gabon and São Tomé and Príncipé (1992–1995).

Bob Woodward, journalist.

➢ NEW YORK

Herbert Allen, investment banker.

John Negroponte, U.S. ambassador to the United Nations (2001–2004).

➢ AFGHANISTAN

Abdul Haq, member of the Northern Alliance; killed by the Taliban in October 2001.

Hamid Karzai, president of Afghanistan (2004–).

Fahim Khan, a Northern Alliance leader.

Ahmed Shah Masood, head of the Northern Alliance; assassinated by al-Qa'ida on September 9, 2001.

Mullah Omar, leader of the Taliban.

Mullah Osmani, commander of the Taliban's Khandahar Corps.

➢ IRAQ

Gen. John Abizaid, Gen. Franks's deputy.

L. Paul "Jerry" Bremer, head of Coalition Provisional Authority (CPA) (2003–2004).

Maj. Gen. Keith Dayton, head of Iraq Survey Group.

Charles Duelfer, former United Nations weapons inspector.

Gen. Tommy Franks, commander of U.S. Central Command (CENTCOM) (2000–2003).

Retired Lt. Gen. Jay Garner, director of Office of Reconstruction and Humanitarian Assistance (ORHA) (March–May 2003).

David Kay, former United Nations weapons inspector.

Maj. Gen. Gene Renuart, Gen. Franks's director of operations.

Lt. Gen. Ricardo Sanchez, commander of Coalition Ground Forces (2003–2004).

➢ AL-QA'IDA

Mohammed Atef, an Egyptian, al-Qa'ida's number three man; killed in an air strike in Afghanistan in 2001.

Mohammed Atta, an Egyptian, ringleader of the 9/11 plot; killed on 9/11.

Kamal Derwish, U.S. citizen, believed to have recruited the "Lackawanna Six" to become al-Qa'ida supporters; killed by a missile strike in Yemen in 2002.

Hambali, an Indonesian, leader of the Jemaah Islamiya, a Sunni extremist organization based in Southeast Asia; captured in Thailand in 2003.

Abu Ali al-Harithi, a Yemeni, mastermind of the USS *Cole* bombing; killed by a missile strike in Yemen in 2002.

Nawaf al-Hazmi, a Saudi, one of the 9/11 hijackers; killed on 9/11.

Usama bin Ladin, a Saudi, the leader of al-Qa'ida; still at large.

Ibn al-Shaykh al-Libi, senior al-Qa'ida paramilitary trainer in Afghanistan; captured by Pakistani authorities in 2002.

Abdel al-Aziz al-Masri, an Egyptian, al-Qa'ida's "nuclear CEO"; reportedly under "house arrest" in Iran.

Abu Khabab al-Masri, an Egyptian, al-Qa'ida WMD expert; killed by a missile strike in Pakistan in 2006.

Khalid al-Mihdhar, a Saudi, one of the 9/11 hijackers; killed on 9/11.

Khalid Sheikh Mohammed (KSM), a Pakistani, the operational planner of 9/11; captured in Pakistan in 2003.

Zacarias Moussaoui, French citizen, involved in the 9/11 plot; sentenced to life imprisonment in 2006.

Abd al-Rahim al-Nashiri, a Saudi, implicated in the USS *Cole* bombing; captured in 2002.

José Padilla, U.S. citizen, believed to be involved in a possible "dirty bomb" plot; arrested in Chicago in 2002.

Ahmad Ressam, an Algerian, involved in a plot to bomb Los Angeles Airport; sentenced to twenty-two years in prison in 2005.

Ramzi bin al-Shibh, a Yemeni, involved in planning the 9/11 attacks; captured in Pakistan in 2002.

Yazid Sufaat, a Malaysian, suspected of providing operational support to 9/11 hijackers and al-Qa'ida, "CEO of anthrax"; arrested by Malaysian authorities in 2001.

Ramzi Yousef, nephew of Khalid Sheikh Mohammed (KSM), involved in planning the 1993 World Trade Center attack; sentenced to life imprisonment in 1996.

Abu Musab al-Zarqawi, a Jordanian, senior al-Qa'ida "associate"; killed in Iraq in 2006.

Ayman al-Zawahiri, an Egyptian, Bin Ladin's deputy; still at large.

Abu Zubaydah, a Saudi, al-Qa'ida operations expert; captured in Pakistan in 2002.

➢ UNITED KINGDOM

Tony Blair, prime minister of the United Kingdom (1997–).

Sir Richard Dearlove, head of the British secret intelligence service, MI-6 (1999–2004).

David Manning, foreign policy advisor to the prime minister (2001–2003).

➢ MIDDLE EAST

King Abdullah, king of Jordan (1999–).

Crown Prince Abdullah, crown prince of Saudi Arabia.

Prince Bandar, Saudi ambassador to the United States (1983–2005).

Samih Battikhi, head of the Jordanian Intelligence Directorate.

King Hussein, king of Jordan (1952–1999).

Saad Khair, head of the Jordanian Intelligence Directorate.

Hosni Mubarak, president of Egypt (1981–).

Prince Naif, Saudi interior minister.

Prince Mohammad bin Naif, Saudi assistant interior minister; Prince Naif's son.

Ali Abdullah Saleh, president of Yemen (1990–).

Gen. Umar Suleiman, head of the Egyptian intelligence service.

➢ ISRAELIS

Ami Ayalon, chief of Shin Bet, the Israeli domestic intelligence service (1996–2000).

Ehud Barak, prime minister of Israel (1999–2001).

Meir Dagan, Netanyahu's counterterrorism advisor, director of the Mossad, the Israeli foreign intelligence service (2002–).

Avi Dichter, chief of Shin Bet, the Israeli domestic intelligence service (2000–2005).

Efraim Halevy, director of the Mossad, the Israeli foreign intelligence service (2000–2002).

Yitzhak Mordechai, defense minister (1996–1999).

Benjamin Netanyahu, prime minister of Israel (1996–1999).

Shimon Peres, prime minister of Israel (1984–1986, 1995–1996).

Yitzhak Rabin, prime minister of Israel (1974–1977, 1992–1995).

Ariel Sharon, prime minister of Israel (2001–2006).

➢ PALESTINIANS

Yasser Arafat, Chairman of the Palestine Liberation Organizaton (PLO) (1969–2004), president of the Palestinian National Authority (1993–2004).
Mohammed Dahlan, chief of security for Gaza.
Amin al-Hindi, chief of the Palestinian external security service.
Jabril Rajoub, chief of security for the West Bank.

➢ LIBYANS

Col. Muammar al-Gadhafi, Libyan leader (1969–).
Saif al-Islam Gadhafi, Colonel Gadhafi's son.
Musa Kusa, head of the Libyan intelligence service.
Fouad Siltni, Libyan diplomat.

➢ IRAQIS

Dr. Iyad Allawi, head of Iraqi National Accord.
Al-Asaaf, Iraqi defector, a former Iraqi major.
Curve Ball, Iraqi defector, a former Iraqi chemical engineer.
Ahmed Chalabi, the head of the Iraqi National Congress.
Saddam Hussein, president of Iraq (1979–2003).
Husayn Kamil, Saddam Hussein's son-in-law.
Gen. Mohammed Abdullah Shawani, chief of Iraqi Special Forces during the Iran-Iraq war.

➢ PAKISTANIS

Gen. Mahmood Ahmed, head of ISI, the Pakistan Inter-Service Intelligence agency (1999–2001).
Aimal Kasi, Pakistani terrorist, killed two CIA employees outside CIA headquarters in 1993; executed in 2002.
Dr. Abdul Qadeer Khan (aka A. Q. Khan), father of the Pakistani nuclear weapons program.
Sultan Bashirrudan Mahmood, former director for nuclear power at Pakistan's Atomic Energy Commission, and

founder of Umma Tameer-e-Nau (UTN), an organization of Pakistani nuclear scientists supporting al-Qa'ida.

Gen. Pervez Musharraf, president of Pakistan (1999–).

➢ FAMILY AND FRIENDS

Stephanie Glakas-Tenet, my wife.

John Michael Tenet, our son.

Evangelia Tenet, my mother.

John Tenet, my father.

Bill Tenet, my brother.

Tommy Glakas, my brother-in-law.

Nick Glakas, my brother-in-law.

Ken Levit, my old friend, who goes back years with me in the Senate.

➢ OTHERS

Manucher Ghorbanifar, Iranian arms dealer, involved in the Iran-Contra affair.

Dr. August Hanning, head of the German intelligence agency, BND (1998–2005).

Jonathan Pollard, U.S. naval intelligence analyst, convicted in 1986 of passing classified information to Israel; currently serving a life sentence.

B. S. A. Tahir, A. Q. Khan's deputy and the Pakistani nuclear weapons network's chief financial officer and money launderer.

PREFACE

Wednesday, September 12, 2001, dawned as the first full day of a world gone mad. Nothing would ever be the same. Early that morning, operating on only a few hours' sleep, I headed out my front door to the armored Ford Expedition that was waiting to carry me to see the president of the United States.

The security outside my home in Washington's Maryland suburbs was tighter than ever before. Arriving at the White House, I saw Secret Service personnel stationed every few feet, all of them brandishing weapons. Clearly visible overhead were fighter aircraft patrolling the skies above the nation's capital. Less than twenty-four hours earlier, America had been attacked by a stateless foreign army. Thousands perished in New York City, at the Pentagon, and in a field in Pennsylvania. At CIA, we had good reason to believe that more attacks might be coming in the hours or days ahead and that 9/11 was just the opening salvo of a multipronged assault on the American mainland.

Days later I made a similar trip to the White House—the exact date is unclear as those days blended together. As I walked beneath the awning that leads to the West Wing, I saw Richard Perle exiting the building just as I was about to enter. Perle is one of the godfathers of the neoconservative movement and, at the time, was head of the Defense Policy Board, an independent advisory group to the secretary of defense. Ours was little more than a passing acquaintance. As the doors closed behind him, we made eye contact and nodded. I had just reached the door myself when Perle turned to me and said, "Iraq has to pay a price for what happened. They bear responsibility."

I was stunned but said nothing. On September 11, I had scanned passenger manifests from the four hijacked airplanes that showed beyond a doubt that al-Qa'ida was behind the attacks. Over the months and years to follow, we would carefully examine the potential of a collaborative role for state sponsors. The intelligence then and now, however, showed no evidence of Iraqi complicity.

At the Secret Service security checkpoint, I looked back at Perle and thought: What the hell is he talking about? Moments later, a second thought came to me: Who has Richard Perle been meeting with in the White House so early in the morning? I never learned the answer to that question.

For better and for worse, the twin topics of terrorism and Iraq would come to define my seven years as Director of Central Intelligence. By the time I stepped down from the job in July 2004, those issues seemed to eclipse all the other work American intelligence had done, and all the other issues we had faced during my tenure. Although I didn't realize it that day, I've since come to think of that brief encounter with Richard Perle as the moment when these two dominant themes in my professional life first intersected.

Growing up in the New York City borough of Queens, the son of working-class immigrants, I never would have imagined I would find myself in such a position. I aspired to a career in government but never gave a moment's thought to a life in the hidden world of intelligence. Yet somehow, through a series of unexpected occupational twists and turns, I found myself in the wilderness of mirrors.

As a career path, intelligence is equal parts thrilling and frustrating, because, by definition, it deals with the unclear, the unknown, and the deliberately hidden. What the enemies of the United States work hard to conceal, the men and women of American intelligence work hard to reveal. Throughout my working life, following the ethos of intelligence, I tried to maintain a low profile—to be little seen or heard among the general public.

When I left government, I felt a need to step back for a little while, to think before I wrote or spoke. Having benefited from time and perspective, I have come to believe that I have an obligation to share some of the things I learned during my years at the helm of American intelligence. I felt I owed it to my family, to my former colleagues, and to history to say what I could about the events I have observed.

This memoir relies on my recollections of a tumultuous period in our nation's life. No such undertaking is completely objective, but it is as honest and as unvarnished as I can make it. There are many things about my tenure as DCI that I am proud of and more than a few things I wish I could do over. Where I, or the organization I led, made mistakes, I say so in these pages. Readers will find no shortage of such admissions. When I point out occasions where our performance was strong, I hope these assertions, too, are given fair consideration. This book reflects how things appeared to me as I found myself literally at the center of the storm.

Where you stand on issues is normally determined by where you sit. And from where I sat, I saw the tidal wave of terrorism building. From where I sat, I also saw a small group of underfunded and lonely warriors swimming against this tide out there all alone, warning, deterring, disrupting, and attempting to destroy a worldwide movement operating in nearly seventy countries and bent on our destruction.

This is the story of how we saw the threat, what we did about it, what was proposed and not done, how our thinking evolved, and why the men and women of the Central Intelligence Agency were ready with a plan of action to respond forcefully to the loss of three thousand American and foreign lives. This is also a story about how we helped disarm a rogue nation of its weapons of mass destruction without firing a shot and how we brought to justice the most dangerous nuclear weapons proliferator the world has ever known. It is a recounting of efforts to bridge historic differences between Israelis and Palestinians and give to diplomats a chance to seek a political solution to an age-old crisis. It also is a cautionary tale of threats still uncountered that would make the attacks of September 11 pale in comparison.

Senior-level people in both the administrations in which I served, Clinton and Bush, tried to do what they saw as best for America. Their results and methods can and should be debated—but not their motives. And when it comes to the U.S. government's handling of Iraq, there are few heroes in Washington, but plenty on the ground in that troubled country. When it comes to the war on terror, though, there are plenty of heroes, in Washington and elsewhere around the world. The same administration that later lost its way on the road to Baghdad performed brilliantly when it came to running down al-Qa'ida in the aftermath of 9/11. CIA undertook an enormous task with great courage and unbelievable dedication. We read too little about these heroes.

With all its burdens and all its pressures, as Director of Central Intelligence, I believe I had the best job in government. The greatest joy for me was the daily interaction with men and women who dared to risk it all every day to protect our nation. I had an opportunity to serve my country and to try to keep it safe in a time of peril. I was not always successful, but I take comfort in knowing that I was in the arena, striving to do what was right. Only in the United States of America can the son of immigrants be given such a privilege. I will always be grateful that John and Evangelia Tenet left their villages in Greece to give me that chance.

PART I

CHAPTER 1

The Towpath

It was like something out of a spy movie.

The date was March 16, 1997, a Sunday. I was at home, on a rare day off, when the phone rang. "Meet me by the C&O Canal, near the Old Angler's Inn in an hour," a voice said, almost in a whisper. "Come alone." That was all. He didn't have to identify himself; he knew I would be there.

The voice belonged to Anthony Lake, who had stepped down as national security advisor two months earlier, when Bill Clinton nominated him to be director of the Central Intelligence Agency. Back in 1992, at the start of the Clinton administration, Tony had made me part of his National Security Council staff. Prior to that I had served as a Senate staffer, and for the previous four years had been staff director of the Senate Select Committee on Intelligence. Over the course of three years on the NSC staff, I had formed a warm personal and professional relationship with Lake and his deputy, Sandy Berger. Then, in May 1995, John Deutch, who was about to become CIA director, tapped me to be his second in command. We had gotten to know each other when Deutch was deputy secretary of defense and had even traveled together once overseas to deal with a sensitive intelligence matter. But now, after only a year and a half in the job, Deutch was leaving CIA, and my friend and former boss Tony Lake had been picked to replace him.

Tony had all the right tools for the job: intelligence, acumen, the confidence of the president, and strength of character. Outsiders who observed Tony when he was national security advisor assumed from his quiet comportment that he was some misplaced mild-mannered professor. Not so. Amid many large egos, Tony was the unchallenged boss at the NSC, a master at process and bureaucratic intrigue. He had observed up close the dysfunctional backbiting that crippled the Carter administration and had worked hard to prevent a repeat performance under Bill Clinton. A rarity in Washington, Tony had no desire to have a high profile, and he emphasized to his staff that we would succeed or fail together as a team. None of us, he stressed, had been elected to the offices we held.

All those attributes made Tony an ideal choice, I thought, to lead CIA. Selfishly, I also knew that his arrival at Langley meant that I would be able to stay on in the deputy's job—a position I was learning to love.

John Deutch—a brilliant, eccentric, and largely misunderstood figure—had an ability to translate his technical expertise into policy in a way few people could. A gregarious bear of a man, he wanted to be respected by the Agency's workforce. But shortly after he arrived at CIA, the Agency's inspector general issued a report criticizing the professionalism of some CIA officers in Guatemala in the 1980s, and John disciplined some of those named. That got him off to a rough start with the workforce. And then things got worse.

His downfall came when he told a reporter for the *New York Times Magazine* that he did not find many first-class intellects at the Agency. "Compared to uniformed officers," the *Times* quoted John as saying, "they certainly are not as competent, or as understanding of what their relative role is and what their responsibilities are." The Central Intelligence Agency is a very emotional place, and after that, John's chances of winning hearts and minds there were pretty much shot. I know he regretted his remarks. It was a valuable lesson that I would put to use later: You have to earn your employees' trust, keep your own counsel, be optimistic, and, as I always said, lead from "the perspective of the glass being always half-full."

John's tumultuous tenure at CIA ended in December 1996

when he abruptly resigned. The conventional wisdom around Washington was that he really wanted to be secretary of defense and that when it became clear that post was not to be his, he left government for good. Whatever the actual reason, after he cleaned out his desk, I became acting director.

I thought I would have to handle the two jobs for only a short while until Lake was confirmed. But four months later, the nomination was still tied up in the Senate. I figured that the delay in Tony's confirmation was behind his request to meet with me, but I had no idea why he had insisted on such an unusual location. His instructions to come alone were especially puzzling. He knew that deputy CIA directors don't go anywhere alone. Since I'd taken the job at the Agency, a heavily armed security detail had been my constant companion. Everywhere I went, I was driven around in a big, black armored SUV with a second follow car full of guys with guns. Threats against senior CIA officials by terrorists and nutcases were very real. In the four months since I had become acting DCI, the security had been ratcheted up even tighter.

Nonetheless, I tried to comply with Tony's request for discretion. I called in the chief of my security detail, Dan O'Connor, and told him that he and I needed to go for a little ride—alone. Dan, known around the Agency as "Doc," for his initials, is a big, genial New York Irishman. He would take a bullet to save my life without hesitation, but he hated the notion of our venturing out without the usual retinue of backups. His duty was to minimize the risk to me, not maximize it. Nonetheless, he drove over to my home, and the two of us headed south toward the Potomac River.

We pulled into the gravel parking lot across from the Old Angler's Inn. From there, with Doc keeping a discreet distance, I set off down a dirt path to the century-and-a-half-old canal that once carried coal from the West to heat Washington's homes. Although it was only mid-March, the parking lot and towpath were crowded with bikers, joggers, walkers, and hikers scrambling along the rocky Billy Goat Trail. Farther downhill, kayakers were pushing off into the churning waters of the Potomac not far from where it comes crashing out of Great Falls.

Memory tells me that a mist was still over the canal that day. Tony was waiting for me, dressed casually in a windbreaker and hiking boots. I was the one who stood out—still in the suit pants and good shirt I had worn to church that morning. I simply hadn't thought to change. We shook hands, and Tony said, "Let's take a walk." I'd been with Tony Lake in tough times, but on this day he had a grim countenance that I had never seen. After a half mile or so, we sat on a bench overlooking the canal.

"I want you to know that I plan to tell the president tomorrow that I am withdrawing my name from consideration as DCI," he said in a measured, flat tone. "It's too hard. They want too much. It's not worth it."

He didn't have to say who "they" were. Tony had been around Washington for a long time. He'd played hardball with the best of them. Now that they had him in their crosshairs, a number of senators were determined to make his confirmation process as difficult as possible. Just how difficult had been driven home to me shortly after he was nominated. I had gone to Capitol Hill to deliver a briefing to members of the Senate Select Committee on Intelligence. After the session, I was pulled aside by Richard C. Shelby, the Alabama Republican who was about to become chairman of the committee.

"George," he drawled, "if you have any dirt on Tony Lake, I sure would like to have it." This brazen remark left me speechless—not a common condition for me. *Doesn't this guy know that Tony is my friend and former boss?* I thought. *What makes him think I would do something like that?*

Others apparently didn't share my reluctance. Soon, issues involving Tony's management of the NSC staff and baseless rumors about personal improprieties arose. The confirmation was clearly in trouble. Still, I believed that, eventually, good sense would prevail.

That day along the towpath, though, Tony told me his heart was no longer in the fight. He had suffered through three days of brutal public hearings and had been forced to endure the worst kind of demagoguery from some of the committee members. Prior to the hearings, Senator Shelby had insisted on, and finally got, administration agreement to allow him to look through the FBI's raw files on Lake. "Raw" means just that—these files contain any allegation ever made against you, no matter how groundless. During the public hearings, Shelby and several of his colleagues took turns attacking the nominee. Democratic senators called it a "trial by ordeal" and a form of "malicious wounding." Even Republican senator John McCain asked Shelby to reconsider his approach—but to no effect.

I'm still convinced that once Shelby had tired of bludgeoning Tony, the votes would have been there, but Tony said that he had heard that Shelby was threatening to ask the FBI for yet another investigation as a delaying tactic. National Security Agency officials told us that Shelby staffers had been asking whether there was derogatory information in their communications intercepts on Lake. NSA rebuffed that fishing expedition, but Tony had had it. Enough was enough. What he told me next stunned me more.

"When I tell the president that I am dropping out, I am going to tell him that he must nominate you to become DCI," he said. To be sure, I was acting DCI, but the prospect of replacing Tony as the nominee had not occurred to me in my wildest imagination. After all, I was just forty-four years old, a relative unknown except within certain bureaucratic intelligence circles. That was one strike against me. Strike two was my health: I had suffered a heart attack fewer than four years earlier.

I can't remember if I replied at all, but my face must have registered the surprise I felt. Tony filled in my silence. "Look, you know the place, you've got the skills, the president likes you, and the Senate will confirm you. Tell me anybody else that can be said about. You'd love the job," he added.

"Yes, but not this way," I answered.

Tears were welling up in my eyes while I processed the mixed emotions I was feeling—shock, uncertainty, sadness, and trepidation. I was like a Broadway understudy who'd just found out that his best pal, the star of the show, had been hit by a bus.

I thought about trying to talk Tony out of withdrawing his nomination, but it was clear that his mind was made up. Then I began expressing doubts about whether I was the right person for the job. Tony was sure that I was, and he didn't want to debate the matter. "Look," he said in his patrician New England tone, "I didn't bring you out here to ask you what you think about my plans. I asked you to come so that I could tell you what I am going to do. I am going to withdraw, and I am going to tell them that they *must* nominate you. It is as simple as that." Tony was worried that President Clinton's instinct would be to go to the mat with Shelby. "He'll want to fight to every last drop of *my* blood," is how he put it. "But that would be terrible for the Agency. CIA needs a director now."

After talking for about a half hour we found our way back to our starting point, shook hands, and headed our separate ways. Back home, I went to the family room, in our basement, to think about what had just transpired. Then, as I always do on tough matters, I asked my wife, Stephanie, for advice. Could I do this job? Should I try? What would it mean for our family? Our child, John Michael, was just finishing up elementary school, a time when a boy needs his dad nearby. As acting DCI, I had had enough of a taste of the job to know that it would eat up my hours. Stephanie has always been my strongest supporter. Over the previous two years, she had come to love the men and women of CIA. Like me, she's also Greek, ready to take virtual strangers under her wing at a moment's notice. The Agency employees and their families had quickly become part of her extended family.

"George, you can do this," she told me. "You *have* to do this, because the Agency needs you. Don't worry about me and John Michael; we will be fine and so will you."

The next afternoon, Monday, March 17, Tony issued a stinging 1,100-word statement about his withdrawal. He said that Washington had gone "haywire," he decried the politicization of CIA, and he said that he hoped for a return to the day when priority would be given "to policy over partisanship" and "to governing over 'gotcha.'" (Nearly a decade later, I'm afraid his wish has not come true.)

On Wednesday morning, I got a call from John Podesta, the deputy chief of staff, telling me that the president would likely nominate me for the DCI job. Like Tony, Podesta didn't seem to be asking me what I thought about the idea. I was invited to come down to the White House to meet with the president.

At the White House, I was led upstairs to the president's personal quarters. There I met with President Clinton, with Lake's successor as national security advisor, Sandy Berger, and with Podesta. The president stayed seated throughout, having recently torn up his knee in a fall at golfer Greg Norman's house in Florida, but there would have barely been time for him to struggle to his feet. We talked briefly, observed the niceties, and then almost before I knew what was happening, presidential staffers were asking that my wife and son be rushed to the White House as soon as possible.

Before long, a pool of White House reporters was called in to hear of the president's intention to nominate me. With my wife and son at my side, I made a brief statement noting my "bittersweet" feelings, since my rise followed the fall of someone I deeply admired, Tony Lake. I promised the president my best efforts and then went back to the job I was already performing.

Thinking back, I find it odd now that there was no job interview. They knew me and what I stood for, of course, but no one asked me what I would do with the intelligence community should I get the job, what changes I might make, or how I intended to repair morale at a place that had experienced four DCIs in the past five years—not to mention two others whose nominations had been withdrawn.

The story of my nomination got big play in the tabloid papers of New York, where I grew up. The headline in one paper called me "The Spy Who Came in from Queens." Enterprising reporters found people from my old neighborhood who had known me for most of my forty-four years. Some explained how surprised they were at my nomination, since, as one person noted, as a child I had a "big mouth" and wasn't known for keeping secrets. Others said they sensed something special about me based on the way I had played stickball thirty-five years earlier. (I was once the Public School 94 doubles stickball champion.)

My favorite quote came from my mom, Evangelia Tenet. Although she had been in this country for forty-five years by that time, the embrace of the Greek American community was so strong that she still got along speaking only broken English. "I have one son in the CIA and one son who is a heart doctor. Not bad, eh?" she told the *Daily News*. Not bad at all, but the real story is my parents, not my brother or me. It is impossible to overstate their influence. Even though I have met scores of presidents, kings, queens, emirs, and potentates, the two people I still admire the most are my mom and dad.

My dad, John Tenet, was his own man since the day he was thrown out of his house at age eleven by an abusive father in Greece. He first traveled to France and found work in a coal mine. There he quickly decided that the mines were not where his future should be, and he made his way to the United States—arriving at Ellis Island just before the Great Depression. He didn't have a nickel in his pocket or a friend in sight. All he knew was that he wanted to be his own boss and take care of his family, and that in America hard work would let him achieve what was unimaginable elsewhere. On that abiding faith alone, he managed to do what so many Greek immigrants did: he opened a diner.

Eventually Dad would become thoroughly American, but his European roots stayed with him. His hero was Charles de Gaulle. I vividly remember April 27, 1960, when my dad took me and my twin brother, Bill, from Queens to Manhattan to see de Gaulle riding in a ticker-tape parade in an open-air limousine. To this day, I can hear Dad shouting, *"Vive la France!"* and see de Gaulle casting his eyes in our direction. I knew I was in the presence of greatness—but, then, I always felt that way when I was around my father.

Dad was a gentle, honest man. He had no formal education, yet he devoured newspapers and was fascinated with world affairs. Our dinner table was the scene of lively debates about politics and news of the old country and of his adopted home. The conversations flowed freely from Greek to English. When Mom and Dad didn't want my brother and me to know what they were saying, they would switch to Albanian.

Dad was the spitting image of Barry Goldwater, so much so that during the 1964 presidential campaign he was often stopped at the Long Island Rail Road platform and asked for his autograph. That says a lot about how times have changed. It seems odd now that New Yorkers would, even for a moment, believe that a presidential candidate might be standing alone waiting for the train from Little Neck to Flushing. Although twenty-three years have passed since his death, I feel Dad's loss as if it happened yesterday.

As arduous as my father's journey to the New World was, my mother's route to America was even more remarkable. She fled what is today southern Albania. Her two brothers were killed by the Communists, and her father, devastated by their murders, died of a heart attack. Alone, Mom somehow managed to make it to the Adriatic coast and board a British submarine after World War II, just as the borders were closing.

Mom made her way first to Rome and then to Athens, and there she might have spent the rest of her life had it not been for one of her uncles, who was in the restaurant business in New York. Uncle Lambros bragged to my dad about his young niece, who was not only beautiful but had recently escaped from a village near where my father was born. Dad must have been enchanted by the tale because in 1952 he flew to Greece, courted Mom for two weeks, and married her. A week later, she arrived in New York to join him in the restaurant business at a place he called the Twentieth Century Diner. She was the baker and he was the chef. It was there, in Queens, with its large Greek American community, that she proudly raised her family.

For an arranged marriage, theirs worked out very well. In another era, with resources and a family behind her, Mom might have gone to college and on to law school. She would have been formidable in a courtroom. My mother has an uncanny ability to read people—private citizens and public figures alike. Mom can spot a liar a mile away. Had I been able to put her to work at CIA, we could have scrapped all our polygraph machines. She is a woman of few words, but her temper is on a hair trigger, especially when anyone tries to make life difficult for her two boys. I tell people—only half kidding—that after dealing with my mom, Yasser Arafat was a piece of cake.

In many ways, I am my father's son. He was a very trusting man, loath to say anything bad about anyone. Many times when I was director of CIA, I would find myself longing for a chance to get Dad's advice on some thorny problem, though he had passed away in 1983. When things got tough, brother Bill would always say, "Just think about what the old man would do." Dad believed in inclusiveness. Keep your friends close and your enemies closer. Sometimes, though, I wish I were more like my mom, who firmly believes that constant confrontation can be cathartic. They were an extraordinary couple. I am thankful every day that their courage and determination brought them to this country.

I thought about my parents' remarkable journey that March Sunday in 1997—a journey that had brought me to that towpath and to this turning point in my life.

CHAPTER 2

The Burning Platform

In a perfect world, I would have been fully prepared for my new job, and the Agency would have had the resources to tackle the growing terrorism menace head-on and across a global frontier. From the lethal 1983 attack on the U.S. Marine barracks in Beirut to the 1988 bombing of Pam Am Flight 103 over Lockerbie, Scotland, to the 1993 World Trade Center bombing to the 1996 attack on another U.S. military barracks, Khobar Towers in Dhahran, Saudi Arabia, we had seen Hezbollah, Hamas, al-Qa'ida, and others at work, and we knew how state sponsors from Libya to Iraq, Iran, and Afghanistan used these killers and suicide bombers in a proxy war against Americans and our friends and interests abroad.

Believe me, there was never any doubt who the enemies were, but in the world we lived in and at the CIA I had inherited, things were never that easy. The CIA of 1997 was not a well-oiled machine with an abundance of resources or an organization that ran with crisp precision. If it had been, plenty of other people would have been vying to lead it. In reality, the job probably fell my way more by default than anything else. One newspaper at the time described me as an "unconventional" choice to run the place. The *New York Times* quoted an anonymous official as saying, "I can't give you a better name" than Tenet or, given the challenges facing the Agency, "even a name at all." At least the *Times* had my name right. Fifteen months earlier my face had been on the cover of *Parade* magazine, along with that of John Deutch. Amusingly, *Parade* identified me for its thirty million plus readers as "David Cohen," who was actually our director of operations at the time.

Perhaps the most critical problem the Agency faced was the lack of continuity in leadership. I was the fifth director in seven years. No company can succeed with that kind of turnover. The view of much of the workforce about edicts from the seventh floor, where the most-senior officials work, was that if you didn't like an order, just wait awhile—the person who gave it would soon be gone.

The problems ran deeper than episodic leadership, though. During the 1990s, the conventional wisdom was that we had won the cold war and it was time to reap the peace dividend. Not only was that assumption wrong—the war was simply evolving from state-run to stateless armies and from intercontinental ballistic missiles (ICBMs) to nuclear manpacks and anthrax vials—but the supposed "peace dividend" was devastating to the spy business at a time when its vitality was most needed. The entire intelligence community, not just CIA, lost billions of dollars in funding. Our workforce was slashed by almost 25 percent. There is no good way to cut an organization's staff by that amount. But there is one incredibly bad way to do it—and that was precisely the method the intelligence community used. They simply stopped recruiting new people. As a result, there was a half decade or so where hardly any new talent was coming in, and many, many experienced hands were going out the door.

When I became deputy DCI in the summer of 1995, we were running two classes a year for new "case officers"—future members of our clandestine service, the men and women who recruit foreign agents to steal secrets. The class in session that summer had a grand total of six future case officers and six "reports officers"—people who don't collect intelligence as much as write up the efforts of their colleagues who do. You can't run a spy service that way. We later learned that, while we were training a handful of case officers each year, al-Qa'ida was training literally thousands of potential terrorists at its camps in Afghanistan, the Sudan, and elsewhere.

Even if we had had the money, the will, and political backing suddenly to ramp up our training program in the mid-1990s, we did not have the infrastructure to support it. Our clandestine training facility had been allowed to deteriorate to an appalling state. Classes were being conducted in dilapidated World War II–era buildings. The housing for our instructors and their families was worse than anything they had to endure when deployed to developing countries. Our best and brightest were not teaching our future officers. Our recruiting program was in shambles, too. Each directorate within the Agency had its own, and there was little or no coordination among them. Of all the telltale signs I tripped over in those first explorations into what was ailing the Agency, the one that stood out the most to me was this: the FBI had more special agents in New York City than CIA had clandestine officers covering the whole world.

It wasn't just the clandestine portion of the Agency that was in bad shape. Our analytic expertise had eroded to an alarming extent. In order to get promoted, analysts who had spent years becoming world-class experts in some critical issue or geographic region had to drop their area of interest and become managers. The Peter Principle is as true in the spy trade as in any other: the best analysts are often not the best managers.

Not surprisingly, morale at the Agency was in the basement. CIA was still reeling from the espionage cases of Aldrich Ames in 1994 and Harold Nicholson in 1996, trusted Agency officers who betrayed the country and their colleagues by selling critical secrets to the Russians. The Agency had also been rocked by false allegations in 1996 that some of its members had been complicit in selling crack cocaine to children in California. The allegations were ludicrous, but even attempting to refute them gave legs to a lurid tale.

Mid- and senior-level officers in the Agency were haunted by the fear of being hauled before Congress or into court and asked to defend their actions. A succession of administrations would tell them that they were expected to take risks and be aggressive. But if something went wrong, Agency officials faced disgrace, dismissal, and financial ruin. Many of those willing to stick it out at CIA rushed to purchase their own "professional liability" insurance. That helped, but the chilling effect of having to do so spread broadly through the organization.

In science and technology, an area where CIA was once a giant, the dot-com revolution was passing us by. Private-sector technology was far outstripping our ability to keep pace with our targets. The information technology tools we were putting in the hands of our officers looked like products of the mid-twentieth century rather than of the approaching twenty-first.

Organizationally, the Agency was a mess as well. There was no chief information officer or chief financial officer. We had no coherent and unified programs of training and education, and our executive board made decisions through a democratic voting process. In a multibillion-dollar organization, "one man, one vote" guarantees lowest-common-denominator solutions—nobody will be truly uncomfortable or unhappy about outcomes. Good leadership, by contrast, demands that some segments of your organization occasionally have to swallow bitter but needed medicine. Organizations such as CIA exist to *defend* democracy, not to practice it.

Overriding all these specific shortcomings, and most damaging, was a lack of an articulated and well-understood strategy for the Agency. We had no coherent, integrated, and measurable long-range plan. To me, that just seemed basic, and so that's where I most focused my energy from day one.

I wish I could tell you that I knew exactly what to do from the start. But I had several advantages. I had been the deputy director for two years. Being the deputy of a large organization in Washington is a great job—nobody knows who you are and nobody cares. And I had used the time to find out all that I could about the insides of the institution, learning about our people and where the best work was being done. The second advantage was the men and women of CIA, the most dedicated, passionate patriots you have ever met in your life. Their work ethic is second to none. The tradition and history of the organization is rich and full of daring and accomplishments. (In fact, there is a memorial wall in our lobby where stars denoting fallen colleagues speak of the ultimate sacrifice.) Change was certainly a necessity, but CIA's history and heritage would provide the foundation upon which to build.

The downside was that now I was no longer the deputy. I couldn't hide behind my boss, and the Agency and the nation couldn't afford for me to be stumbling my way up the learning curve. You might think that I had been preparing for this job for two decades, ever since I first went to work as a Senate staffer, but in fact a series of staff jobs does not prepare you for executive leadership. Certainly I knew the substance of the work, but leading a large, multifaceted organization with many lines of business, especially in more than one hundred countries overseas, is a lot different from running a relatively small congressional committee staff. I spent plenty of sleepless nights wondering, given the monumental task before me, if I was up to the job. No previous experience had prepared me to run a large organization. I was no Jack Welch and I knew it.

I knew one thing that needed to be done, however: restoring humanity to the organization. The obligation of leaders is to listen and care for all their people, and not just those in the most skilled of occupations. A long time ago, in the Twentieth Century Diner, I had learned from my dad that if you took care of people, they would take care of you. And at CIA, if men and women believed that you cared about them and about their families, there was nothing they would not do for you.

Throw your arms around an employee, ask him about his family, send someone a note about an ailing mom, walk around and talk to real people doing their great work, make them all feel that they are part of something special—from the kitchen staff to the cleaning crew to the crusty seasoned operations officer you share a cigar with on the office balcony at the end of the day. Show them that you care—and when you have to kick them in the butt, they will understand that it is not personal, but rather about doing the job right for the country.

If you looked at the organization and dissected its business lines, the men and women of our clandestine service, the spies, would be our fighter pilots. Our analysts resembled a large college faculty; our scientists and engineers were the geeks who made everything work. Our security officers, logisticians, communications officers, and disguise specialists were the men and women who allowed us to be fast, agile, and responsive. They needed to feel special because they were, and they needed to be united with a common purpose, a mission statement—to protect America and its families—that tugged at their hearts.

The first thing I did was build a leadership team that all these people would trust. I brought in very few outsiders. The message I wanted to send to the workforce was that the talent to help us get where we needed to go was already among us. To stress the importance of our relationship with the military, I picked Lt. Gen. John Gordon, USAF, to be my deputy. To head up the Directorate of Operations—the Agency's clandestine service—I lured out of retirement a legendary officer named Jack Downing. Jack had served in Moscow and Beijing, and was a skilled linguist. His very presence on the team conveyed the notion that we were getting back to the basics of uncovering secrets to protect the nation.

As head of our analytic unit, the Directorate of Intelligence, I installed John McLaughlin, to whom I (only half jokingly) referred as the smartest man in America. A highly respected analyst, John was renowned for the precision, rigor, and honesty that our tradecraft required. No coincidence, perhaps, he is also a world-class magician. His nickname, Merlin, suggests both his vocational and avocational talents.

For executive director, I picked Dave Carey, the former head of the Agency's Crime and Narcotics Center, and I retained Dick Calder, a much-esteemed member of the clandestine service, as head of the Directorate of Administration. In every case, I was going for talent, but I also wanted everyone in house to understand that our core functions were going to be run by people who had walked the walk before.

One person I did bring in from the outside was A. B. "Buzzy" Krongard. He had been the CEO of the investment banking firm Alex. Brown. That's heady territory, with salaries and perks to match. If Buzzy hadn't been so ready to serve his nation in a time of great need, I never could have recruited him as a special advisor. His mission was to gather the data and assemble the metrics about all of our business processes that would allow us to make the changes critical to the Agency's survival. He brought business savvy to an organization that seemed to pride itself on its unbusinesslike methods. Prior to Buzzy's arrival, the Agency was a "data-free zone." We didn't know where the money was going; we didn't know why people joined our Agency or why they left. All that would change with Buzzy's expert help.

I also gleaned from the outside someone to head our Office of Public Affairs. For years the Agency's PR strategy was to proudly say "No comment" about virtually everything. Trouble was, we had long ago stopped functioning in a "no comment" environment. The media demanded responses, and when they didn't get any, they assumed you had something to hide, even when, as with us, hiding things was part of your job description. To remedy the matter, I brought in Bill Harlow, an experienced communications professional who had worked in the comparatively media-friendly (and media-savvy) press operations at the Pentagon and White House. (I should note that despite Bill's best efforts to get me to do a Sunday talk show, I had a seven-year unblemished record of almost never speaking to a television camera. It was my belief that a sitting DCI should maintain a low public profile and leave the "talking head" role to others.)

With the leadership team in place, in August 1997, we were meeting at one of the Agency's clandestine facilities not that far from Washington when someone said we were standing on a "burning platform." If we didn't work quickly to extinguish the blaze, the organization and all of us in it would sink into the sea. The term "burning platform" stuck—probably because it was so metaphorically accurate and because it reminded us every day of just how much was at stake. So we set out to learn how other organizations in disarray had transformed themselves. By the spring of 1998 we had a plan in place—a document we called the "Strategic Direction." A key part of the document envisioned what kind of officers we would need to have at the Agency in the year 2010. We looked at the skills they would need to possess, their languages, academic backgrounds, and so on. For five decades, CIA officers had been modeling themselves on the swashbuckling, mostly Ivy League–educated heroes of "Wild Bill" Donovan's wartime Office of Strategic Services. Brains still count, and a little panache is always useful, but if CIA was going to be able to do its job in a seventh and eighth decade, we had to take into account the new world in which our people would operate.

It took us nearly eight months of soul-searching to develop this plan for the future. On May 6, 1998, I stood up in front of five hundred Agency employees in our igloo-shaped auditorium known as "the Bubble" to talk about the burning platform and what we were going to do about it. Thousands of other employees watched me on closed-circuit television. Many of them were justifiably skeptical of what they were hearing. After all, they had seen so many other leadership teams come and go. How did they know I wasn't just the flavor of the month?

I tried to grab their attention by driving home how serious our problems were. CIA had recently celebrated its fiftieth anniversary, but unless we performed some sustained miracles, I said, the Agency was unlikely to be relevant by the time it reached its sixtieth birthday. I told them that, God and the president willing, I was going to be around for the long haul. There was no other job I wanted and no place I would rather be. The statement seemed necessary on my part, but I was stunned when it inspired a thunderous ovation. The reaction, for certain, was not about me. More than anything else, the applause spoke to how desperately the place wanted and needed stability.

I continued, promising that the days of trying to do more with less were over. The things we were proposing were going to cost money, but I assured them that they shouldn't worry about that part. My job was to get the necessary funding, and I pledged to try my damnedest to do so. I didn't entirely succeed, but I made myself a royal pain in the ass trying. I begged for large increases in intelligence funding and obtained modest "plus ups"—small increases in our budgetary top line. We reallocated significant portions of our budget to counterterrorism. The budget for CT, as it is called, went up more than 50 percent from 1997 until just before 9/11—at a time when most other accounts were shrinking. In the fall of 1998, I asked the administration for a budget increase of more than two billion dollars annually for the entire intelligence community over the next five years. Alas, only a small portion of that increase was granted.

So strongly did I believe that we were desperately short of needed resources that I went around my own chain of command. Although I was a cabinet officer in the Clinton administration, I struck up a relationship with then Republican Speaker of the House Newt Gingrich, who was a strong believer in the fact that the intelligence community needed more support. To his credit, Gingrich pushed through Congress a supplemental funding bill in the 1999 fiscal year that provided for the first time a significant increase in our baseline funding. My off-the-books alliance with the House Speaker alienated some members of President Clinton's team. Although the president was generally supportive of our mission, resources simply were not forthcoming. My only regret is that much of the money in the 1999 supplemental was for one year only, and was not continued in the years immediately following.

Perhaps the most important message I had for the CIA workforce that morning was that we were going back to the basics of our core mission. From now on, we would emphasize blocking and tackling. Everything must support and empower the most important part of our business, the pointy end of the spear: espionage, stealing secrets, and what we call "all-source analysis."

Before I stepped away from the lectern in the Bubble that day, I promised we would rebuild our field strength, increase the number of our operations officers, and augment the number of stations and bases. Those promises were kept. Over the next six years we upped the number of our stations and bases by close to 30 percent, in some cases reversing decisions made a few years before to draw down, and in others opening new facilities in countries that had only recently come into being.

The cornerstone of our business is people—analysts, field officers, managers, technicians, and, yes, spies. And no part of the Agency had been more neglected in the downsizing after the collapse of the Soviet Union than our human capital. The first thing we did was commit ourselves to establishing a centralized recruiting office on par with the finest in private industry. To bring in the best talent, we got back on college campuses, launched a national advertising campaign, and ensured that we had hiring bonuses in place for the skill areas we needed most—anywhere from thirty to fifty thousand dollars for scientists, engineers, information technology specialists, and people with unique language skills—serious money for serious needs. Some of the things we did might sound routine for the private sector, but I can guarantee you that they were revolutionary for a government intelligence agency. Traditionally, CIA recruits had to wait in a kind of limbo while we ran security checks on them. No more. We started making conditional offers of employment on the spot, and we gave recruits a paycheck while they were awaiting clearance. To be sure, this method increased our risk calculus. Today about 40 percent of all Agency employees have been there five years or fewer, barely time to get to know someone. But the simple fact is that the old standards and practices weren't getting the job done.

How did all this pay off? By 2004, 138,000 people were applying for a little more than two thousand Agency jobs. This wasn't just the result of increased interest in our business after 9/11—we also experienced a steep climb in résumés received throughout the late 1990s and in 2000 and 2001. Our corporate attrition rate was 4 percent, remarkably low for any major organization. A survey of nine thousand engineering and science students at eighty-six universities named the Agency the top government organization to work for and the fifth best employer overall—in front of companies like Pfizer, Disney, and Johnson & Johnson. And the *Black Collegian* magazine named CIA one of the best places for young African Americans to work—twenty-seventh on a list of fifty companies, ahead of such giants as AT&T, GM, Ford, and PepsiCo.

This second item was especially gratifying to me because I had made it a priority to enhance the Agency's record on diversity. Forget for a moment the ethical reasons for diversity. More than any other entity, the intelligence community has a business need to have its workforce reflect a broad cross section of our populace. We needed demographic diversity and diversity of thought. If all of our employees looked like me, we would never be able to penetrate our toughest targets around the world. The critical decision was to stop treating diversity as a compliance issue and to treat it as a central business imperative.

This issue vividly came to light early in my tenure when I attended a meeting in the Bubble called by some of our African American employees. Those were several of the most eye-opening hours I spent during my time at CIA. One after another, black employees rose to tell disturbing stories of how over the years they had been disrespected and treated as second-class citizens at the Agency. I vowed then and there that we would fix the problem, and I did everything in my power to make good on that promise. We built a program inside CIA that guaranteed that everyone would be afforded the opportunity to advance and grow—the only standard that mattered was excellence. Concurrently, we put in place a program to ensure that every man and woman would have the training and education opportunities to advance. These were not just words; they included metrics and performance reviews of all our major components, and accountability for leaders who did not get the message.

As we were rebuilding CIA, we recognized that our training and education programs, just like recruiting, had been allowed to function independently without an integrated set of common values. So we made a major investment in creating "CIA University." Today all CIA training takes place under one roof, in ten different schools: schools for operational and analytic tradecraft, foreign languages, business, and support information technology, and, most important, a leadership academy where all levels of managers are taught how to lead change and take care of their people.

Just before leaving office in 2004, I testified on Capitol Hill about our clandestine services. That year, we were graduating the largest class of clandestine officers in our history. Since 1997 we had deployed a thousand operations officers in the field. The numbers were great, I said, but nonetheless it would take another five years before our clandestine service was where it needed to be. This shouldn't have been a surprise. When you have had a decade of neglect, it takes you at least as long to recover. No matter how bright the people you recruit, you cannot give them instant experience. Basic training takes about a year. Add in another year, or maybe two, for language school. Then the fledgling officers have to go out in the field and learn by doing. No one showed up at his first station instantly productive.

We also set about improving our second major function: analysis. We changed the dynamic that encouraged top-notch analysts to pursue managerial posts so they could rise up the status ladder. Instead, we created a career path for people who wanted to gain deep analytic expertise. Now such people can go to the top of the pay scale and even be paid more than their managers as long as they enhance their skills and remain productive.

When I first became DCI, I was handed a plan that had been in development for some time to completely overhaul the way we compensated our people. I set it aside because I knew instinctively that with the organization in so much disarray, the workforce would fixate on that and nothing else. We had more important work to do. Five years later, at the urging of Buzzy Krongard, when we judged the institution was healthy enough, we moved to implement a performance-based pay system. We needed a system that would create incentives for valued officers to take on the highest challenges, one that would encourage them to stay and help struggling colleagues to improve. The new system was structured so that it rewarded taking time off day-to-day duties to acquire critical skills. The plan was initially greeted with great cynicism, but we launched a large communication program to educate and make changes based on employee input.

Time and again, I told employees that senior leaders like me were only stewards for a short period of time. The workers, not the drive-through bosses, had to own the institution and take ideas and implement them on the local level.

I'm convinced that the plan could have produced an invaluable boost to morale, but unfortunately, until the day I retired, Congress refused me the authority to implement it across the enterprise. We were allowed instead to conduct only a pilot program affecting thirteen hundred support personnel, and that was a resounding success. The employees knew what they had to do and managers were held accountable. Even more regrettably, the leadership team that followed ours scrapped the plan entirely. In their eyes, the plan suffered from the "not invented here" syndrome. In addition, the new team didn't have the credibility or the will to drive home the sales pitch to the workforce. Still, not implementing the plan Agency-wide was a terrible mistake.

As limited as our human resources were when I took over as DCI in 1997, our technological capacity might have been even worse. Once, CIA was the place to go to achieve technological feats that couldn't have been managed anywhere else—like the creation of the U2 spy plane. But time and technology had passed us by. The private sector was infinitely more agile than were we in adapting the latest technologies. The then head of our Science and Technology Directorate, Ruth David, and her deputy, Joanne Isham, came to me with a bold plan. We had to find a way to harness the brilliance of young innovators in the IT industry. To them, we were their fathers: stiff, buttoned up, wearing suits. They wanted nothing to do with us. We needed to bridge that generation gap.

We decided to use our limited dollars to leverage technology developed elsewhere. In 1999 we chartered a private, independent, nonprofit corporation called In-Q-Tel. A hybrid organization, In-Q-Tel blends research and development models from corporate venture capital funds, businesses, nonprofits, and government. While we pay the bills, In-Q-Tel is independent of CIA. CIA identifies pressing problems, and In-Q-Tel provides the technology to address them. The In-Q-Tel alliance has put the Agency back at the leading edge of technology, a frontier we never should have retreated from in the first place. This highly unusual collaboration between government and the private sector enabled CIA to take advantage of the technology that Las Vegas uses to identify corrupt card players and apply it to link analysis for terrorists, and to adapt the technology that online booksellers use and convert it to scour millions of pages of documents looking for unexpected results.

If you were to ask me how far we came in the effort to transform CIA, I would say we built the foundation and first four floors of a seven-story building. We were far from perfect, and the world never stood still for a minute. After 9/11, making organizational changes had to be calibrated to allow men and women both to perform their mission and to continue the transformation. In the real time of the real world we operated in, the onslaught of threats and crises never abated as we tried to remake the institution. We couldn't afford pit stops. We were changing the tires as the race car was careening around the curves at 180 miles an hour. The mission had to come first. Buzzy Krongard used to say, "Country, mission, CIA, family, and self." That was the CIA I knew.

The job of being DCI was really two jobs—running both CIA and also the larger intelligence community, sixteen diverse agencies. One of the criticisms of not only me but of all my predecessors is that we focused on CIA to the exclusion of the fifteen other parts of the intelligence community. But when I arrived at a badly damaged CIA and intelligence community, I believed first and foremost that it was essential to rebuild the director's base, CIA. If the central pillar of American intelligence was wobbly, all else would be extremely difficult. Rebuilding and transforming CIA, I believed, would give me leverage to use recruitment, training, education, and diversity achievements at CIA to drive similar gains in the rest of the intelligence community.

The resource shortfalls that plagued CIA were shared by the entire community. Despite what might have been seen as a CIA-centric focus, my highest budget priority was to restore the capabilities of the National Security Agency, which by the mid- to late 1990s was in serious jeopardy.

It was in this period that we began to make investments across the community in capabilities that would serve us so well after 9/11. While the money never showed up in the early years, we were preparing for the future.

My plan all along was to get CIA healthy while laying the foundation to do the same with the intelligence community. We made progress, but looming international crises would not wait for us to complete the task.

CHAPTER 3

Shot Out of a Cannon

Jack Devine, a very able clandestine service officer who was acting deputy director of operations during the John Deutch era, once said to me, "George, somebody is going to fire a bullet today in northern Iraq, and you are going to find out where it landed two years from now." As I was to learn, truer words were seldom spoken. So many things were going on in such disparate venues and coming at me from so many angles that it was impossible to keep track of everything. Too often, what seemed trivial at the moment would grow to huge significance, while what seemed hugely significant would disappear into the background noise. A predictable life this was not.

On a typical day as DCI, I felt pretty much as if I had been shot out of a cannon. People were always queued up wanting my undivided attention on dozens of unrelated matters. I bounced from meeting to meeting, with people thrusting thick briefing books into my hands and snatching them away almost before I'd had a chance to digest the first page.

My growing responsibilities even caused my space at home to shrink. Stephanie, John Michael, and I lived in a modest house in suburban Maryland that we had bought ten years before I became DCI. Now that I had the job, we had to give up a portion of our basement so that a security command post and classified document vault could be built. Inevitably the security detail became part of the family—and ours were wonderful, dedicated people—but even so, having armed men and women living in your basement takes some getting used to.

My workday actually began at about ten o'clock the previous night. That's when a printer in the basement command post would start to hum with the first draft of the next day's intelligence briefing for the president. The President's Daily Brief (PDB), or "the book," as we called it, was our most important product. Most nights I would spend an hour or so reviewing the draft articles comprising the PDB, then call the PDB night editor with suggestions on needed changes and areas that required greater explanation. Sometimes, I spiked items that weren't ready for prime time.

By 5:45 in the morning I'd be awake, and usually around 6:15 or 6:30 I would head out the door and jump in the armored SUV idling in the driveway. Waiting in the vehicle in addition to the driver would be an armed security officer riding shotgun and a briefer ready to hand me the completed PDB, a stack of raw intelligence reports that the briefer had plucked from the overnight intake of secrets, and something guaranteed to sour my mood: a thick compilation of news clippings from the morning papers—the overnight leaks. In many cases, staying on top of the news was nearly as important as staying current on the incoming intelligence. In both administrations that I worked for, what was in the news would often drive the policy makers' agenda. That was often the first thing they wanted to talk about.

The two secure telephones in the car were in constant use, with the people from the CIA operations center providing updates and with calls from my staff asking for decisions, relaying messages from the White House, and telling me of constant schedule changes. It was sometimes hard to hear the scrambled communications over the phones because of the competing radio transmissions between my vehicle, a chase car, and members of my security detail pre-positioned at wherever my first stop would be.

During the Clinton years, if I had no early morning appointments downtown, our convoy would cross the Potomac on the Beltway, then head down the George Washington Parkway to headquarters at Langley. Others were doing the actual briefing of the president then. Once George W. Bush came into office and made it apparent that he wanted me on hand personally when he was briefed, we would weave in and out of traffic all the way to the White House. The darting about was both for security reasons and because of the need to get where we were going quickly.

Traditionally, VIPs being ferried around Washington sit in the right rear seat of their official vehicle. I used to enjoy encouraging new briefers to take that spot, calling it my "lucky seat." Halfway to our destination, I would casually mention that the "lucky seat" was also the location that terrorists target with their rocket-propelled grenades.

En route downtown during the Bush administration, my briefer would walk me through the final version of the PDB, a series of short, one- or two-page articles printed on heavy paper and contained in a leather binder. The president's briefer, a different CIA analyst from the one who rode in the car with me, would be waiting in an office we had in the Old Executive Office Building (OEOB), directly across from the White House. Wilma Hall, a White House institution who had served under a half dozen or so presidents, ran my hideaway office and was a comforting anchor in a sea of confusion. There the president's briefer and I would huddle over "the book," trying to divine what questions the president might ask and often calling out to the Agency to contact subject-matter experts to get more data before showtime. Initially our office was in Room 345, looking out over Pennsylvania Avenue. (After 9/11 we were transferred to a room, away from the street, to minimize the potential effects of a terrorist bomb.)

The president's briefer traveled wherever the commander in chief went, updated him, took direction on additional information the president wanted to see, and reported back to me six days a week. It's a killer job. You are up all night preparing for the next day's briefing and up most of the next day preparing for the day after that. The compensation for the awful hours is a chance to witness history up close and personal, the chance of a lifetime.

Usually, after a year in the position, briefers would be rotated to a new job in order to preserve their sanity and, in some cases, marriages.

All around Washington, other CIA briefers were doing the same thing—meeting with their principals, from the vice president and secretaries of state and defense, to a handful of others privileged to receive the PDB. Those briefers would quickly report back to headquarters any significant reactions they got, and often those reactions would give us an early warning of what we might hear coming out of the Oval Office a few minutes later. Official Washington is like a spiderweb. Press it anywhere and the reverberations can be felt throughout the whole structure.

Around 8:00 A.M., the briefer and I would go across the street to the West Wing of the White House and troop up the back staircase to the Oval Office. The actual briefing would generally take between thirty and forty-five minutes—an hour when things were really busy. The vice president, Dick Cheney; Condoleezza Rice, then national security advisor; and Andy Card, the president's chief of staff, always sat in unless they were out of town. The briefer would usually "tee up the piece," explaining each PDB article's background or context, and then hand each item to the president to read. Often there would be additional material to flesh out the story—the nitty-gritty on how we had stolen the secrets contained in the item, and the like. Everyone loves a good spy story. More important, it was an opportunity to pull back the curtain, to talk to the president about a sensitive source or a collection method. The written items were generally short, and the president would read them carefully. Sometimes he would start tossing out questions before getting to the bottom line—a practice that would cause others in the room to start doing so as well. This interactive process was something I welcomed.

My role was to provide color commentary and to provide the larger context. Since I had been around for a while, I could often give some of the historical underpinnings for why other governments were acting as they were. After 9/11, at the conclusion of the PDB briefings, we would be joined by the attorney general, John Ashcroft; FBI director Robert Mueller; and the secretary of homeland security, Tom Ridge, to go over a matrix of recent terrorist threats, weighing their validity and discussing what we were each trying to do to thwart them. By 9:00 A.M. we were generally done with this process. Also, post-9/11, the morning show was followed three days a week—Mondays, Wednesdays, and Fridays—by a "Principals Committee" meeting in the Situation Room, one floor below the Oval Office. The national security advisor would chair these meetings, except when the president chose to attend.

With luck I might be able to head to my office by 10:00 A.M. During the twenty minute ride to headquarters, I usually got in four or five calls using the SUV's secure, scrambled, and sometimes over-scrambled phone system.

When I reached my office, Dottie Hanson, my longtime special assistant, would have a list of calls on my desk that required my attention and another list of Agency and intelligence community people who had been bugging her for "just ten minutes" of my time. Dottie had to change my schedule three or four times a day, almost always beginning in the evenings—that's when things began to settle down in the other offices around town, especially the one at 1600 Pennsylvania Avenue. I had no particular sense that she was doing this. I just went where I was pointed and consulted my "daybook"—an artful compilation of research papers, backgrounders, and biographic information that my staff prepared daily—before I arrived at the office. Dottie knew the building well; she had been with CIA for more than forty years. Indispensable and loyal, she was a good judge of character and always gave valuable advice. People sometimes joked with me, Who really ran the Agency? Let me clear that up right now: it was Dottie.

Being responsible for CIA alone would have been a big enough job, but as DCI, I was also accountable for the rest of the intelligence community. That meant trying to monitor fifteen other agencies, including what the National Security Agency was up to, not easy with a place that generated thousands of intelligence reports on intercepted communications, called "signals intelligence," each week. I also had to concern myself with the work of another agency, now known as the National Geospatial-Intelligence Agency, which was cranking out hundreds of dispatches daily that tried to interpret what they were seeing from satellite reconnaissance photos. And I had to trust that somewhere in the organization people were marrying these products up—providing the "all-source analysis" that attempts to assemble a big picture.

I wasn't in the job long before I realized that there was precious little time for me to step back and say, "What does all this mean?" So I directed my "issue managers"—people who had responsibility for specific geographic regions or subjects—to send me a memo every two weeks summarizing the latest developments within their areas of responsibility, and to tell me what worried them the most. Even if the issue was not on the front burner today, it might be within months. I needed a baseline. With so much swirling around and through me, I often felt as if I were trying to watch eight television shows at once.

Another big part of the DCI's role was to maintain contact with the heads of foreign intelligence services. I met with visiting senior security officials from just about every country imaginable. Most countries had multiple intelligence services, and so I would need to be in touch with various sets of people from the same country. I would meet with both the Israeli Mossad and Shin Bet, for example, or the British MI-5 and MI-6. Mossad is the CIA equivalent; Shin Bet, the Israeli internal security service. MI-5 handles internal security in the United Kingdom, while MI-6 is the foreign intelligence service. Occasionally, a delegation from one service would be cooling its heels in one waiting room while we were trying to move a group from that country's rival nation out the other door. Traffic jams were to be avoided at all costs.

These weren't social visits. There were briefing books to study before each meeting, telling me what the group wanted from us and what we wanted from them. Sometimes we were seeking insights on threats from their region, but very often our visitors carried with them detailed requests for information, training, or financial assistance that needed to be dealt with. Visiting delegations often brought with them ceremonial gifts. Some were small tokens; others, touching and beautiful artifacts. With rare exceptions, I would accept on behalf of the U.S. government, and sometimes the gift would then end up auctioned off or stored. Any gift that was going to be placed on display at the Agency first had to be x-rayed to ensure that it was not bugged with listening devices.

These meetings were often held at the cost of other pressing matters, but these vital relationships needed careful tending if it ever became necessary to call in the chits from our side. After 9/11, the time invested in such meetings paid off in willing partners ready to help us in a common cause when so much was on the line.

Responding to the requests (and sometimes demands) of Congress was an equally large part of the job. I participated in hundreds of closed-door hearings and briefings during my tenure, not just for our two oversight committees but also before a half dozen other committees that thought they were owed a piece of my time. As a former Hill staffer, I understood the need to tend to Congress. It is important work. I believe in thorough and thoughtful oversight; it distinguishes this country from all other countries in the world. But I occasionally found myself wishing committees had focused more of their time on the long-term needs of U.S. intelligence rather than responding to the news of the day.

When I was back at Langley, the afternoons were invariably packed with meetings, briefings, and the occasional pop-up crisis. I hated being tethered to the office and would sneak away as much as possible to drop in unannounced in offices around the 250-plus-acre headquarters compound. Early in my tenure, one Friday afternoon, I wandered into an office in the bowels of the headquarters building where two female employees were in the middle of a conversation that I had apparently interrupted. "Hi, howya doing? What are you working on?" I asked. One of the pair, a crusty veteran of the organization, stared at me for a second, then said, "I hope you don't mind my asking, but who the hell are you?" I chose that moment to pop an unlit cigar in my mouth—something I was known to do at the time. The woman's eyes got wide, her face turned red, and she said, "Oh my God, you're him, aren't you?"

Although I had spent most of my professional life on Capitol Hill, I increasingly found myself most comfortable on the other side of the world. Out in the desert, or in Jerusalem or Ramallah, Riyadh or Islamabad, I got along just fine. Maybe I'd gone native and didn't realize it.

At least 90 percent of the trips I made overseas during my seven years as DCI were to the Middle East or to the border nations of Central and South Asia. I went often, and I kept going back, to build the personal relationships that might at some point yield a breakthrough.

You need to put capital in these countries' banks—including, in my case, the capital of your own time—respect their sovereignty, and as a normal practice, refrain from sticking your finger in their chests. It is important to deal with them honestly and fairly and have them learn over a period of time that they can trust your word. A key piece of this is absolute patience. It takes time to develop a relationship as a trusted partner.

This wasn't Henry Kissinger's brand of highbrow shuttle diplomacy. This was some hybrid of intelligence work and diplomacy practiced by the son of Greek immigrants. The closer I am to my ancestral Mediterranean, the more at home I feel. For some reason, whether talking to crowned heads of state or streetwise security officials risen improbably to power in the cauldron of Middle East politics, my style seemed to work.

I'm reminded particularly of a trip in the spring of 2000 to

Georgia. We flew into the capital about midday, did our business there, and then retreated to a dacha, or country house, where the Georgians had insisted on hosting a party for us. The dinner got under way promptly at seven that evening. There must have been at least fifty of us seated at a very long table, with the Georgians on one side, Americans on the other, and a contingent of Georgian singers clustered down at one end. The "singers," in this case, were far more adept at drinking than song. One fireplug of a vocalist—maybe five feet five, with a barrel chest, like a sawed-off Rich Armitage—began the night with two fifths of Johnny Walker Black in front of him. Three hours later both were empty.

I had never been to a Georgian dinner before, but I had been briefed enough on the customs to know that the host is called the *tamada*, who is also the master of ceremonies and leads the toasts. Sure enough, we no sooner sat down than the tamada popped to his feet and toasted me with a glass of sweet Georgian wine. When he was through, I naturally rose and returned the favor, and with that, I figured we were done with the formalities and could get down to our meal. No way. A few minutes later the host popped up again, went to the wall behind him, and pulled down a big, hollowed-out antler. Then he picked up a bottle of wine, poured half of it into the antler, toasted me again, and chugged the antler dry. Well, there was an antler behind me, so I got up and did the same, and when I sat back down, it was 7:12 P.M. and I was officially pie-eyed. Let me stress that this was not typical of my condition before, after, or during work. But sometimes when you are trying to bond with foreign counterparts, you have to bend to local customs.

In any case, I had a long dinner ahead of me and many, many more toasts to come, a number of them led by the increasingly boisterous covey of professional drinker-singers at the end of the table.

It was maybe two hours into the party when I heard the Georgians across the table from us talking in derogatory terms about the Russians. By then, I was deeply into the spirit of the evening, so I leaned over to Dave Carey, CIA's number three man at the time, who was sitting next to me, and whispered, "Ah, to hell with the Russians!" Unfortunately, what I meant to come out in a whisper instead came out at a hundred decibels, to the great delight of the Georgians, who jumped up and start applauding and toasting me still more.

Just about then, the Georgians decided to teach us how to do "the chair dance," a local custom that goes like this: you turn your chair around; sit on it backward, and to the beat of music, you and your chair bounce around the table. By then, I'm certain, the CIA security detail that was watching all this through a window from an adjoining room was thinking, "We have *got* to get the DCI out of there. Nothing good is going to come of this." In fact, of course, something good did come of it. That kind of bonding experience is worth its weight in gold in that part of the world.

The next morning, though, arriving back at the airport for our flight on to Uzbekistan, it was hard to think of anything more than my pounding head. That's about all I was doing when a senior Georgian official came up to me and said, "We have bad news. The Russians have denied you flight clearance to get to your next stop." We always wondered whether the Russians had had the Georgians' dacha "wired" and taken offense at my impromptu remark from the night before.

The Georgians, at least, showed us a good time. Relations with Moscow were always strained at best or weird at worst. Maybe it's the residue of the cold war or the incompleteness of Russia's transformation to a democratic society, but the same lack of connection dogged the one visit I made to Moscow, to meet with the head of the FSB, the federal security service of the Russian Federation. We convened at FSB headquarters, atop the notorious Lubyanka prison, a portion of which has now been turned into a KGB museum. Substantive issues (which, for reasons of security, I can't get into) were on the table, but we never got close to addressing them. First, our hosts offered us a tour of the

American section of the prison museum, which includes, among other artifacts, the silencer- and poison needle–equipped pistol that Gary Powers carried when his U2 spy plane was shot down over the USSR in 1960. We declined—we weren't there to play tourist—so our hosts hurried us off to a very elaborate restaurant for dinner, and that's when things really got weird.

Waiting at the top of the stairs at the Praha Restaurant entrance was a very tall, voluptuous blond woman. At her side in attendance were two dwarfs, neither much more than three feet tall. As we reached the top of the steps, our hostess turned, the dwarfs turned with her, each taking one of her hands, and the three of them then paraded side by side down a long hall, leading us into the restaurant proper.

You would think that a meal that started so, um, uniquely might at least have led to a little conviviality, but that wasn't the case. Finally, out of other gambits, I did what I often do when the going gets rough at such gatherings; I asked John McLaughlin to perform his famous money trick. So John took out a thousand-ruble note, went through his extraordinary mumbo jumbo and fancy prestidigitation, and, presto, when he opened his hands again, it was a hundred-thousand-ruble note. "How do you think we get our money?" he said to the FSB director, Nikolai Kovalev, with an absolutely straight face. By then, the look on Kovalev's face was priceless. I could just see him thinking, "Ronald Reagan said he was going to spend us into oblivion with the Strategic Defense Initiative, and now this man McLaughlin has just manufactured money for them. We'll never beat them!"

John once performed the same trick for Carlos Menem, then president of a debt-strapped Argentina. A week later, we received word that upon reflection, Menem wanted to make John his finance minister.

Some places I almost didn't get back from. In 1996, when I was still the deputy DCI, we were halfway across the Atlantic Ocean, returning from a trip to Croatia. Suddenly we heard a hissing sound from the front of the plane, and shortly thereafter a wide-eyed young military steward walked into the cabin. His name was Daniel, and previously he had proudly told us that this was his very first "VIP" flight. Now he came back tightly gripping an emergency manual, told us we had an "in-flight" emergency, and ordered us to don our life vests. "Why?" we asked. "What emergency?"

The plane's exterior windshield had cracked, and the interior windshield was in danger of breaking, too, he explained, which would cause immediate depressurization of the cabin. That event, he said, would force the jet to "land on water." Daniel went on to say that "when" that happened, we would have a minute and twenty seconds to exit the plane and get in the inflatable life raft.

"Don't you mean *if* that happens?" I asked.

One of our traveling party, a division chief with nearly four decades of CIA service under his belt, looked at Daniel and said, "Son, I was born in the 1930s. I can't do anything in a minute and twenty seconds." He reached for a beer to fortify himself against the cold Atlantic waters.

As our plane was limping toward Gander, Newfoundland, Daniel came back to tell us that while the good news was the windshield was still holding, the bad news was that it appeared that our landing gear would not come down. Eventually the gear was lowered, and we made a safe landing, passing through a cordon of fire trucks and crash vehicles. Daniel and the crew of the air force plane performed magnificently, but I suspect he will never forget his first VIP flight. I know I won't.

When you are the DCI, you never get away from the job. Either you travel with it or it travels with you. In my seven years as DCI, I made seventy-seven trips to thirty-three countries, about one trip a month on average. Saudi Arabia was one of my most frequent stops; I went there nine times, a clear indication of the importance of the U.S.-Saudi relationship. Domestically I traveled less often, although I visited our clandestine training facility regularly. But it's the times I was supposed to be away from the job—the rare vacations—that I remember best.

In September 1997, I took Stephanie and John Michael to Bethany Beach, Delaware, for a quiet weekend. We were on the beach, pretending to be a normal American family, when I was summoned by my security detail to take a frantic incoming phone call from the head of the Jordanian intelligence service. He told me that the Jordanians had just captured a group of Israeli intelligence officials as they attempted to assassinate Khaled Mish'al, the head of the Damascus office of Hamas, by injecting a lethal poison in his ear. The attempt had been carried out in broad daylight in downtown Amman, the Jordanian capital. Two members of the Israeli hit team had been apprehended, and six others reportedly had taken refuge in the Israeli embassy. Mish'al was hovering on the brink of death. King Hussein, who had been enormously helpful in the Middle East peace process, was understandably furious. Meanwhile, Jordanian officials were screaming at the Israelis to get an antidote that might save Mish'al's life.

I had had a lot of experiences by then, but nothing in my training or background had prepared me for what to do when someone comes up to you on the beach to tell you that some friends of yours have just botched an assassination attempt using a poison. That's the way the job was, though—full of surprises, few of them pleasant.

I don't want to imply that every day was stomach churning or worse than the one before it. There were triumphant moments, nights I would go home feeling on top of the world. One of the most memorable came in the aftermath of one of the worst days in the Agency's history.

On January 25, 1993, Aimal Kasi, a lone Pakistani gunman armed with an AK-47, walked up to the main entrance to CIA headquarters and shot five people waiting to enter the compound. Dr. Lansing Bennett, a sixty-six-year-old Agency physician, and Frank Darling, twenty-eight, a communications specialist, were brutally murdered while doing the most mundane daily chore—driving to work. Darling's wife, Judy Becker Darling, also an Agency employee at the time, was sitting beside her husband and watched in horror as Kasi coolly walked among the cars stacked up at a stoplight and randomly singled out a few occupants to die. Amazingly, Kasi simply walked away in the ensuing chaos. Recovering the car he had stashed, he drove to his apartment, where he left his weapon, and then headed to Dulles International Airport for a flight back to Pakistan.

A massive international manhunt was mounted with a combination of investigative expertise, physical daring, and a generous application of reward money. Finally, four and a half years later, in 1998, Kasi—or a man who we suspected was Kasi—was lured to Dera Ghazi Khan, a dusty town in central Pakistan, with promises of being able to buy Russian goods in Afghanistan and sell them at a premium across the border in Pakistan. While he waited for the deal to go through, the suspect stayed in a three-dollar-a-night rooming house. That's where we determined to run him down.

I remember as if it were yesterday standing in the Global Response Center (GRC) on the sixth floor of our headquarters building listening to the radio traffic coming back as a joint FBI-CIA team dressed in local garb entered the dingy hotel in the middle of the night, kicked down the door, and wrestled a startled bearded man to the floor. We waited anxiously while the team cuffed its prisoner and quickly forced his fingers onto an inkpad to obtain positive identification. Then one of the team members in Pakistan called out, "Red Zulu, Red Zulu!" and a guy standing near me shouted, "We got him! He's our man!" As cheers went up in the GRC and the backslapping and high fives began, I allowed myself to light a rare victory cigar. Apparently, in the excitement, it fell on the floor. I know this because for years afterward a piece of burned carpet hung framed on the GRC wall.

A few days later, several of my top aides and I went out to

Dulles to watch Kasi being brought to justice. From a building the FBI controlled at the end of a runway, we followed the aircraft bearing the shackled terrorist as it made its slow approach. I couldn't help but wonder at that moment what must have been going through Kasi's mind. Four and a half years earlier he'd flown out of this same airport thinking that he had gotten away with murder. He hadn't. Standing side by side with our FBI colleagues in stony silence as Kasi disembarked, I felt I was representing the thousands of Agency men and women who had been praying and working for this moment to arrive.

The next day, I invited the FBI agents and CIA officers who had participated in arresting Kasi to come out to the Agency headquarters and bask in the applause and thanks of a grateful Agency workforce—an emotional moment that no one present that day will ever forget. You often hear about rivalry between the FBI and CIA. Some of those stories are true. But in this case there was an outpouring of respect, pride, and gratitude, not to mention hugs and tears. As the crowd filed out at the end of the ceremony, Bruce Springsteen's "Born in the U.S.A." boomed out of the auditorium's speaker system.

After his capture, Kasi said that he had conducted the shooting because he was upset with U.S. policy in the Middle East and Iraq. In a letter sent from his jail cell to a reporter, he said that his hope had been to kill the CIA director, at the time Jim Woolsey, or Woolsey's predecessor, Bob Gates. In fact, just a few weeks before the attack outside CIA, a man with a rifle was spotted in the woods behind Gates's home. That person was never captured, but the possibility of being personally targeted was something that all of us who succeeded Gates lived with constantly. As for Kasi, almost a decade would pass before he was finally executed in a Jarratt, Virginia, prison, on November 14, 2002.

There were many moments like Aimal Kasi's capture, times when all the hours, all the risks, all the planning, would be rewarded. Some I can't write about at all. Otherwise, sources would get compromised, channels closed down, and lives lost.

Unfortunately, when you run a place like CIA, it's the low lights that stand out in the media—the mistakes, the goofs, the gaffes—the things everyone can see and no one, it seems, can resist commenting on. For many of those, I would like to turn back the clock and erase them. Some, I can't stop remembering.

On May 11, 1998, the Indian government conducted underground tests of three nuclear devices. It followed up a couple days later with tests of two more. Within two weeks, Pakistan responded with its own tests. We knew that both countries had nuclear desires, intent, and capabilities, and we knew the risks all too well. The India-Pakistan border is one of the most contentious in the world, maybe even more contentious than the border that divides Israel and the Palestinians, and the region is one of the world's most populated. Unleashing nuclear weapons on the subcontinent could kill literally millions. That said, the timing of the tests caught us by surprise.

The morning that the world learned of the first Indian tests, I received a call from our Senate oversight chairman, Richard Shelby. Not surprisingly, he asked me what had happened. One of my habits is to be plainspoken, maybe too much. "Senator, we didn't have a clue," I told him. Within minutes, Shelby was on CNN, calling the miss a "colossal intelligence failure." Was it a failure? No doubt. "Colossal" is in the eye of the beholder.

The very same day, I got a call from my boss, President Clinton. "George," he said, "I want you to know I have full faith and trust in you. You're doing a damn good job—don't worry." For a forty-five-year-old guy in the middle of his first major crisis as DCI to have the president of the United States pick up the phone and reassure him like that was a great morale booster. Afterward, I said to myself, okay, forget Shelby. The only guy that matters has just checked off. Let's go find out what went wrong here and see what we can do to prevent a similar event in the future. And so I asked the former vice chairman of the Joint Chiefs of Staff, Adm. David Jeremiah, to lead a team to examine how and why we had missed the boat so badly. A month later, the results were in.

Jeremiah's team confirmed that the identification of the Indian nuclear test preparations was a difficult intelligence-collection and analytical problem. The Indian program was not derived from the U.S., Chinese, Russian, or French programs, but was indigenously developed and thus harder to detect. Three years earlier, in 1995, we had learned about similar test preparations and strongly urged the Indians to stop. They had, but in confronting them we had given them a road map for how to deceive us in the future. This time, only a limited number of senior Indian officials were aware of the planned tests.

The field of expectation had changed, too, and we were perhaps slow to catch up with it. Back in the days when our adversary was the Soviet Union, we were not expected to predict or prevent weapons tests. In almost every case, the only way we ever knew about the location of a new Soviet test site was by detecting a test after the fact. If the intelligence community subsequently could tell policy makers how big a test was, this was considered a success. Now we were expected to predict and prevent tests in non-superpower nations. Adding to the challenge in this instance was the fact that our limited overhead satellite collection capability was stretched thin, in large part because some of it had been diverted from the Indian subcontinent to focus on Iraq and the protection of U.S. airmen patrolling the no-fly zones around Baghdad.

One major conclusion of the Jeremiah report was that both the U.S. intelligence and policy communities had an underlying mind-set that Indian government officials would behave as ours behaved. We did not sufficiently accept that Indian politicians might do what they had openly promised—conduct a nuclear test, as the incoming ruling party had said it would. The lesson learned is that sometimes intentions do not reside in secret—they are out there for all to see and hear. What we believe to be implausible often has nothing to do with how a foreign culture might act. We would learn this in a different way years later with regard to Iraq. We thought it implausible that someone like Saddam would risk the destruction of his regime over noncompliance with UN resolutions. What we did not account for was the mind-set never to show weakness in a very dangerous neighborhood—particularly in regard to a growing Iranian military capability. Relying on secrets by themselves, divorced from deep knowledge of cultural mind-sets and history, will take you only so far.

A year later, my job was at risk again, this time maybe for better reasons. In early May of 1999, on the eve of leaving for London for one of our regular conferences with our British Commonwealth counterparts, my then executive assistant, Michael Morell, called me in the middle of the night. Mike had just been contacted by CIA's operations center after it had received a call from Gen. Wesley Clark, the commander of U.S. forces in the Balkans. Clark's question: "Why did the CIA tell me to bomb the Chinese embassy in Belgrade?" In retrospect I should have fired back a note asking why the "no strike" databases for which General Clark's command was responsible weren't up to date as required. If they had been, the tragedy might have been avoided. That doesn't excuse our mistake, however.

A check of the newswires showed that the Chinese government was indeed saying that its embassy in Belgrade had just been bombed by U.S. aircraft. For a few hours we thought that it was simply a matter of an errant bomb or a missile veering from its intended target. Tragic, but these things happen in war. I was airborne, en route to London, when we started getting word that air force bombers had hit precisely what they were aiming at and that they had indeed used targeting data supplied by CIA. Three people had been killed in the strike, which did substantial damage to the building, and more than twenty were injured. I still had no idea why the targeting data had been so faulty, but because it was obvious this was going to become an international incident, I asked my deputy at the time, Air Force Gen. John Gordon, to get to the bottom of the situation as soon as possible. Unnamed Pentagon officials were already rushing to the phones to absolve their department of any blame, while telling the media that the mistake rested with the Agency's use of faulty maps. But that was only part of the story.

During the course of the brief air war in the Balkans, CIA had provided intelligence on scores of military-selected targets. Soon, though, the Pentagon started to run out of militarily significant sites to hit and asked the Agency to suggest targets that we wanted to see destroyed. The very first one offered up was what we believed to be the Yugoslav Federal Directorate of Supply and Procurement (FDSP), a military warehouse involved in shipping missile parts to rogue nations such as Libya and Iraq. Unfortunately, the warehouse had been mis-plotted on maps not intended for the creation of strike packages. In fact, we had given the Pentagon the coordinates of the Chinese embassy. The warehouse was about three hundred meters away. After the bum information was passed to the Pentagon, several fail-safe mechanisms collapsed at their end. The military was supposed to keep up-to-date "no strike" databases that warned aircraft away from hospitals, schools, churches, mosques, and places like embassies. But that database had been neglected.

One of our officers, not involved in nominating targets, happened to notice in passing the warehouse plotting and raised questions about it. He remembered seeing information a few years earlier that the supply building was located a block away from the location identified. Showing great initiative, this officer telephoned the Department of Defense Task Force in Naples three days before the bombing to say that he thought the FDSP headquarters building was a block away from the identified location. Nonetheless, on May 7, the officer found to his surprise that the building was on the target list for bombing that night; he again phoned Naples. The aircraft was already en route to the target. Later, military officials in Europe would say that they believed the CIA officer was trying to convey that while the building might not be the supply *headquarters*, it was still a legitimate target. Recollections differ of exactly what was said, but the one certainty is that no one up or down the line knew that the facility in question was the Chinese embassy.

Not long after my plane touched down in the United Kingdom on the day after the bombing, I got a phone call from President Clinton's national security advisor, Sandy Berger. "You better get back here right away," Sandy said. "I'm trying to save your job." With that, I turned around and headed home to face the music. As I soon found out, the "bad map" story was already the butt of many jokes on late-night television and in editorial cartoons. We found no humor in it since three Chinese intelligence officers had died as a result of our and the Pentagon's combined mistake.

Inevitably, great pressure was being brought to bear on the White House for heads to roll over the issue, and mine seemed a likely candidate. If someone was going to plead my case, I was glad to have Sandy Berger doing it. I had worked closely with Sandy at the NSC before becoming deputy director. Sandy had one overriding concern always: to protect the president. An embarrassing screwup like this one—one based on a lack of focus and inattention to detail—was exactly the kind of thing he hated to see. But the two of us also spoke the same language. Sandy was very direct; he would have fit in well in the Queens neighborhood where I grew up. Most important, you always knew where you stood with Sandy. If he was hopping mad at you, you were going to hear it directly from him, not learn about it through blind quotes in some newspaper column.

When I got to the White House on my return from London, Sandy was true to form. He let me know directly just how displeased he was with CIA's performance on the embassy targeting, but he saved my job. To my relief, President Clinton rejected the calls for me to be held personally accountable for the incident.

Deputy Defense Secretary John Hamre and I were hauled before Congress to try to explain how such an egregious error could have occurred. Hamre was candid and faced up to his share of the responsibility. The general view from the Pentagon, however, was that "stuff" happens in war, and they weren't going to hold anyone in DOD responsible for their share of the blame.

Nearly a year after the bombing incident, our CIA Accountability Board determined that several Agency officers involved in identifying the proposed bombing target had failed to take necessary and prudent steps to ensure that the appropriate site was struck. Several people received written or oral reprimands. One retired military officer who was working for the Agency as a contractor, the person most responsible for the misplaced targeting, had his contract terminated and was essentially fired. I supported the dismissal, but I regret it today. Yes, his performance was flawed, but there were others up the chain of command who should have borne more responsibility, and the complete absence of accountability from the Pentagon for its part in the incident meant that this man was the single recipient of censure. That wasn't right, and unfortunately, it wasn't the last time on my watch that CIA would take sole responsibility for errors in which other agencies shared the blame.

The Accountability Board wasn't the last that we would hear about the incident. In the days just before the war in Iraq started, in March of 2003, one of my senior officers from the Directorate of Operations came up to me with a smile and said, "Hey, boss, you aren't going to believe this. We just got an urgent back-channel message from the Chinese intelligence service." He paused for effect, having gotten my attention.

"So, what did they say?" I asked.

"They sent us the geographic coordinates for their embassy in Baghdad and said they hoped it was accurately listed in all the Pentagon's databases."

The Chinese embassy bombing was not my worst day as DCI before 9/11. That sad distinction goes to April 20, 2001. I was working late that Friday evening when reports started dribbling in about an incident that had happened earlier that day in a remote region of Peru. We had been involved there in a highly classified program helping the Peruvian air force interdict flights suspected of carrying illicit drugs bound for the United States. The "Airbridge Denial Program," as it was known, used civilian aircraft under contract to CIA to pass on actionable intelligence to the Peruvians. Americans weren't firing on suspected drug planes; Peruvians were.

As far as I was concerned, this was an important mission and a good example of just how widely our resources were spread across the globe. In the mid-nineties the United States had detected more than 400 narcotics flights leaving Peru annually carrying an estimated 310 metric tons of semi-refined cocaine. Over the previous five years, we had put a big dent in that. With our help, the Peruvians had forced down or shot down thirty-eight suspected drug flights and had probably discouraged many more. On this day, though, the program went horribly wrong.

James Bowers and his wife, Veronica, were evangelical Baptists who had been carrying out missionary work in the Peruvian Amazon region for several years. They worked to bring educational, medical, and other assistance to a remote area—truly "God's work." The Bowers had recently adopted a baby girl in the United States, whom they had named Charity. They needed a residence visa from the Peruvian government for the baby to remain in Peru. Kevin Donaldson, a fellow member of their missionary group, agreed to fly them in a single-engine floatplane to Islandia, a Peruvian town near the tri-border area of Peru, Colombia, and Brazil. From there, the Bowers family traveled to a nearby town, where the necessary paperwork could be completed.

On the return flight, the missionaries' floatplane flew a course following the Amazon River, and in keeping with local practice, the pilot tried to remain within sight of the waterway in case he needed to make an emergency landing. The problem was that their flight path also made them appear to be "an aircraft of interest" to the U.S. and Peruvian planes looking for drug traffickers, although the aircraft took no evasive action. After finding no flight plan on file for the aircraft, observers upgraded the small floatplane to "suspect" status. From there, the tragedy built upon itself. The Peruvian aircrew didn't follow agreed-upon procedures. The Americans lacked adequate Spanish skills to communicate with their counterparts. When the private aircraft did not respond to radio calls, a Peruvian fighter fired on it. Veronica Bowers, thirty-five, and seven-month-old Charity were killed in the incident.

We quickly obtained audio recordings of the cockpit-to-cockpit communications (and miscommunications) and subsequently got some video of the downed floatplane's surviving passengers trying to save themselves in the Amazon. The sounds and images of the incident haunt me to this day. On the audio recording you could hear the contract Agency aircrew question their Peruvian counterparts before the jet opened fire. The American crew working under contract for CIA kept asking their counterparts if they were "sure" that those in the plane in question were "bad guys" or *"banditos."* They tried to restrain the Peruvians, with no effect. It was clear, from listening to the tapes, that Americans and Peruvians were talking past one another, unable to understand what they were hearing. Toward the end of the tape, pilot Kevin Donaldson can be heard screaming, "They're killing us, they're killing us!" In broken Spanish, the Agency contractor aircrew shouted for the Peruvians to stop. *"No mas, no mas!"* But it was too late for Veronica and her baby. I'll never forget the end of the tape, with the Agency aircrew simply sighing and saying, "God!"

CHAPTER 4

Waging Peace

CIA Director George J. Tenet told President Clinton last month that he would find it difficult to remain as director were convicted Israeli spy Jonathan Jay Pollard released as part of a Middle East peace agreement, according to sources.

—*Washington Post*, November 11, 1998

Difficult" is the wrong word. "Impossible" is closer, but even that doesn't do the situation justice. Here is what happened in mid-October 1998 at the Wye Plantation Conference Center, a beautiful 1,100-acre estate along the Wye River, on the Eastern Shore of Maryland. The story itself, though, begins three years earlier, with a brutal murder.

The November 1995 assassination of Israeli prime minister Yitzhak Rabin—by an Israeli opposed to the peace process, less than two years after Rabin had shared the Nobel Peace Prize with his foreign minister, Shimon Peres, and Yasser Arafat—had a profound effect not only on Rabin's countrymen but on the Palestinians. The Israelis were accustomed to Palestinians cheering on the rooftops whenever disaster struck across the border. Not this time. Rabin's murder sparked an outpouring of genuine emotion among the Palestinians, and with it, the entire Israeli perception of their neighbors began to change. Peres had been handed Rabin's job, his legacy, and his momentum, and for a few months, peace seemed not just conceivable between the Israelis and the Palestinians but genuinely possible.

Then, beginning in late February 1996, came a wave of suicide bombings—four in nine days that left more than sixty dead—engineered by the militant Islamic group Hamas. Arafat, who had been elected to the presidency of the Palestinian Authority that January, reacted with surprising speed, arresting scores of militants, including the man suspected of recruiting the suicide bombers, and raiding more than two dozen Islamic organizations and institutions thought to lend financial and other support to Hamas.

To us at CIA, it was evident that Arafat had been surprised by the violence. Hamas was stronger than he realized, strong enough to threaten his power. The bombings had done more than derail the peace process—that was an old story in the Middle East. This time they had called into question the whole structure of the process and the premises on which it was built.

It is difficult to overstate the importance of Middle East peace. The issue transcends humanitarian concerns to stop the violence and suffering. And it is even more important than the desire to eliminate a root cause for much of the global terrorism that plagues our world. The best hopes and the worst fears of the planet are invested in that relatively small patch of earth.

In March 1996, desperate to restart negotiations, a high-level U.S. delegation flew out to the Middle East to meet with leaders there. Aboard were Bill Clinton, still finishing up his first term in office and with a reelection campaign in the offing; Dennis Ross, Clinton's special envoy to the region, with ambassadorial status; my then boss, John Deutch; and others. In flight, Dennis would later tell me, Clinton asked a very simple question: What do we have to do to save this? And out of that was born the Summit of the Peacemakers, held that spring at the Egyptian Red Sea resort of Sharm el-Sheikh. The idea of the summit was to demonstrate unmistakably to the Israelis what had been so evident before Hamas went on its killing spree—that they were not alone. The Palestinians were threatened by the same things Israelis were; they, too, condemned these kinds of violent acts.

Clinton and the others didn't stop there. On that same flight, a second realization was born: that without simultaneous progress on security issues, the political process alone was never going to bring peace to the Middle East. Every deal in the world could be struck with all the goodwill imaginable, but unless the Palestinian and Israeli security forces were in constant communication and working to achieve mutually beneficial goals, Hamas, or some similar group, would always be able to destroy what the politicians had created. The Israelis wanted to know that terrorists would not be given safe harbor by the Palestinians. The Palestinians, in return, wanted to know that their people would not be crushed by an oppressive Israeli security apparatus.

Clinton and Ross agreed on the principle but said that someone would have to be in charge of making the security arrangements, and Deutch apparently said, "I know just the guy for the job." It turns out that I was the guy. Security was the key. You can talk about sovereignty, borders, elections, territory, and the rest all day long, but unless the two sides feel safe, then nothing else matters.

To be honest, I wasn't enamored of suddenly finding myself in the middle of all this. At one level, it was a natural fit. CIA already had significant ties to the security forces in both Palestine and Israel, and we had aided and abetted plenty of negotiations. But our job in those instances was to provide behind-the-scenes input and insight to the actual negotiators, not to sit at the table ourselves. This new plan called for taking on a quasi-diplomatic role in what was largely a political process, and initially that struck me as inappropriate for someone in my position.

There was no way that my new role wouldn't become very public, very soon. The DCI had volunteered me, and the president had agreed. Under those circumstances, I couldn't say no. But I made it absolutely clear from the beginning that we wouldn't be mediators or umpires. That was a policy maker's job, and at CIA, we don't make policy; we implement it. As I saw our role, it was to be an honest broker, someone both sides could turn to and both sides could trust. The more the Palestinians and Israelis initiated dialogue on their own, the less we were in the middle, the better off everyone would be.

"Listen," a Palestinian negotiator said to me one day after a grueling session, "we know you have a close and strategic relationship with the Israelis that we will never be able to re-create with you. All we ask is that you be fair." That's a principle to live by in the Middle East, and it was our gold standard from start to finish.

In early March 1996, just days before the Peacemakers summit convened at Sharm el-Sheikh, and in the first real exercise of my new duties, I flew to Israel with some of our top people to begin trying to forge common ground between the Israeli and Palestinian intelligence services. And sure enough, the story went public before my plane touched down.

Citing anonymous sources, the *Jerusalem Post* reported on March 10 that "the American delegation was headed by deputy CIA director George Tenet." In the *New York Times*, Tim Weiner wrote that "official meetings between an American intelligence official of Mr. Tenet's rank and his Palestinian counterpart may be unprecedented."

I can't say if that's so, but the emphasis on security issues as a parallel track with the political issues—the recognition that without security there could be no peace process—was unique, at least in my experience. Dennis Ross, the lead American negotiator at Sharm el-Sheikh, made the same point forcefully to Yasser Arafat. "The peace process is over unless you do something on the security issue. And you can't fake it—it has to be real," as Dennis later recounted his conversation with the Palestinian chairman. The message got through. The bombings had already convinced Arafat of the threat Hamas posed to him, personally and politically. Once Dennis had helped him understand that we stood ready to help and that ours was an offer he couldn't refuse, Arafat told Bill Clinton that he was willing to engage in talks with the Israelis, and the peace process was once again up and running. Sort of.

As so often happens with these things, life and other concerns intervened. The Wye River summit that was meant to be the second leg of an ongoing process kept getting put off. Like everything in the Middle East, except the onset of violence, it took longer than expected. When the conference was finally held, in October 1998, more than two years after the Sharm el-Sheikh gathering, I had been DCI for fifteen months.

Dennis tried to set the table for Wye by meeting beforehand with Mohammed Dahlan, the Palestinian security chief, on the beach in Gaza. Dennis's message was essentially what he had said to Arafat two years earlier: The Palestinians had to be ready to make concessions to the Israelis on the security front. They needed to accommodate Israel's concerns in unprecedented ways. Then he went on to list what those were going to be. Dahlan's response was predictable. No, he could never agree to that. He would look like a quisling, and on and on. Fine, Dennis told him, we'll change the words, but we cannot alter the substance. Dahlan said yes to that—he really didn't have a choice—but Dennis was still uneasy. Without a security proposal, he would have no leverage on Bibi Netanyahu, who had succeeded Shimon Peres as prime minister in the spring of 1996, and without leverage, nothing was going to get done.

When Dennis returned, he asked me to fly out to the Middle East and help the Palestinians develop a specific security plan that they would then bring with them to Wye—an insurance policy, of sorts, that the leverage would be there when he needed it. And thus I found myself, only days before a summit was to begin, locked into the secure Sensitive Compartmented Information Facility—or SCIF, as it's known—at the U.S. consulate in Jerusalem with Mohammed Dahlan; Jabril Rajoub, chief of the Palestinian Security Service on the West Bank; and Amin al-Hindi, leader of the Palestinian General Intelligence Service.

These men, who would be my counterparts in countless meetings in the years to come, shared some traits. Several of them spoke decent Hebrew, an artifact of having spent years as prisoners in Israeli jails. They also were competitors among themselves. It was sometimes hard to know where their official talking points stopped and their personal agendas started. I was accustomed to politicians with egos and agendas, however, and I struck up warm personal relations with them all. Perhaps it was my Greek ancestry, but I was used to people speaking emotionally, with lots of arm-waving and raised voices. Dahlan in particular was prone to launching into histrionic rants about slights real and perceived that had been visited on his people. Of course, he always had a purpose in mind.

My goal, as per instructions, was to move beyond all that and get on paper the specific concessions that the Palestinians were prepared to make and implement. Their goal, it soon became apparent, was to do anything but.

At first I figured they were just fundamentally disorganized and incapable of doing graphs and opening up Microsoft Word so they could start writing things down. Before long, though, I came to realize the Palestinians were simply concerned that anything they put on paper had a high possibility of getting leaked to the Israelis, and from the Israelis to the media, before anyone ever got to Wye. That would mean trouble in their own communities for having made concessions, but from their point of view, it also was imprudent for them to commit to anything, on paper or in face-to-face negotiations, before they had seen the color of Israeli money and knew what reciprocal concessions the Israelis were willing to make.

Four or five hours of hard jawboning didn't budge them from their position. They had no intention of showing their cards in advance, or even sitting at the table. I left the consulate that day uncertain what, if anything, the Palestinians might show up with, but at least they understood we were serious about getting the job done.

My second appointment was more successful, or so it seemed at the time. Dennis had also asked me to meet with Ami Ayalon, chief of Shin Bet, the internal Israeli intelligence service. Dennis worried that Netanyahu was for political reasons going to demand security requirements that went beyond any reasonable standards. A retired Israeli navy admiral, Ami was a real straight shooter—and we could count on him not to play games. For our get-together, he was accompanied by one of his deputies, Israel Hassoon.

In our first Israeli-American meeting at the U.S. consulate, I saw hopeful signs. If Ami said the Israelis were prepared to negotiate security issues in good faith, and if he believed that the concessions we were urging the Palestinians to make would be acceptable to Israel, then Wye might really prove to be a turning point. That's basically what Ami told me when I saw him—a good omen, except that he also told me he was not going to be a part of the Israeli delegation at Wye. Dennis later theorized that Netanyahu wanted to leave him at home because Ami, like Rabin, just couldn't lie. Physically, both men were incapable of it. You can't play team poker when your partner can't put on a poker face. For his part, Ami explained that he didn't want to get involved in what was sure to become political theater. I shared that feeling, but it was strange to think of negotiating security arrangements without the chief Israeli security official in the room.

By October 15, 1998, when everyone had gathered at Wye River, Ami Ayalon seemed to be about the only person who wasn't there or on his way. Benjamin Netanyahu and Yasser Arafat headed their delegations, of course, but the second tier were also key players in the peace process. Abu Ala, Abu Mazen, Saeb Erakat, Jabril Rajoub, and Mohammed Dahlan were there along with Arafat.

In addition to Ariel Sharon, the Israelis had Shlomo Yanai, the chief military planner, and Meir Dagan, Netanyahu's counterterrorism advisor; Gen. Mike Herzog, head of the Israeli Defense Forces' strategic planning division; and Gen. Amos Gilead, a superb intelligence officer. Israel Hassoon showed up to represent Shin Bet, and he ended up being one of the unsung heroes of the entire affair.

In addition to the president, the U.S. team included Sandy Berger; Secretary of State Madeleine Albright; Dennis Ross; Martin Indyk, assistant secretary of state for Near East; Stan

Moskowitz, one of CIA's senior officers in the Middle East; and Gemal Helal, the State Department interpreter. Vice President Gore showed up on Sunday afternoon for a few hours to add his presence as well.

Naturally, such a distinguished assemblage gathered for so momentous a purpose attracted a huge opening press conference, held in one of the large meeting rooms. I chose to sit upstairs and wait it out. Although I'd grown accustomed to my role in these negotiations, I still wasn't comfortable with such a public display.

I was on hand, though, later in the week, for what to me was the most emotional moment of the entire event. At President Clinton's urging, King Hussein and Queen Noor of Jordan flew in from the Mayo Clinic, where the king was being treated for cancer. The king gave a poignant speech, urging both sides to listen to each other and be prepared to make concessions to the greater goal of regional peace. That alone would have been riveting, but the fact that the king had made this effort while he was so clearly struggling for his life—he'd lost a great deal of weight and all his hair, even his eyebrows, to chemotherapy—bathed the moment in emotion and heroism.

But this was Bill Clinton's show from the beginning. The president was someone who loved to try to solve big problems, and they don't get much bigger than this one. But there was more to it than that, more to it even than regional security and humanitarian concerns. Finding a solution to the Israeli-Palestinian issue might have had a major impact on the conditions that promote Middle Eastern terrorism. Clinton understood what was ultimately at stake, and he had been working his entire presidency on finding a solution.

As he always did, he had read up extensively on the issues. It was incredible how much detail he was steeped in and how easily he could call it back up. And he had no intention of letting this meeting fail, however long it took. Late at night, sometimes at two or three in the morning, you could hear Clinton's helicopter lifting off for the White House, where he would work until dawn on budget issues. In the mornings would come the thump-thump-thump of the helicopter returning. I have no idea when he slept, or how long he might have gone without sleep altogether. We arrived at the Wye Center on a Friday expecting to be heading home no later than the end of the next Monday. By Tuesday, with no end in sight, I had to start scrambling for clean clothes.

My part of the operation didn't exactly race along either; in fact, it was a roadblock in the entire process. Without a security arrangement, the political side of the equation was never going to fall in place; and without something hard and fast on paper, from both sides, we were never going to get there.

As I had a few days earlier in Jerusalem, Dennis spent hours that first Saturday trying to get the Palestinians to commit to the plan we had laid out for them. Meanwhile, the Israelis sat and stewed, waiting for Dennis to give them equal time. By the first joint session late that afternoon, Netanyahu and his gang had worked themselves into a fine and deeply suspicious lather, and I still had nothing concrete to show them from the other side.

The script had called for this to be a very small meeting, just a few principals from either side. I was going to walk in, say something like "Here's the security piece, just waiting to be signed, sealed, and delivered." Instead, eight or nine people showed up from both camps—the room was jammed—and Netanyahu was having none of it.

"Look," he said, "as much as we like and trust you, we haven't seen the substance of the security plan. You have seen it, but we haven't seen anything. So how are we supposed to know? This is our security, not yours." I couldn't argue with that. He was right, so I told him, "Bibi, I will work this. We will go and do it." And that became my life, day and night, for the next five days.

Odd memories persist from that time. I can recall chatting with Meir Dagan, the Israeli counterterrorism advisor, during a break in the negotiations. I asked if he knew Gen. Amin al-Hindi, the head of the Palestinian external security service. Meir looked straight at me and said, "I know Amin al-Hindi. I chased him around the West Bank for two years trying to put a bullet in his head." "Well," I told him, returning his smile, "he is on the other side of the room. You could end the whole thing right now. Just go over and pop him." Happily, he understood that my suggestion was a rhetorical one only.

During another break, at a time when I was deeply frustrated by the fact that no progress was being made, I ran into Mohammed Dahlan. "Let's go to the game room," he said to me. "I'm going to teach you how to play Israeli jailhouse pool." "What is that?" I asked. "What are the rules?" "Oh," he told me, "it's simple. The guy who gets the most balls in loses." So for the next hour and a half, in the elegant Wye Plantation game room, the two of us worked our way around the pool table, doing everything we could not to get a ball close to a pocket. I never asked Mohammed exactly what lesson I was to take away from the experience, but it seemed to be a metaphor for the entire peace process. I think the game of pool was his way of showing me that committing on the security side would shift the pressure to the political arrangements, but neither the Palestinians nor the Israelis were anxious to get there.

Dahlan could be problematic himself. In the Palestinian manner—and the Israeli one, for that matter—he was prone to long tirades. That's where the Shin Bet representative, Israel Hassoon, came in so handy. When Dahlan was just about to go round the bend, Hassoon—in a singsong voice that steadily but slowly increased in volume—would begin to repeat and stretch out Dahlan's familiar name: Abu Fahdi, Abooo Faaaahdi, Aboooooooo Faaaaaaaaaahdiiiiii. Then Hassoon would speak to Mohammed in Arabic in a hushed tone, and suddenly we were back on track. The effect was often amazing, but the entire process was, too. Every word, every gesture, every parry and feint and thrust sometimes seemed to have been scripted thousands of years earlier.

One evening, desperate to escape our confinement, Stan Moskowitz and I sneaked off to town to watch a Yankees–Cleveland Indians game in the American League Championships. We clandestinely "exfiltrated" ourselves off the plantation and went to a nearby hotel, where other CIA folks supporting the negotiations were staying. When we got there, I called Madeleine Albright. "Where are you?" she asked. "You cannot leave! Please come and get me . . ."

About that time, I received a handwritten note from my son, John Michael, then eleven years old. He'd scrawled on a card, "Hey, Dad, what's up? How have you been? I know how hard it must be trying to get them to sign a peace treaty. Just pray to God to help you because he is the only one who knows the answere [*sic*]. Have a good time. Get them to make peace, and come home soon. Love, John Michael." I remember showing the note to Abu Allah, and he asked me for a copy.

At the negotiations, while others used armored limousines and large security details to get to meetings, Stan Moskowitz decided that he and I should ride Schwinn bikes between "Palestinian-land" and "Israeli-land," as we called the large homes where the delegations were staying. He declared it much more efficient and fun.

During one ride, as the secretary of state's motorcade blew by us, Stan leaned over and asked, "How much if I can get Madeleine Albright on a bike?" We almost got Arafat to ride one. We were like two kids from Queens and the Bronx on the way to a stickball game, leaving skid marks everywhere as we approached solemn meetings. (Sadly, Stan's death in the summer of 2006 robbed us of a great intellect and a passionate proponent of peace in the Middle East. I miss him still.)

While I was in Jerusalem, the Palestinians had reached out to the Israelis with a specific work plan for the city of Ramallah. Now the Israelis expected the Palestinians to work out a detailed plan for the rest of the territories under their control and commit to laying down a specific ninety-day security plan that would operate indefinitely into the future.

At the opening trilateral session, Shlomo Yanai, dedicated to his country and a pragmatic and thoughtful man, stated that it was essential that Israel know this was a work plan and that it was being implemented. More than anything else, Yanai and the Israelis needed something tangible so that the Israelis would have real confidence that steps were being taken. Predictably, perhaps, Mohammed Dahlan expressed what would be a familiar refrain at Wye: that this Israeli requirement was humiliating and unfair. He said that dealing with Israelis was always some kind of a test and that passing one exam always led to another.

The opening discussion illuminated the crux of the problem. For the Palestinians, concessions and action plans against the military and civilian infrastructure of Hamas had enormous political implications. The lack of trust, and the possibility of leaks, would potentially cast Dahlan as an Israeli lackey. The high theater for everyone's benefit on Dahlan's part was not lost on the Israelis, and in particular Israel Hassoon, who understood Dahlan's dilemma. Yet not lost on us either was the absolute requirement for the Palestinians to act and ultimately be held accountable for what they did or did not do.

This is where CIA came in. We were the one entity both sides could trust. But there had to be a work plan and there had to be measurable time lines for bilateral cooperation to have a chance. Incrementally, though, progress did get made. By Wednesday morning, after nearly five days of head-butting, we finally had a draft agreement nearly in place, and that was when the Israelis decided to play hardball. They put their bags outside, signaling that they were going home. The security draft wasn't acceptable, they were suggesting. Without that nothing would get done, so why stick around?

Dennis Ross, for one, wasn't impressed. "Okay, call them on it," he told Madeleine Albright. "Ask them what time they want to leave. We'll make all the arrangements." Dennis's belief was that when people put their bags out, they don't intend to leave. If they did, Netanyahu would be the loser, not the Palestinians—the one who walked away from a historic opportunity for peace.

I wasn't so sure, so I went in search of Yitzhak Mordechai, the Israeli defense minister, who had arrived at Wye less than a day earlier. Mordechai was a serious man who instinctively distrusted the showmanship of populist politicians. Madeleine Albright had told me to ignore the luggage, but I could barely get into the building without tripping over it, so I called out to some sheepish-looking Israelis standing nearby, "What's with the bags? You guys going somewhere?" Then I found Mordechai and asked him to take a walk. "Here's where we are and what we have," I told him, and went on to lay out the security negotiations to date. "Look," he said, "I'll go talk to them. I will get us to 'yes.'" And with that, the bags returned to the rooms and we got back to business. Perhaps the Israelis were just doing some sort of scripted, good cop/bad cop routine, but whatever the backdrop, it worked. Mordechai was critical to the final stages of the security negotiations, including some concessions that put us over the top.

At last, on Wednesday, October 21, at a 6:00 P.M. meeting, a deal was reached.

Days of negotiation followed. In the end the Israelis agreed that a thirty-day plan would be developed in the field jointly between Palestinian and Shin Bet officials, that it would be coordinated within seven days with Chief of Staff Mofaz and Director Ayalon, that all Palestinian entities would have to adhere to the plan—an important point for the Israelis, as Dahlan could not speak for the West Bank—and that cooperation would be continuous. Finally, CIA agreed to host biweekly trilateral meetings to assess implementation, enhance communication, and help the two sides overcome obstacles. I then asked Defense Minister Mordechai if the agreement meant that the security file was closed. The defense minister said yes.

The Israelis were taking an enormous risk, betting that the Palestinians would fulfill their obligations. Mordechai had been indispensable in selling the agreement to his political leadership. The Palestinians needed our help in building their security. We agreed to do so. But in return, I said to them, "At the end of the day, only one thing matters—performance. The credibility of the CIA is on the line. There will be no second chances." We all seemed to be on the same page on security issues, but there was one final matter yet to be resolved: Jonathan Pollard.

Jonathan Pollard had been convicted in 1986 on one count of passing top-secret material to the Israelis while working as a navy intelligence analyst. He was then (and still is) serving a life sentence at a federal prison in Butner, North Carolina. Many people in the intelligence community believed that Pollard hadn't been motivated by love of Israel alone. There were indications that he offered to spy for other countries as well. But many Israelis considered Pollard to be a soldier, and this was the Israeli ethos—leave nobody on the battlefield. It was understandable on one level, but I was still shocked to hear Pollard's name arise in the middle of these negotiations. We were there to broker peace, not to pardon people who had sold out their country.

Martin Indyk recalls that Pollard came up at the first meeting President Clinton had with Netanyahu at Wye. I was not at that meeting. After the session, according to Martin, Sandy Berger asked the president whether Bibi had raised the issue of Pollard. The president said yes, and that he had told Bibi he would deal with that at the end.

On Tuesday evening, the president had asked Dennis Ross how important Pollard was to Bibi. Dennis felt that Pollard could be released but that he should be saved for the final negotiations—some months or years ahead. Ross told Clinton he thought he could get this deal without Pollard.

On Thursday, Sandy Berger called a session that included me, Dennis Ross, Madeleine Albright, and some others, and that's when Sandy dropped what for me was a bombshell. "You need to be aware of the fact that Netanyahu has put Pollard on the table," he said.

"No," I responded. "You're wrong. Pollard is not on the table." And with that I got up and walked out of the room. Sandy followed me out. "This is ridiculous," I told him. "Pollard has nothing to do with what we are doing here."

"Look," he said, "the president hasn't agreed to anything, but I promise to give you a shot at the president if the Israelis put this back on the table."

I talked the matter over with Stan Moskowitz, who was just as alarmed as I was about the possibility of the Israelis using our legitimate desire for peace to spring Pollard. Then I stewed over it myself for a few hours until I knew what I had to do. I'd just negotiated the security arrangement. If Pollard were included in the final package, no one at Langley would believe I hadn't had a hand in that, too. In the margins, the deal would reward a U.S. citizen who spied on his own country, and once word of that got out (and that would take a nanosecond or two), I would be effectively through as CIA director. What's more, I should be. I would have no moral capital left with my troops. Better to go out on my own, first, especially when I felt so strongly about the issue.

Finally, I called Stephanie, to be certain I was doing the right thing.

"You're right," she told me after I had explained the situation and told her I was going to resign if the president wouldn't hold the line. "Stick to your guns."

About midnight that Thursday, Madeleine came up to me and said, "If you're going to say anything to the president about Pollard, now is the time to say it."

"Why?" I asked, but she just repeated herself.

"If you've got something to say, say it now."

Madeleine was absolutely critical here; she knew a terrible deal when she saw one and she knew that releasing Pollard would put me in an impossible position. As soon as Madeleine was gone, I cornered Sandy and told him I needed to see the president alone.

"What do you want to talk with him about?" he asked. Sandy sounded agitated, but that might have been the strain of the summit, not my request. Everyone's nerves were getting a little raw by then.

"Pollard," I told him.

Within the hour, I was led into a back room where the president was waiting—just the two of us alone. I'd seen Bill Clinton plenty of times by then, in Cabinet meetings—although I attended only those that dealt with national security—at Camp David, during the Peacemakers summit at Sharm el-Sheikh, and other places. We had a good professional relationship, but nothing had prepared me for this. I was flying solo now.

"Mr. President," I began, "I just need to make you aware of something. We've done a security agreement here that I think is important. As a result, I think the negotiations may succeed, but if Pollard is released, I will no longer be the Director of Central Intelligence in the morning. This is an issue that has nothing to do with this set of negotiations."

I can be an emotional guy. But I was very calm at this moment, very matter-of-fact. I knew what had to be done. "I've worked very hard to restore morale at the agency," I continued. "I think our efforts are paying off, but I also just negotiated this security agreement. Everyone knows that. If a spy is let out as a consequence of these negotiations, I will never be able to lead my building." I went on to say that other people needed to be consulted here—the attorney general, for example—but the bottom line, I said, "is that it's just the wrong thing to do. I just want you to know that I appreciate the fact that you've allowed me to serve and I appreciate the opportunity you've given me, but I won't be your CIA director in the morning."

When I was through, the president thanked me, and I walked out of the room uncertain whether I would still have a job come morning.

The talks meanwhile continued through the night—the president really was indefatigable. My part in the deal-making was officially done, even if my stake in the deal was, at least to me, larger than ever.

At six that morning, Stan and I were sitting in a small room off the main negotiating area with some of the Israeli and Palestinian participants, including Bibi Netanyahu and Mohammed Dahlan, when the president came walking in with Arafat and led him to Netanyahu so they could shake hands and seal the deal. After a round of congratulations, everyone began filing out of the room.

Stan and I were the last ones remaining when Dahlan turned at the door and said, "There will be one more thing."

No, we told him, it was done. Didn't he see the handshake?

"You wait," he said. "The Israelis always want one more thing."

That, of course, is exactly what the Israelis say about the Palestinians, but in this instance, Dahlan was correct.

When we walked into the big dayroom next door, Netanyahu was sitting in the corner, in an obvious funk, with Clinton talking to him. Finally, the president came over to us and said, "We have a problem. Netanyahu still wants Pollard."

Dennis Ross would later tell me that he and the president went off to the bathroom to have a private conference after Netanyahu had tossed in the Pollard monkey wrench again. According to Dennis, he asked the president if he had promised Pollard to the Israelis. Clinton said no, but reading between the lines, Dennis believes that the president had all but walked up to that point.

"You don't have a choice," Dennis remembers telling the president. "If you *promised* Bibi you would release Pollard, then you have to release him. But this agreement is too good for Bibi to give up. Hang tough, and we will get a deal."

According to Indyk, the president met with Netanyahu one more time and told him that he would not be able to give him Pollard because the Director of Central Intelligence would resign. Netanyahu said in that case the deal is off. As we would soon learn, the story had already leaked and the Israeli press was reporting that Netanyahu would be bringing Pollard home with him on the plane when he left for Israel. Martin remembers an

Israeli journalist calling him and asking if it was true that Pollard was going to be released. No way, Martin said.

Somewhere in this same time frame, Yitzhak Mordechai broke ranks to come over and sit next to me. "You know," he said, "we really must have Pollard."

"Mr. Minister," I answered, "with all due respect, it's inappropriate. Let's flip sides here. Put yourself in my position, and I think you'll see that this is just not something I will ever agree to. If the decision is made over my head, there's nothing I can do about it, but there is no budge from my position."

I went back to my room after that. Clearly, I had become a more prominent player in these negotiations than I had ever expected. I knew I was right but, still, I felt uncomfortable.

I wasn't alone for long. I was barely settled in before John Podesta, Clinton's chief of staff, called. John was not pushing, just delivering a message. "The vice president asked me to phone you," he began. "Do you know how important this agreement is?"

"Yes, I know it's very important."

"Well, the Israelis won't sign unless they get Pollard."

"John," I told him, "this agreement is in their interest. They will sign it. Do not give them Pollard." Just so there could be no misunderstanding, I repeated my position. "If you give them Pollard, I'm done, but you don't have to. They will sign this agreement because it is in their interest. Just hold fast."

I was confident that my position on Pollard was the correct one—but that didn't stop me from feeling an enormous amount of self-imposed pressure. What if I am the reason this whole peace process collapses? I thought. I took a stroll with Dennis Ross along Wye's boardwalk and told him that I didn't think I had any choice other than to adopt the stance I was taking but that I was really worried about becoming a human roadblock to peace. "Don't worry," Dennis said to me. "In the end we will get the deal."

News of Pollard's supposed release spread quickly outward from the Israeli media. Before long the White House started getting heat from all kinds of people, including then House Speaker Newt Gingrich, who called the president to oppose Pollard's release. This cemented the president's determination not to release Pollard.

I do know that when Stan Moskowitz and I next saw the president, he clearly had made up his mind. Instead of sidestepping the subject, he put his arm around Stan, looked at me, and said, "Why don't we swap Stan for Pollard?" he joked.

And of course the Israelis did do the deal, just as Dennis and I were convinced they would. This was a game of chicken; Netanyahu and company were holding out to the last minute to see if we would blink. The Palestinians signed on, too. The Wye River Memorandum, as the final agreement became known, was as much in their interests as it was in the Israelis', and for a precious short time, we could congratulate ourselves on a job well done.

I passed up the Wye River signing ceremony that Friday afternoon in the East Room of the White House. I didn't think it was any more appropriate for the Chief Spy to be seen there than it would have been for me to show up for the photo session at the start of the negotiations.

The day after the signing ceremony, Stephanie and I had a private lunch with King Hussein and Queen Noor at the house they kept on River Road in Potomac, not far from my own house but a thousand real estate zones removed. "I'm really proud of what you did in that negotiation," the king told me. But for me, it was the king who deserved congratulations. His appearance at the negotiations had been heroic, given his failing health. King Hussein died three and a half months later. About a month before the king died, I had flown to see him at the Mayo Clinic. Stephanie had given me some holy oil from the Church of the Nativity in Bethlehem with instructions to pass it on to Queen Noor and let her know that we were praying for a miracle. Before he died, the king went to the effort of sending Stephanie a touching letter of thanks for her gesture.

When I was with King Hussein, I always felt that I was in the presence of wisdom and history, and yet when I met him for the first time, at his own palace, he had come up to the car I arrived in, opened the door himself, and said to me, "Good morning, sir. It's good to meet you." For a guy from Queens, having a king call him sir made quite an impression. I was forty-two years old then, new to my work, a rookie in the presence of a legend. In the years since, I've often wondered what impact his wisdom would have had in helping all of us avert the mess we find ourselves in today.

A few months after Wye, the *New York Times* came out with a story that all but quoted my conversation with the president at Wye, including my promise that I would resign if Pollard walked. I was in the middle of one of Washington's great dining experiences, at L'Auberge Chez François, in Great Falls, Virginia, hosting a raucous dinner with a bunch of visiting Australian intelligence officials, when someone called from Langley to say that the White House wanted me to deny the *Times* story. "No," I remember saying. I told my spokesman, Bill Harlow, to simply say, "No comment."

Was this the peace to end all peace? Hardly. It was only a beginning, but the Palestinians were ready to act in a way they had not acted before. As a result of security cooperation, the instances of terrorism from 1996 to 1999 plummeted. The two parties deserve the lion's share of the credit, but CIA officers were critical to building and opening lines of communication. And the United States was diplomatically engaged as well. As Stan Moskowitz had said, CIA was nurturing trust with the Palestinians. Our diplomats were pushing Arafat, and he trusted us because they were also pushing the Israelis. Counterterrorism worked because security and diplomacy were joined at the hip. What CIA's role provided to our government was a basis to help intervene in the coming years, to give the political process the oxygen it needed to keep breathing.

CHAPTER 5

Beyond Wye

Between the close of the Wye summit in October 1998 and the end of September 2000, no terrorist attacks occurred inside Green Line Israel—an interlude in the violence that seems almost impossible to conceive of today. Then, on September 28, 2000, Ariel Sharon, the leader of Israel's opposition Likud Party, visited the Temple Mount in Old Jerusalem, home to the remains of ancient Jewish temples, as well as the Dome of the Rock and the al Aqsa mosques, and maybe the most contentious piece of real estate known to man.

Sharon's announced purpose was to look into complaints by Israeli archaeologists that Muslims were vandalizing the site, but he arrived flanked by a thousand Israeli soldiers and policemen, on the day after an Israeli army sergeant had been killed in a terrorist attack. About a day later, the Second Intifada began, and the peace process was effectively in shambles. Over the next half decade, roughly 950 Israelis would be killed, more than half of those in Israel proper and many in gruesome suicide bombings. Through the end of 2005, some 3,200 Palestinians would die.

It wasn't for want of trying that the Middle East peace process collapsed. I participated in three more major pushes for peace in the Middle East during the Clinton administration: the epic Camp David summit that got under way on July 11, 2000, and ran virtually nonstop for two weeks; the follow-up meeting in Paris that began October 4, 2000, less than a week after peace was shattered yet again by the outbreak of the Second Intifada; and the October 16–17 summit at Sharm el-Sheikh, co-chaired by Clinton and Egyptian president Hosni Mubarak.

The security arrangement we had hammered out at Wye River was always the foundation for these meetings and helped both sides to understand what reciprocal security really meant. The Palestinians and the Israelis created joint operation centers and began training people who could help enforce the peace and ensure compliance with the agreements. All the while, we were working to increase the Palestinians' operational capabilities to give them more credibility in the eyes of the Israelis so they could take action against terrorists in their midst. And for a critical two years what we had put together at Wye, and the work we had done to implement it, actually worked, perhaps not by the letter of the agreement but at least in spirit.

At CIA, we had taken on a public role with which many of us inside the building and many on Capitol Hill and elsewhere were distinctly uncomfortable. At a personal level, we all had poured vast amounts of energy into the challenge. Sitting in a room with Palestinians and Israelis isn't like sitting in a room with corporate department heads or even divorce lawyers. For starters, I knew, absolutely knew, that for the first three or four hours, we initially would have to listen to exactly what we had heard at previous meetings—a litany of grievances. That was the given, and we had no choice but to take it, knowing that at any moment maybe 40 percent of what we were hearing simply wasn't true. It was also a given that somewhere in the middle of the session there would be a family argument so heated that we feared that both parties were going to come to blows. That's just the way things were. The Israelis and Palestinians yell and scream at each other. There's nothing in the least Anglo-Saxon that happens in these negotiations.

I was on standby for the July 2000 Camp David summit. The security issues were not uppermost in the discussions at first. The talks had moved on to other issues and involved new players, at least on the Israeli side. Netanyahu was gone, replaced by Barak. Arafat, though, was still in charge on the Palestinian side and, as ever, was difficult, if not impossible to move. The principals involved had almost no leverage where he was concerned. Madeleine Albright had a love-hate relationship with the chairman that, by then, was more on the "hate" than the "love" side of the line. President Clinton might have moved him, but Arafat confounded even Clinton's best efforts.

On the surface, it was stunning just how much the Israelis were prepared to sacrifice in the name of reaching some sort of lasting accord and hard to understand why Arafat could say no. Yet CIA's assessment in advance of the summit was that while Barak was coming to Camp David to conclude a framework agreement for a permanent settlement, Arafat had no such intention. Arafat believed that he had a firm commitment from Barak to turn over three Arab villages near Jerusalem. When by mid-May it became clear that he was not going to get the villages anytime soon, Arafat concluded that he could not trust Barak to deliver on his promises. Barak's argument that his tenuous situation at home required him to preserve his political capital for the final-status talks rather than spend it on a series of interim steps did not hold water with Arafat. The chairman had come to the summit because he did not want to insult President Clinton. But without a return of villages and Israeli flexibility, he would wait out the current effort.

Ten days into the talks my standby status changed. A worried Madeleine Albright called and asked if I would come up to Camp David on the afternoon of July 22 to try to persuade Arafat to negotiate on the basis of Barak's plan. Geoff O'Connell, who was Stan Moskowitz's successor, and I huddled with a despairing Albright and the peace team in her cabin. She told us that the negotiations had more or less collapsed after that famous photograph of Barak and Arafat urging each other to go first as they entered the president's cabin. In fact, neither Arafat nor Barak had met with each other since. Albright asked me to visit the chairman and try to persuade him to come back to the table.

I went to Arafat's cabin and told him that the Israelis would never again extend an olive branch like this. I reminded him of how much the president had done to move the peace process forward. "Now," I said, "you have to come back to the table." I asked him directly if he was willing to negotiate. If not, it was time for everyone to go home. To my surprise, the chairman immediately agreed, saying that he was ready to consider anything the president put before him. The whole conversation lasted about fifteen minutes, and we were shortly back at Albright's cabin.

Obviously expecting the worst, the secretary was stunned but energized by the news. She ordered us back to Arafat's cabin and had the State Department's top Arabic interpreter, Gemal Helal, accompany us to ensure there was no communication problem. Back we went, and Arafat again pledged to negotiate, but this time with an important caveat: he could never compromise Jerusalem's status. He went on at considerable length about the Armenian community, its desire to be part of a Palestinian state, and the need to bring an Armenian representative to Camp David immediately to participate in the talks. In retrospect, he was laying down a marker that would allow him to say no.

The rest of the day was spent shuttling between the Palestinians and the Israelis. We felt we were close to a deal on most of the security issues. Albright hosted a dinner that evening and invited both Arafat and Barak. To our surprise, Barak refused to attend. We later learned that he had retreated to his cabin shortly after the first day of the talks and had not come out since, except for solitary walks.

After a few hours' sleep, we returned to Camp David and took part in a long round of bilateral and multilateral security discussions. The president was expected back at Camp David around 3:30 P.M. Albright ordered us all to meet and pull together what we would tell him. Shortly before the meeting, we got together with Mohammed Dahlan and Shlomo Yanai, who had been hammering out the details of a security agreement. There were six issues: early warning, air space, emergency deployment, demilitarization, counterterrorism, and the Jordan Valley. Both Dahlan and Yanai told us that their discussions were going well, and they outlined their proposed solutions. While there were some minor differences, they were confident that they could resolve them before meeting with the president. I relayed that back to Albright, and she gave the president a very encouraging report after he arrived.

The president convened the negotiating session and, to my surprise, remained in the chair leading the effort until the meeting ended in the middle of the night. He began by telling the group, "We have a lot to do. Let's go through the agenda as quickly as possible. Where there is agreement, we will move on and concentrate on where there is disagreement. Everyone should operate on the basis of two assumptions:

— No one is bound by anything they say without a comprehensive agreement.
— Let's assume that we can ultimately reach a deal on who controls what territory."

Shlomo Yanai opened the discussion by reviewing Israel's need for early warning sites on Palestinian territory from which they could detect border intrusions. Yanai outlined a proposal for setting up three early warning sites. Yanai's proposal closely matched what he and Dahlan had told us was acceptable earlier in the afternoon. Clearly anticipating a positive Palestinian response, Yanai turned the floor over to Dahlan.

Dahlan opened by complaining that all the agenda items were Israeli. He told us that the Palestinians had their own requests. They would not raise them now, but he reassured us that he thought the Israelis were capable of meeting them. Dahlan then stated, "We said we *understood* the Israeli need for early warning sites. We did not say that we *agreed* with them." Uh, oh, I thought, something had happened in the three or four hours since our meeting with Yanai and Dahlan.

The rest of the session followed that script. Yanai would propose a solution, and Dahlan would object. The president did a magnificent job trying to bridge gaps and come up with creative ideas to resolve differences. When we broke to get some sleep, I thought we were close to an agreement. Again, I made the long drive to Washington, but shortly after getting to bed, I was summoned back to Camp David. By the time I arrived, the talks had collapsed. Eventually the parties went home empty-handed.

In October 2000 the various parties reconvened in Paris. By then the Intifada, the Palestinian uprising, was a week old, and we were trying to come up with something dramatic to stem the violence. Madeleine shocked me by turning my way early in the meeting and saying, "You take this over." Reluctantly, I did. I mentally ran through my talking points, and in fairly short order we came up with ten steps that needed to be taken—ten steps that both sides agreed on, a big breakthrough. While Dennis Ross went off to summarize the ten steps and get them on paper, Arafat left to visit French president Jacques Chirac, and everything started to go wrong again.

With Chirac, the Palestinian chairman seized on the most controversial of the ten points—an investigation into the causes of the Intifada. In our meeting, both sides had accepted an American-led tribunal, with input from the European Union, but Arafat pressed Chirac for an international court, a show trial with a stacked jury that Israel would never agree to. Chirac backed Arafat, and we were at stalemate yet again.

Barak didn't even bother to appear a little over a week later at Sharm el-Sheikh for the summit hosted by Clinton and Mubarak. Egypt occupies a unique position in the Middle East. The Saudis make the same claim, for cogent reasons, but Cairo, not Riyadh or Medina or Mecca, is the intellectual capital of Islam. Egypt is a nation of some seventy-five million people, three times the population of Saudi Arabia, with a gross domestic product four times the size of Syria. That alone would make it important, but like Saudi Arabia, it also sits at a crossroads of international terrorism. The Muslim Brotherhood was born in Egypt; Anwar Sadat was assassinated there. Egypt, allied with other Arab countries, has fought four wars against Israel, in 1948, in 1967, in 1968–1970, and again in 1973. It's still the country that Palestinians most look to, however forlornly, as their protector.

Umar Suleiman has been head of the Egyptian intelligence service for many years. A general as well as an intelligence chief, Umar is tall and regal looking, a very powerful man, very deliberate in his speech. He's also tough and engaging. In a world filled with shadows, he is straight up and down. Umar has also done as much behind the scenes as anyone else I can think of to try to bring peace between the Palestinians and the Israelis. That was true when the United States was still engaged in the process. It's even more so now that we are long gone from it. When nobody was trying to go see Hamas, when nobody was talking to the Palestinians, when nobody was talking to the Israelis, when nobody was pushing forward with innovative ideas to try to get people talking to each other, Umar was on the ground taking risks.

I didn't know Hosni Mubarak as well, but he has been one of our most reliable partners in fighting terrorism and in trying to bring peace to the Middle East. Ours wasn't a peer-to-peer relationship. He was a very important historical figure. He had been president of Egypt since 1981, following Sadat's murder. He barely escaped assassination himself in 1995, while in Ethiopia, four years later, he escaped death again when he was nicked by an assailant's knife. He has a tremendous amount of wisdom, but although a serious man, he also had a lighter side. The October 2000 summit at Sharm el-Sheikh was an example. Umar Suleiman and I had spent the entire day locked in a room with the Palestinians and the Israelis trying to strike a security bargain. When we were through, I went off to brief Yasser Arafat on the details, while Mubarak drowsily took a seat in the corner of the room. Arafat had a way in these circumstances of looking at me as if I were speaking in an incomprehensible foreign language. This was typical of him; he was buying time to think things through. But on this occasion, the situation was not business as usual. From the corner of my eye, I saw Hosni Mubarak, the president of Egypt, host of the conference, and the closest thing Palestine had to a guarantor, looking at me and Arafat and twirling his finger beside his head, the universal symbol for "This guy you're talking to is nuts!" I went on with the briefing—I am a trained professional, after all—but it wasn't easy, especially when Mubarak dissolved into quiet laughter over his little gag.

Trust with Arafat was always problematic. Particularly in the last year of the Clinton administration, he saw how desperately the American president wanted peace—for humanitarian and strategic reasons, and to establish a legacy. Arafat always wanted one more thing, and one more thing was never enough because what he really wanted was for the peace process to be ever-active and eternally unresolved. Keeping the process going gave Arafat leverage. Walking up to the edge of agreeing and then backing away made him a central player on the world stage. It stamped him as legitimate. His own people would see him splashed all over CNN. And he loved having CIA right in the middle of negotiations. In the Middle East, CIA is a powerful talisman. He got what he could from us, and from that point on gave little back.

When the Bush administration came to power, they did not hold Arafat in high regard. The Clinton team had made him a central part of the peace process. Yet Arafat could never get the deal done. Therefore—and it was a view I supported—there would be no more letting him in the front door. No more conveying the image of him as a global player. No more reward for behavior that led us nowhere.

As the administrations changed, my role, and that of CIA, in the negotiations between the Palestinians and the Israelis changed, too. The Bush administration also had a more traditional, and perhaps more appropriate, view regarding CIA's involvement. They clearly weren't comfortable with the Agency's filling the semi-diplomatic function we had taken on over the last few years. They wanted to bring it under their own roof. I did, however, make one last effort at the administration's behest. In early June 2001, I flew out to Amman, Cairo, and Tel Aviv. I don't think the Bush people expected much to come out of my trip—to them, it was more like a duty call—but after a week of intense negotiations and constant shuttling from capital to capital, we managed to produce what became known as the Tenet Security Work Plan, a very clear, very straightforward timetable that laid out the steps both sides had agreed to take to strengthen the security framework.

And that, too, like so much else, was never implemented. Dennis Ross was gone by then. There was no attempt to replace him with someone else whose job was to think about this issue day and night, and thus there was very little push on the political side. Colin Powell had flown out in late June to try to get something moving politically, but despite his best efforts he was unable to succeed. Once more, we had edged up to a workable cease-fire, and once more, it had withered and died before it could ever take root. In the absence of a political process, this was inevitable. Soon afterward, I made a determination that there was no role for us to play anymore. As I always saw it, our part in the process was to be an honest broker, but after June 2001, there was nothing left to broker honestly. Better to retreat, protect our institution, liaison with both the Israelis and the Palestinians, report accurately and honestly to all sides what was happening on the ground—the classic work of an intelligence agency—and step back out of the light.

Or so we thought. During the spring of 2002, CIA found itself in the middle of one other highly public crisis. On April 2, some two hundred Palestinians, about fifty armed, broke into the Church of the Nativity, one of the holiest places in all of Christendom, while fleeing an Israeli Defense Force incursion into Bethlehem. The site is administered by a coalition of clerics from the Armenian, Roman Catholic, and Greek Orthodox churches and is built over what Christians believe to be Christ's birthplace. Barricading themselves in the Church, the Palestinians presented a terrible dilemma to the Israelis in what would turn out to be a very lengthy standoff. Many of the clergymen who worked at the site remained inside as "voluntary hostages," hoping that their presence might deter bloodshed.

Early on, the Israelis called on CIA's senior man in the region, Geoff O'Connell, and asked him to intercede with the Palestinians to help end the standoff. What made the situation especially dicey was that some Palestinian officials would have dearly loved for the Israelis to overreact, damage the holy site, perhaps kill the monks along with the terrorists, and stir up international outrage.

Geoff contacted a senior Palestinian official. Within a couple days they came up with a plan. The Israelis had given Geoff their bottom-line negotiating position—a handful of the most wanted men holed up in the church would either have to go on trial or be immediately exiled from Israeli- or Palestinian-controlled territory. With difficulty, Geoff got the Palestinians to agree to the exile arrangement. Then the Israeli side had a change of heart. Shin Bet officials apologetically told O'Connell that they were unable to complete a deal that they had previously proposed. To make matters worse, the Israelis asked CIA to back off and let European negotiators try to bring the situation to closure. Back off we did.

Over the ensuing several weeks Israeli snipers killed or wounded not only several Palestinians but also church workers who were mistaken for terrorists. The Israelis also cut off food entering the site. Before long, conditions inside were rapidly deteriorating.

After three weeks of getting nowhere, the Israelis came back to Geoff and said, "Look, we really need you to get involved again. We can't let this drag on much longer."

So Geoff reengaged with the senior Palestinian official while CIA officers entered the church and made direct contact with some of the Palestinians taking refuge there. Although Geoff briefed the Europeans at every step along the way, they were still unhappy that we were once again involved, supplanting their efforts. The Europeans had been dealing with the families of the men under siege in the church, failing to recognize that the real decision making was not with them but with Yasser Arafat and the Palestinian National Authority.

After much back-and-forth, O'Connell struck a deal once again. It looked like a happy ending. The Israelis started taking down the barricades around the church, but then it was Arafat's turn to renege. This situation exemplified the difficulties involved in bringing peace to the Middle East. Finally Arafat agreed to most of the elements of the deal, but there was still one sticking point: the weapons the Palestinians had taken into the church with them.

The Israelis quite naturally didn't want the Palestinians to leave heavily armed, just as they had arrived. But Arafat insisted that the Israelis could not have the weapons. Our theory was that he didn't want Israeli forensics to later show that these same weapons had been used in terrorist attacks. That would have handed Israel a PR victory.

Once again, O'Connell came to the rescue with an idea. "We'll throw the weapons in the sea!" he announced. For a while everyone thought that was a splendid solution. But once again the deal came undone. The Israelis wanted the weapons thrown in the Mediterranean, and the Palestinians wanted them thrown in the Dead Sea, closer to their territory. You can't make this stuff up.

Finally, O'Connell came up with plan B, or maybe it was plan C—the United States would take control of the weapons and hold them in perpetuity. All the negotiators present agreed, but it was up to their seniors to bless the concept. Geoff called me and had me track down Arafat. I reached the chairman in Egypt and congratulated him on the deal.

"Deal? What deal? I know nothing of a deal," he blustered, in typical fashion. In the end we convinced all sides that this was as good an arrangement as they were going to get, and after thirty-eight days, control of the Church of the Nativity was returned to its rightful owners.

I only wish our broader involvement in the peace process had met with similar success. Yet, however much I regret the outcome, I wouldn't have foregone the process itself. In all our dealings with the Israelis and Palestinians, we negotiated in good faith. When Israel asked us to back off, we backed off. When the Palestinians needed their hand held, we held it. Ultimately, I told both sides, the United States can't want peace in their region more than they do.

Once you got involved in the peace process, it was difficult not to be totally consumed by it. We had very deep bonds with the Israelis, who were like us in so many ways. The relationships that we developed with their intelligence professionals were deep and meaningful ones. They became personal. Dany Yatom, Efraim Halevy, and Avi Dichter would become lifelong friends. These were people I could rely on. These were people we could talk to. We had common motives and concerns.

At the same time, it was hard not to develop affection for the Palestinians. I understood that they wanted to put themselves in a better place. Politics and historical animosities were not things that the security talks alone were going to overcome. But my view was that if there was some way we could improve the lives of these long-suffering people, we should try it. Yes, it was an emotional environment. But there was enormous talent and potential on both sides. There was great possibility. It was never a matter of being pro-Israeli or pro-Palestinian. I was pro–both sides.

It is clear that both parties bear ultimate responsibility for the success or failure of the process. We cannot tell the Israeli prime minister what his security needs are. We cannot tell the Palestinian prime minister what his security needs are. But the United States, during this period and on this issue, occupied a special role. And that worked, not only to security and moral benefit but also to the benefit of the world at large.

Although our strategy was focused first and foremost on the Israelis and Palestinians, there were other dividends. It gave us greater legitimacy in the Arab world because we showed dignity and respect in dealing with the Palestinian people. It allowed us to show the Arab street that we cared about an issue that the Islamists and the terrorists used as a mobilizing grievance. Because we were seen as fair, doors opened for us. Not just with intelligence chiefs throughout the region, but also with heads of state, so that when we really needed their help, they would be there for us. That time was coming soon. Almost always, that last impenetrable barrier to peace had the same name: Arafat.

CHAPTER 6
Arafat

The one constant in the Middle East during my time in office was Yasser Arafat. From his first appearance on the cover of *Time* magazine in 1968 through his final years confined by the Israelis to his headquarters in Ramallah, until his death in December 2004 in Paris, Arafat was the face—for good and bad—of the Palestinian struggle.

His own security chiefs knew his limitations. Often they recognized the need for change; they understood there was no accountability built into the system. But it was clear to me that they would never break ranks with the Old Man, as we often urged.

Arafat was a hero of the revolution, the leader of his people. The one hard and unavoidable fact was that the peace process could not succeed without him, and he did not want it to succeed in any way acceptable to Israel or the United States. There were many times in the negotiating room that we all hoped he would disappear. Yet the moment he was out the door, we seemed to talk about no one else.

The Israelis knew Arafat. They knew him better than anyone else in the world, and the debate would always be: Who is he? Does he have a strategy? I was having a long discussion about this one night with Shlomo Yanai, then head of military planning for the Israeli Defense Forces. Shlomo is an old tanker; he'd been badly burned in one of the battles. A strategic thinker, he is someone whom I came to rely on for his integrity and forthrightness.

After much back-and-forth, he finally said, "Answer the following question: Is Arafat Moses or is he Ben Gurion?" Then he answered himself: "He's Moses. He will never do the deal. He will never sign an agreement. He will never compromise his position because he wants to take his people to the Promised Land. The Promised Land for Arafat is Jerusalem, and he will never concede." It was as insightful an analysis of Yasser Arafat as I'd ever heard.

Though the United States had long ago established relations with Arafat, it would be misleading to characterize them as friendly. After all, it was Arafat's organization that was involved in many terrorist acts in the 1970s and 1980s. Although he shared a Nobel Peace Prize in 1994 with Shimon Peres and Yitzhak Rabin, Arafat turned his back six years later on the best peace offer Palestine might get in our lifetime. There were times the man drove me nuts, and times when I wanted to hug him. He was far and away the most complicated person with whom I ever dealt. I never knew which Arafat was going to show up, but I always knew that whichever one did, there would always be a story to tell afterward.

One of the first times I met him was at a dinner at the Greek Orthodox archbishop's residence in Bethlehem. I was still the deputy DCI then, but I sat next to Arafat, underneath a painting of the Last Supper, in a room packed with guns. I remember looking at that painting, looking at my plate, contemplating all the religious tension that already colored everything in Bethlehem, and thinking, It's over. I'm done for. This might be my last supper, too.

The Palestinian on my right was someone I had never met before. Halfway through the meal, I turned to him and said, "So, what did you do before this?"

"I was in an Israeli prison for seventeen years," he answered.

"Why did you go to prison?" I asked.

"I blew up an Israeli school bus," he replied in a matter-of-fact manner.

This is going to be different, I recall thinking. You're not in Kansas anymore.

Arafat was very solicitous of me during the meal and even took food off his plate and moved it to mine, saying he was worried I didn't have enough to eat. After dinner I happened to mention that I was Greek Orthodox, and with that news Arafat warmed up even more. Apparently he had some affinity for the Greeks.

All of a sudden Arafat started rolling out gifts, insisting on photos, the whole grand host thing. In the years to come, he would get angry with me, or we would go at each other, but it never became personal between us and that moment of connection never faded. I would walk into Arafat's headquarters and there would be forty or fifty people all talking at the same time, yelling, laughing, telling lies to each other because they didn't want to hurt anyone's feelings by telling the whole truth, and I would think to myself, This is just like the Greeks I knew growing up in Queens.

The truth is, I love the Israelis—their passion for life, what they've done to stand up for themselves, and what they've done in establishing their state—but I bonded with the Palestinians as well. And Yasser Arafat was part of that. I couldn't keep myself from liking him. "Friend" is always an odd word when you hold a job like DCI. Maybe "marriage of convenience" is more accurate, but that doesn't fully capture how I felt about Yasser Arafat, either.

There were all the eccentricities, the unpredictability, the constant theater. To truly set Arafat off, all you had to do was say the word "Kuwaitis," and he would be gone. "Ah, the Kuwaitis, they can go to hell," he would say, "but not with my money!" I never knew what they had done to offend him; maybe he had an account frozen in some Kuwaiti bank. But he was never going to forgive or forget.

We used to have a pool among ourselves whenever we went to see Arafat over how long it would take him to say, "I'm still suffering," a constant refrain with him. We'd each pick a time and put our money down. Since I was generally leading the conversation from our side, I would keep a close eye on my watch and then, at just the right moment, ask him, "Oh, Chairman Arafat, how are you?" The answer: he was still suffering, always.

I remember the time the Israelis sent some low-level emissary along, someone we had never heard of. Arafat took one look at him, pulled himself up in high dudgeon, and shouted, "Can you believe they sent this *boy coffee* to see *me*?" We guessed he meant "coffee boy."

There was also the time we were at the U.S. ambassador's residence in Paris—in October 2000, at another conference negotiating a peace that we would never achieve—when Boogie Ya'alon, the deputy chief of general staff for the Israelis, called Arafat *rais*, which means "president." In front of Madeleine Albright and the delegations from all sides, Arafat went into a sudden rage. "You will call me *General* Arafat! I was the greatest general in the Egyptian army!" I didn't even know he was in the Egyptian army, much less a general or a great one. But I wasn't about to correct him.

Initially, the Bush administration wanted me to stay out of the peace process business and leave things in the hands of the diplomats. That was fine with me. But on June 1, 2001, there was a horrendous terrorist attack on a Tel Aviv disco called the Dolphinarium. Twenty-one young Israelis, mostly Russian immigrants, were killed by a suicide bomber. The carnage shocked the Israelis, and it appeared that the already ugly atmosphere in the region was about to get even uglier.

So, a few days later, I was dispatched to the area to see what could be done to revive peace efforts, trying to construct a workable security agreement that might allow the political process to go forward.

We were in the Israeli cabinet room, right outside Ariel Sharon's office, putting the final touches on a possible pact, when the Israelis began demanding a side agreement, some sort of cover they could hide behind if things went sour or, more likely, could leak to the press to sabotage the whole process.

"No sides," I told them.

"No deal," they said.

Over eight days, we shuttled between the two sides and put together what our team believed to be a fair proposal, resurrecting and enhancing old ideas and generating new ones that would have required tough actions against their own people.

The pact, called a "work plan," was a detailed list of specific steps that would lead to resumed security cooperation, enforce strict compliance to a cease-fire, suppress terrorism, and redeploy the Israeli Defense Forces to positions they had held eight months earlier. Among other things, it called for an immediate halt to hostilities, the arrest of terrorists by the Palestinians, an easing of travel restrictions imposed by the Israelis, and a pullback of Israeli troops. Eventually, after a cooling-off interval, the plan envisioned implementation of peacemaking suggestions laid out in April 2001 by the Mitchell Commission, a five-member fact-finding body led by former senator George Mitchell that looked into the causes of and possible solutions to the Intifada.

By the evening of June 11, our work was done and we convened one last trilateral meeting to make a final appeal for acceptance. I said, "Frankly, we are out of time. More innocent Palestinian and Israeli civilians continue to die. Israeli children who died last week were not soldiers carrying weapons. The three Palestinian women who died yesterday were not engaged in terror or violence. Courage and risk to stop all violence against your respective peoples must start tonight. There must be a return to normal life for the Israeli and Palestinian people. All these things can happen. They must happen. They will happen if you live up to your obligations in the work plan we have presented. But these words must be followed by actions that are embodied in the paper I have presented. The Palestinians must apprehend terrorists and provide transparency into their actions. The Israelis must not attack innocent Palestinian civilians. But in truth, I cannot feel this more than you. And Geoff O'Connell cannot preside over meetings that only result in words. I will not let him do this. We want to help you. Allow us to do that tonight by responding affirmatively so that we can begin tomorrow."

The next morning, the Israelis said yes. Then began the long wait for an answer from Arafat.

I traveled to Jerusalem, where I saw Arafat's principal advisors—Saeb Erakat, Mohammed Dahlan, Jabril Rajoub, and some others—around noon, told them the Israelis had agreed to the terms we had all been hashing over, and gave them until four o'clock to sign on as well. When my deadline passed with no response, I told my people at our hotel in Tel Aviv to tell the airplane crew to get ready and then to put our luggage out on the street. I had learned something from Bibi at Wye.

Then I called the Palestinians to say I was going home—no harm, no foul, but I wasn't hanging around to see what would happen. I was in the hotel dining room, preparing to leave, when I got a call from my friend Saad Khair, Jordan's intelligence chief, saying that if I went back to see Arafat, he'd give me the deal. Umar Suleiman followed that with another call; Mubarak also wanted me to go see Arafat. Jabril Rajoub chimed in as well: "Come back. The old man will sign."

So I went back up the hill to Ramallah.

Israeli security and military officials provided an escort from our hotel, but as always, they had to drop us off several hundred yards from Arafat's door, a no-man's-land of sorts that separates Israel from the Palestinian Authority. For that trip, my party and I climbed into our armored vehicles, with my security detail in the front and back cars and us in the middle. Going to see Arafat was often eventful. On a similar trip two days earlier, just as we entered Palestinian territory, we pulled around a curve and found a pickup truck blocking the road with its hood up and two Palestinians standing alongside. The setting was a textbook scenario for an ambush or assassination. What's more, two Israeli settlers had been killed in the area earlier that day when they inadvertently wandered into the wrong zone.

As my staff shouted at the Palestinian truckers and they shouted back, I wondered if we were going to be added to the day's death toll. After about thirty seconds of this, our Suburbans blasted over rocks that lined the side of the road and careened into Arafat's compound. Thankfully, this trip was less eventful.

When we arrrived, Arafat wasn't at the door to greet me as he usually was—a bad sign. The expression on his face when I got inside augured even worse: the same look my mother used to give me when she was really, really angry.

Arafat continued glowering for a while, and then said, "I have to have a side agreement with you about this agreement."

"No," I told him. "Sharon wanted one, too, and I told him he couldn't have one. I'm going to treat both sides equally. Besides," I said, "you are going to leak it to the press and ruin the deal."

When I was through, he looked at me, smiled, and said, "That's right." Almost immediately he said, "Okay, no side agreement. But I want to write you a letter."

"Mr. Chairman," I answered, "I think the cease-fire agreement you have is important and fair—but, I cannot want it more than you do. If you do not want to take the deal as is, I will go home. And I do not want a letter!"

Arafat continued to insist on a letter. After spending five minutes going round and round over it, Geoff O'Connell said, "If the chairman wants to write you a letter, he can write you a letter. After all, he is the president of the Palestinian people."

Of course, Geoff was right. At that moment, it looked like Arafat wanted to kiss him. I wanted to throttle him. I knew we had just guaranteed several hours more of painful dithering.

There were just three Americans in the room: Geoff, John Brennan, one of my most senior advisors, and me. Arafat had only two aides with him, and they began discussing what might go into the letter. With each draft paragraph, Arafat would retreat to the next room, where he had twenty or thirty advisors sitting. I heard lots of shouting.

"What's going on?" I asked John Brennan, who speaks Arabic.

"Nothing good," he told me.

While Arafat's side was yelling at each other, I got on the phone and updated Bill Burns, the very able assistant secretary of state for the Middle East, and Jonathan Schwartz, a senior State

Department lawyer who helped ensure that nothing I agreed to was inconsistent with U.S. policy or other agreements we were party to.

We negotiated three paragraphs this way. Finally, I thought we were done. After the third paragraph of the letter had been completed, Arafat walked in and said, "I want one more thing." I objected; the bazaar was closed.

We were in the middle of one of these exchanges when a burst of automatic weapons fire rocked the headquarters. After a quick exchange of furtive glances between the chairman and his lieutenants, Arafat and his aides said in virtual unison, "Celebrations. Don't worry. No danger. People are celebrating something." Earlier in the day, effigies of Bill Burns and me had been burned in the streets of Ramallah.

At last, around two in the morning, we were done, or seemingly done. Arafat sent the three-paragraph letter out to be typed, leaving me alone in his office with John and Geoff. By then, my back was killing me, so I lay down flat on the floor. That's where I was when the chairman walked in, saw me, said, "Oh, I do this for my back when it hurts as well," and proceeded to lie down next to me and started talking, with his nose about two centimeters away from mine. I could see Brennan and O'Connell thinking, Oh, great! Get off the floor before the cameras show up!

Finally, the freshly typed letter, sealed in an envelope, appeared at the door and was handed to John Brennan. I didn't trust Arafat's typist and kept trying to make eye contact with Brennan to silently signal him to open the letter and read it.

He was as exhausted as I and wasn't getting my message. So I finally blurted out, "John, open the damn letter and read it!" He did and found it to be what we were expecting, except that my name was misspelled. Arafat wanted to have the letter retyped, shouting at his staff about the error and insisting that the next version include the salutation "My dear beloved Director Tenet."

That was the last thing I wanted to take back to Washington, especially after we had been whispering sweet nothings on the floor a short while before, so I insisted that we take the letter as it was and head back to our hotel. As we jumped into our vehicles I called Steve Hadley at the NSC to report what we had done and then called Stephanie to let her know that after an arduous eight days, I'd soon be heading home.

En route to Tel Aviv, we also learned that a Greek Orthodox monk had been killed on the West Bank that evening. Sadly, people get killed all the time in the Middle East, but my conspiratorial mind caused me to wonder if this was intended as a message to me.

The next day we hosted a trilateral meeting not far from the Dolphinarium disco itself. Unaware of the excess drama, President Bush called me the next day from Air Force One to offer congratulations. But as so often happened with the Palestinians and Israelis, the political side of the equation didn't keep pace. A little more than a week later, the whole deal came apart, another roadside ruin on the bumpy path to peace.

I was among the last senior American officials to see Arafat alive, in Ramallah in 2002. He was a disheveled figure by that point, isolated from his people, indeed virtually imprisoned in his headquarters by Israeli tanks. By title, though, he was still the leader of Palestine, so I went over to urge him to reform his security services—put them in a unitary chain of command, appoint a minister in charge, and so on. Again, he didn't greet me at the door. This time, he didn't dare. This was a much more somber man, a much sadder occasion. Looking at him, I couldn't escape the feeling that all this—the tanks, the sandbags—was such a waste. There was so much talent among the Palestinians. There were so many similarities between them and the Israelis. And for a very special moment in time, everybody in government—the Palestinian government, the Israeli one, and our own—trusted CIA enough on security issues that we really might have been able to make a difference.

That time had passed, though. The window had closed. Sad as it was, we were just going through the motions. Arafat, I'm sure, knew it. He would never lead his people to the Promised Land; he couldn't even walk out the front door. In fact, he was neither Moses nor Ben Gurion.

PART II

CHAPTER 7

Gathering Storm

The attacks of 9/11 so dominate the national consciousness that it can be hard to recall that there was a time, not that long ago, when terrorism in general and the war on terror in particular seemed remote from our lives. For most Americans prior to 9/11, terrorism was something that happened "over there." Yes, it would periodically leap into the headlines—for example, when the Marine barracks and U.S. embassy in Lebanon were bombed in the early 1980s—but almost as quickly the issue would recede.

For me, terrorism was a dominant theme not just during my seven years as DCI but also during my tenure as Deputy CIA Director before that. I don't claim any special prescience. But you simply could not sit where I did and read what passed across my desk on a daily basis and be anything other than scared to death about what it portended.

Beneath the surface of the Islamic fundamentalist world, hatred for the West kept building and building for countless reasons. We could see it approaching. We could see those who were trying to harness this mindless animosity and bend it to their own purposes. And we struggled mightily every day to find ways to defuse or deflect the coming explosion.

The struggle didn't begin with me. Looking for new techniques to force our own bureaucracy to focus on specific looming intelligence threats, in 1996 then DCI John Deutch drew down from the limited funds in our tight intelligence budget and, as an experiment, set up what we called "virtual stations." The idea was to create stateside units that would act as if they were an overseas operation. They would be housed separately, away from our headquarters compound, and staffed with a small number of people, both analysts and operations officers, who would focus on a single issue.

As it turns out, only one such station was ever established. The issue we selected for our test case was called "Terrorist Financial Links." The unit kept the acronym TFL for a short while, but before long it morphed into something even more focused.

The then-obscure name "Usama bin Ladin" kept cropping up in the intelligence traffic. Bin Ladin was the only son of the tenth wife of a wealthy Saudi construction magnate. The Agency spotted Bin Ladin's tracks in the early 1990s in connection with funding other terrorist movements. They didn't know exactly what this Saudi exile living in Sudan was up to, but they knew it was not good. As early as 1993, two years before I came over to CIA, the Agency had declared Bin Ladin to be a significant financial backer of Islamic terrorist movements. We knew he was funding paramilitary training of Arab religious militants in such far-flung places as Bosnia, Egypt, Kashmir, Jordan, Tunisia, Algeria, and Yemen.

UBL, as we came to call him, was just one of many examples of the disturbing trend in terror. Longtime threat Hezbollah, Hamas, Egyptian Islamic Jihad, and dozens of other disaffected groups competed with him for attention, but by the middle of the decade, UBL was front and center on the Agency's radar screen. In March of 1995, for example, Pakistani investigators reported that Ramzi Yousef, the mastermind of the 1993 World Trade Center bombing, who had just been captured in Islamabad, had spent a great deal of time in recent years at a Bin Ladin–funded guest house in Peshawar.

Before long, the TFL virtual station became the "Bin Ladin issue station." It also soon carried the code name "Alec Station." The unit's first leader, Mike Scheuer, named it after his son.

The plan from the beginning was that this "virtual station" would run for two years, after which time the experiment would be evaluated and its functions folded into the larger Counterterrorist Center under which it fell. As it turned out, the unit operated for almost a decade.

It was in Afghanistan in the late eighties, during the war to expel the Soviets, that UBL first made contacts with many of the Islamic extremists who would later form the foundation of what was to become al-Qa'ida—Arabic for "the base." In a media interview in 1988, UBL told of a Soviet mortar shell that had once landed at his feet. When it failed to explode, he said, he knew he had a sign from God that he should battle all foes of Islam. Not long afterward he began using his personal fortune to train and equip militant "Afghan Arabs" for a holy war, or jihad, that would go beyond Afghanistan and eventually reach around the world. (Internet-based conspiracy theorists keep alive the rumor that Bin Ladin had somehow worked for the CIA during the Afghan-Soviet war or had more informal contacts with American officials during that time. Let me state categorically that CIA had no contact with Bin Ladin during the Soviet's Afghan misadventure.)

UBL returned to Saudi Arabia after the Soviets were driven out of Afghanistan in 1989, but the Saudis already had enough trouble with fundamentalist extremists, and Bin Ladin soon ran afoul of his own government despite the prominence of his family. Saudi Arabia's close cooperation with the United States during Operations Desert Shield and Desert Storm, particularly the fact that American troops were allowed on Saudi soil, fueled Bin Ladin's hatred of the West and further estranged him from the Saudi rulers. In 1991 the Saudis were thrilled to see him decamp for Sudan.

In Khartoum, UBL found a much warmer reception and began to occupy more and more of our attention. The country's leader, Hassan al-Turabi, invited him to help organize resistance to Christian separatists in southern Sudan and to build a network of companies that would later serve as fronts for Bin Ladin's worldwide terrorist network. Simultaneously UBL was providing financial help for militant organizations around the Middle East as well as setting up outposts where paramilitary training was provided to jihadists from all over the Muslim world.

Initially, we believed Bin Ladin was principally a financier, and in January 1996 we described him as such, but Alec Station was quickly putting together a picture of someone who was more than a Saudi dilettante with deep pockets and a hatred for the West. UBL, we were learning, was an engine of evil.

Unfortunately, the U.S. embassy in Khartoum was shuttered in early 1996 due to a deteriorating security environment and threats to U.S. officials. In retrospect, that was a mistake—we lost a valuable window into the burgeoning terrorist environment there as a result. But if the intelligence gathering got harder, it nonetheless went on.

In Sudan, Bin Ladin opened several businesses in which he employed veterans of the Afghan war against the Soviets. Many of these men would later become al-Qa'ida operatives. The businesses were quite successful and served to multiply Bin Ladin's already considerable wealth. More worrisome, though, was the increasing evidence that UBL had begun to plan and direct operations himself.

By 1996 we knew that Bin Ladin was more than a financier. An al-Qa'ida defector told us that UBL was the head of a worldwide terrorist organization with a board of directors that would include the likes of Ayman al-Zawahiri and that he wanted to strike the United States on our soil. We learned that al-Qa'ida had attempted to acquire material that could be used to develop chemical, biological, radiological, or nuclear weapons capability. He had gone so far as to hire an Egyptian physicist to work on nuclear and chemical projects in Sudan. At al-Qa'ida camps there, his operatives experimented on methods for delivering poisonous gases that could be fired at U.S. troops in Saudi Arabia.

The defector also told us that Bin Ladin had sent some of his people to Somalia three years earlier to advise the Somali warlord Mohammed Farrah Aideed, who at the time was attacking American forces working in support of Operation Restore Hope, a 1992–1993 U.S. humanitarian aid effort to deal with famine and chaos in Somalia. In fact, the Somalia experience played a significant role in Bin Ladin's perception of the United States. He has said publicly that the U.S. withdrawal from Somalia demonstrated that Americans were soft and that the United States was a paper tiger that could be defeated more easily than the Soviets had been in Afghanistan. (That perception contributed to his surprise five years later, when CIA, operating with U.S. Special Forces, arrived on the ground in Afghanistan so swiftly after 9/11 and, with the help of Afghan surrogates, so effectively destroyed his sanctuary.)

When the United States started putting pressure on the Sudanese to expel Bin Ladin, he became a burden to his hosts. But the question of where he might go was a problem. The Saudis had stripped him of his citizenship in 1994 and certainly didn't want him coming back to the kingdom. Press reports and the Internet rumor mill continue to contend that the Sudanese had offered to extradite UBL to the United States, but I am unaware of anything to substantiate that.

What we do know for certain is that on May 19, 1996, UBL left Sudan, apparently of his own accord, and relocated to Afghanistan. In many ways this was the worst-case scenario for us. Afghanistan at the time was in the midst of extraordinarily chaotic fighting—even by Afghan standards—that would soon leave the country in the hands of the Taliban, a brutal, backward band of fanatics. Inevitably, UBL was quick to form an alliance with Mullah Omar and the Taliban rulers who had seized control of the country, and arguably, for the first time in history, we had something that was not "state-sponsored terrorism" but rather a state sponsored by a terrorist group.

Very soon, dark warning signs were spilling out of Afghanistan. The British newspaper the *Independent* published an article in July 1996 quoting UBL as saying that the killing of Americans at Khobar Towers the previous month was the beginning of a war between Muslims and the United States. The next month,

August, UBL joined other radical Muslims in promulgating a "fatwa," or religious edict, announcing a "Declaration of War" and blessing attacks against Western military targets on the Arab Peninsula.

After 9/11 some senior government officials contended that they were surprised at the size and nature of the attacks. Perhaps so, but they shouldn't have been. We had been warning about the threat at every opportunity. As the red flags multiplied on the horizon in the years before, we tried our best to call attention to them. In 1995 we published a National Intelligence Estimate called "The Foreign Terrorist Threat in the United States." It warned of the threat from radical Islamists and their enhanced ability "to operate in the United States." The Estimate judged that the most likely targets of a terrorist attack would be "national symbols such as the White House and the Capitol and symbols of U.S. capitalism such as Wall Street." The report said that U.S. civil aviation was an especially vulnerable and attractive target.

In 1997 another National Intelligence Estimate, the coordinated judgments of the entire intelligence community, stressed that "Civil aviation remains a particularly attractive target for terrorist attacks." We know that the message was received. The White House Commission on Aviation Safety and Security, chaired by Vice President Al Gore, said in its report that "the Federal Bureau of Investigation, the Central Intelligence Agency, and other intelligence sources have been warning that the threat of terrorism is changing." The report went on to stress that the danger was "no longer just an overseas threat from foreign terrorists. People and places in the United States have joined the list of targets."

In open public testimony in February 1997, I told Congress, "Even as our counterterrorism efforts are improving, international groups are expanding their networks, improving their skills and sophistication, and working to stage more spectacular attacks." In January 1998, at another open hearing, I stressed that "the threat to U.S. interests and citizens worldwide remains high . . . moreover, there has been a trend toward increasing lethality of attacks, especially against civilian targets. . . . A confluence of recent developments increases the risk that individuals or groups will attack U.S. interests."

As if to reemphasize my point, a month later Bin Ladin issued another fatwa, this one stating that all Muslims had the religious duty to "kill Americans and their allies, both civilians and military," worldwide. UBL followed up that pronouncement with a media interview in which he explained that all Americans were legitimate targets because they paid taxes to the U.S. government.

A PDB briefing prepared for President Clinton on December 4, 1998, was titled, "Bin Ladin Preparing to Hijack US Aircraft and Other Attacks." Between April 1, 2001, and September 11, 2001, as many as 105 daily intelligence summaries were produced by the FAA for airline industry leaders. These reports were based on information received from the intelligence community. Almost half of these mentioned al-Qa'ida, Usama bin Ladin, or both.

Unfortunately, even when our warnings were heard, little was done domestically to protect the United States against the threat. To cite two obvious and tragic failures, only after 9/11 were cockpit doors hardened and passengers forbidden from carrying box cutters aboard U.S. commercial airliners.

In combating terror it was necessary to work closely with foreign allies. None would ultimately have to step up more than the Saudis.

I had many memorable meetings with the Saudis over the years. In the spring of 1998 the Saudis foiled a plot by Abd al-Rahim al-Nashiri—head of al-Qa'ida operations in the Arabian Peninsula and the mastermind of the attack against the USS *Cole*—to smuggle four Sagger antitank missiles from Yemen into Saudi Arabia.

Vice President Gore was scheduled to visit Saudi Arabia a week or so after the seizure. We would have expected the Saudis to pass this information to us immediately.

John Brennan, at the time our senior liaison to the Saudis, confronted the Saudi head of intelligence, Prince Turki, about the lapse, but Turki professed ignorance. Brennan suggested I make a quick trip to Saudi Arabia to underscore the importance of sharing such information.

I went to see the crown prince's brother, the interior minister, Prince Naif, who oversaw the Mabahith, the Saudi internal intelligence service.

My "audience" with him took place in a grand receiving room in one of Naif's opulent Riyadh palaces, with scores of Saudi officials observing from chairs lining the perimeter of the hall.

Naif opened, as I recall, with an interminable soliloquy recounting the history of the U.S.-Saudi "special" relationship, including how the Saudis would never, ever keep security-related information from their U.S. allies, despite American unwillingness to share important information with Riyadh. After a while, I had had enough.

John McLaughlin and Brennan were by my side. I was struggling to be diplomatic, but they could see the frustration building.

There was a joke around the office calling me "the subliminal man." It was based on a *Saturday Night Live* skit in which one of the comedians, Kevin Nealon, would say normal things like "How are you, madam?" and then quickly and quietly mutter something different under his breath, such as "You miserable twit." The staff knew that when I was being oh so polite, I was probably thinking something else. McLaughlin wrote a note and passed it to Brennan. "The DCI is about to go 'subliminal.'" He was right.

I scooted my chair forward toward Naif and, without thinking and with no intention of being disrespectful, put my hand on his knee, something you are never supposed to do with royalty.

I said, "Your Royal Highness, what do you think it will look like if someday I have to tell the *Washington Post* that you held out data that might have helped us track down al-Qa'ida murderers, perhaps even plotters who want to assassinate our vice president?"

I don't remember the response of the crowd in general, although Brennan tells me that you could practically feel the air being sucked out of the room as the Saudis simultaneously gasped for breath at the sight of my touching such a powerful royal patella, but I do remember Naif's reaction—what looked to be a prolonged state of shock, with his eyes continuously shifting back and forth between my face and my hand on his knee.

I let him go at last, but I assured him that I would be back the next week, and every week after that if necessary, to ensure that the flow of terrorism-related information between U.S. and Saudi officials was timely and unencumbered.

Crown Prince Abdullah was decisive in breaking the log jam. Within a week of my visit, Brennan was given a comprehensive written report on the entire Sagger missile episode.

During the latter part of 1998, I was aggressively seeking additional resources from our government to fight terrorism. Twice, on November 5, 1998, and October 15, 1999, I wrote personal letters to President Clinton seeking a major increase in our funding. For the most part I succeeded in annoying the administration for which I worked but did not loosen any significant purse strings. In the aftermath of 9/11, politicians from both parties claimed heroism after the fact, saying they had encouraged the DCI to spend more money on terrorism. No, they didn't—at least not in any consistent or coherent way. Neither they nor the 9/11 Commission ever understood that you do not simply snap your fingers and throw resources at one problem while your overall capabilities are in such bad shape.

You can't toss spies at al-Qa'ida when you don't have them, especially when you lack the recruiting and training infrastructure to get them and grow them. You don't simply tell NSA to give you more signals intelligence when their capabilities are crumbling and they are "going deaf"—unable to monitor critical voice communications. Nor could you ignore the need to replace costly, aging imagery satellites without which the country would lose much of its reconnaissance capability, essentially "going blind."

The fact is that by the mid- to late 1990s American intelligence was in Chapter 11, and neither Congress nor the executive branch did much about it. Their attitude was that we could surge ahead when necessary to deal with challenges like terrorism. They provided neither the sustained funding required to deal with terrorism nor the resources needed to enable the recovery of U.S. intelligence with the speed required. Nevertheless, while having to do more with less, we made a conscious decision to invest in future capabilities—not to go deaf, dumb, or blind—that allowed us to stay steps ahead of our adversaries. When money flowed to us after 9/11, we were ready to accelerate our efforts. While our budget declined by 10 percent over the decade, we quadrupled the resources devoted to counterterrorism while investments in other national priorities either remained flat or declined. We did this for the most part by robbing Peter to pay Paul. Still, we never had enough people.

While we were trying to restore our capabilities, the world did not stand still. Nobody relieved us of the burdens of dealing with two wars in the Balkans, tensions in South Asia, China's military buildup, the threat to Taiwan, or the threats posed by North Korea, Iran, or Iraq. The strain was enormous.

The challenge was not just resources but attitude. The policy of the U.S. government at the time was to treat terrorism as a law-enforcement problem. The Justice Department devoted considerable effort to gathering evidence that could be used in court to bring Islamic militants to trial on charges of conspiracy to commit murder if—and it was a *big* if—we could even capture them. At the Agency, we believed that the terrorists sitting around campfires in Afghanistan were probably not losing much sleep over the doings of some U.S. district court—unless, that is, they were planning how to bomb the courthouse itself.

Case in point: Bin Ladin was indicted in June 1998 on charges of plotting to murder U.S. soldiers in Yemen six years earlier. Five months later, he was indicted again, this time in the East African embassy bombings. I can't imagine this fazed him in the least since he was living comfortably in his Afghan sanctuary.

Beyond legal action, there are two other tracks that a country can follow to go after a threat like Bin Ladin. It can attempt to use overt military force or the clandestine capabilities of its intelligence services in a "covert action." The Clinton administration tried both methods. The requirements to make each of these methods successful and the rules under which they are conducted are very different.

If we had been able to provide timely and reliable information about where UBL was at a given moment, *and* precisely where he was going to be a number of hours hence, while simultaneously assuring policy makers that an attack could be conducted without endangering many innocent women and children, the administration would have ordered the use of military force.

Although there were a number of opportunities, we could never get over the critical hurdle of being able to corroborate Bin Ladin's whereabouts, beyond the single thread of data provided by Afghan tribal sources. Policy makers wanted more. I understood their dilemma. As much as we all wanted Bin Ladin dead, the use of force by a superpower requires information, discipline, and time. We rarely had the information in sufficient quantities or the time to evaluate and act on it.

The use of covert action is quite different from the use of overt military power. Almost all of the "authorities" President Clinton provided to us with regard to Bin Ladin were predicated on the planning of a capture operation. It was understood that in the context of such an operation, Bin Ladin would resist and might be killed in the ensuing battle. But the context was almost always to attempt to capture him first. This was the way people up and down the CIA chain of command understood the president's orders. My own understanding of that constraint was deepened in a meeting I had with Attorney General Janet Reno. She made it clear to me and to Geoff O'Connell, the then head of CTC, that she would view an attempt simply to kill Bin Ladin as illegal. Legal guidance by the attorney general matters.

The review of covert-action proposals was very carefully handled. Each time these authorities were updated they showed a deep concern for proportionality and the minimization of loss of life. There was even greater sensitivity shown when the use of surrogates to carry out our will was contemplated.

After 9/11, some policy makers asked rhetorically why I wouldn't have wanted to kill Usama bin Ladin with covert action when I had tried to do so with cruise missiles. This was a completely misleading argument. Our country has appropriately always viewed the secret activities of CIA far differently from the overt use of military force. Despite what they might have said subsequently, everyone understood the differences at the time. Almost every authority granted to CIA prior to 9/11 made it clear that just going out and assassinating UBL would not have been permissible or acceptable.

In the aftermath of 9/11, everyone has become fixated on the word "kill," as if anything but the most vigorous pursuit of the term prior to 9/11 represented some form of risk aversion. It is easy to adopt such a stance after a tragedy like 9/11, but it was simply not the legal or political reality that we operated under prior to that day.

From my perspective, this is a largely pointless debate. Policy makers can sign some covert authorities and lull themselves into thinking that they have done their jobs. But in the absence of hard intelligence—in this case regarding Bin Ladin and the al-Qa'ida leadership structure operating inside Afghanistan—covert action is a fool's game, an illusory silver bullet. With numerous fleeting opportunities to act militarily, and additional authorities being provided, I came to understand that we were putting the cart before the horse. While in the aftermath of 9/11 some would reflect on this period and say that CIA was either risk averse or incompetent to execute the authorities provided by the president,

I understood something else: we had to increase our odds by engaging in old-fashioned espionage inside the Afghan sanctuary. We needed more intelligence, not just about Bin Ladin but about his entire leadership structure inside Afghanistan. That is precisely what we would set out to do. There is one other thing I learned: Ultimately, no matter how hard we worked inside Afghanistan, real increases in the quality of the data acquired there would ultimately occur only when we finally disrupted al-Qa'ida's environment through direct action, forcing them up out of their comfort zone, putting them on the run, and causing them to make mistakes. Action begets intelligence. As one Special Operations commander told the 9/11 Commission, "You give me the action and I will give you the intelligence."

Over time, the covert-action authorities granted to us by the Clinton administration were modified—for example, to give us the ability to work with groups such as the Northern Alliance to collect intelligence, but not to use the Alliance to take lethal action against Bin Ladin and al-Qa'ida.

We could press ahead on collecting information about Bin Ladin and other terrorists. We could work with foreign intelligence services to disrupt their efforts and throw them off their stride, in the same way a beat cop might keep vagrants moving along. Our Counterterrorism Center worked hard to develop better human sources in Afghanistan so that we would have improved windows into what UBL was planning and where he was. But we were not in the freelance assassination business—that's for the movies, not the complicated real world that CIA operates in.

There were a number of opportunities to use military action against Bin Ladin, but these opportunities were fleeting, and tough decisions would have to have been made in narrow windows of time. My job was to assess objectively whether the data we had, often only from a single source, could ever get policy makers above a 50 or 60 percent confidence level so they could launch cruise missiles in the next thirty minutes. It never did.

Was this good enough for them? It was not. It was understandable, in the aftermath of 9/11, when everyone's risk calculus had changed, that people became more aggressive with regard to taking action. I know my officers wanted to be more aggressive, but my job was to lay down what we knew, accurately and objectively. I tried to do so, without a trace of advocacy. My own frustration was that, as much as we all wanted Bin Ladin dead, we didn't have enough information to give policy makers the confidence they required to pull the trigger.

Hindsight is perfect, of course, and it is easy to say now that launching a major covert action against the Taliban sooner might have made a difference before 9/11. But policy makers across two administrations had reasons to be cautious. They had legitimate concerns about the impact such a plan might have on the stability of the neighboring Pakistani government. Actions in the region could have had unintended consequences regarding the tenuous Indian-Pakistani situation. It may also have been impossible to launch a major assault against the Taliban without Pakistani concurrence. Two administrations may have waited too long to act. The Taliban and their Afghan surrogates were allowed to remain too comfortable in their sanctuary. Had we been authorized to shake them from their complacency, we might have produced the intelligence that could have averted the coming disaster. I just do not know.

One step we did take in light of our expanded authorities was to work with members of an Afghan tribe that had helped us in 1997 in our search for the murderer Aimal Kasi. The tribe provided some very good tracking data on Bin Ladin. On a number of occasions they were able to relay to us information on where UBL had recently been. Prudently, he moved around a lot, most often between Khandahar and a walled compound outside of town called Tarnak Farms.

During the spring of 1998, the first of what would become several plans to try to capture Bin Ladin emerged. The idea was for our surrogates to snatch him in Afghanistan and allow us to bring him back to the United States, if possible, to face trial. Counterterrorist Center officers developed a plan where members of the tribe would be used to break into the Tarnak Farm compound, breaching its ten-foot walls. UBL had several wives there, so exactly where he would be found was mostly a matter of guessing which wife he had decided to grace with his company on any given evening, but we had a pretty good idea which houses inside the compound those wives were most likely to be found.

If the tribe had been able to find UBL and spirit him away, they were going to literally roll him up in a rug, take him to the desert, and hide him away, perhaps for a lengthy period, until the United States could stealthily get an aircraft in to "exfiltrate" him (remove him from Afghanistan clandestinely) so that he could face justice in the United States.

Clearly, this was a plan with a lot of "ifs" and "maybes," including the questions of whether UBL would even be there at the time and, if so, whether tribal forces could get past his protection and locate the house he was in before he fled. Several practice runs seemed to convince the plan's proponents that it had, at best, a 40 percent chance of succeeding. Others thought the odds considerably worse. From our point of view, trying to effect a capture and having UBL die in a shoot-out was perfectly acceptable, but we couldn't simply have our surrogates burst in, guns blazing, and hope for the best. That sort of "kill 'em all and let God sort 'em out" approach might have had a lot of appeal after the massacres of 9/11, but 1998 was a different environment, legally and otherwise. Naturally the tribal leaders thought we were crazy when we tried to explain to them the concepts of restraint and rule of law. Such legal niceties are foreign to Afghans.

Mike Scheuer, the head of Alec Station, was strongly in favor of going ahead with the operation. I took his recommendation very seriously, but six senior CIA officers stood in the chain of command between Mike and me. Most of them were seasoned operations officers, while Mike was an analyst not trained in conducting paramilitary operations. Every one of the senior operations officers above Mike recommended against undertaking the operation. They believed the chances of success were too low and the chances of killing innocent women and children were too high. Geoff O'Connell told me that it was the "best plan we had" but that "it simply wasn't good enough." Revisionist historians will tell you that the U.S. Special Operations Command evaluated the plan and pronounced it a good one. If the plan had been carried out by the Special Operations Command, it might have worked. But no one in the U.S. government authorized us to use elite American troops. Instead we had to rely on a largely untested group of tribal Afghans to conduct the mission.

I had only limited confidence in the tribals. They were good at passing information regarding Bin Ladin's alleged location, but frankly, there were serious concerns about their operational capability. In the end, I made the decision not to go ahead with the plan. I believed it would have been irresponsible of me, knowing of the opposition the plan engendered among my most senior operations officers, to have passed it on to the president's desk. It didn't take long, though, for that decision to be thrown back in my face.

On Friday, August 7, 1998, about two months after I pulled the plug on the Tarnak Farms operations, the phone at my bedside started ringing sometime before 5:00 A.M. These late-night and early-morning calls were a normal occurrence by then, but there was nothing regular about this one. The senior duty officer in the Agency's Operation Center was on the line. "Bombs have just gone off at our embassies in Nairobi, Kenya, and Dar es Salaam, Tanzania," he said. "The damage is massive; the death toll will be high." High turned out to be an understatement, at least by pre-9/11 terms. There were 240 people killed and some 4,000 wounded in the two attacks. As I dressed and headed to the office, the status of U.S. officials at both sites was still uncertain. It quickly became clear that the embassy bombings were indeed the work of al-Qa'ida.

A day or so later, I paid a visit to Alec Station, which by this time had been moved back into CIA headquarters. That's where one of Scheuer's subordinates, quivering with emotion, confronted me about my Tarnak Farms decision. "If you had allowed us to go ahead with our operation," she said, "those people might still be alive!"

It was a tough moment. Of course I had some self-doubt. But the fact is that al-Qa'ida operations are planned years in advance. We later learned that they first cased the Nairobi embassy more than four years earlier. A Bin Ladin snatch in June would not have stopped either bombing. But given the emotion of the moment, I let the analyst vent and just walked away.

This act demanded some sort of retaliation. Working with the Pentagon, we assembled a list of al-Qa'ida–related targets that might be struck. One of the difficulties of fighting a terrorist opponent is the paucity of targets susceptible to the application of military force. I recall no discussion of sending in the 82nd Airborne or the like to put U.S. boots on the ground in Afghanistan, but in mid-August, as we were searching for ways to respond, we received a godsend: signals intelligence revealed that a meeting would be held by Bin Ladin. We were accustomed to getting intelligence about where UBL *had been*. This was a rarity: intelligence predicting where he was *going to be*.

In tightly held discussions within the NSC, we determined not only to go after Bin Ladin in Afghanistan but also to demonstrate that we were prepared to go after his organization worldwide. On our list of potential targets were businesses in Sudan and elsewhere in which he had been involved. These businesses not only were part of the terrorist financial network but also had possible connections with al-Qa'ida attempts to obtain chemical and biological weapons. But while attacking the terrorist summit meeting in Khost was a "no brainer," the other targets were a matter of considerable debate.

The phone at my bedside rang again early on the morning of August 20. This time it was President Clinton calling from Martha's Vineyard, where he was vacationing and trying to ride out the Monica Lewinsky storm. I never saw any evidence that Clinton's personal problems distracted him from focusing on his official duties. Perhaps they circumscribed the range of actions he could take—he was, after all, losing political capital by the hour—but they certainly didn't seem to do so in this case. The president wanted to talk about the potential targets, especially a tannery that Bin Ladin owned in Sudan and the al-Shifa pharmaceutical factory in Khartoum with which he was involved and which we believed was somehow implicated in the production of chemical agents. A spoonful of clandestinely acquired soil collected from outside the factory gate had shown trace amounts of O-ethyl methylphosphonothioic acid, or EMPTA, a chemical precursor for the deadly VX chemical agent. In the end, the president decided to drop the tannery from the target list. There were too many chances for collateral damage with too small a payoff. But the factory at al-Shifa and the camp at Khost were to be struck by cruise missiles.

I understood why the administration favored cruise missiles. They didn't require putting pilots at risk, and they carried none of the burden or baggage of inserting combat troops. But in hindsight, I'm not certain at the time we fully comprehended the missiles' limitations. The slow-flying missiles are a good choice for taking out fixed targets such as pharmaceutical factories but are far less ideally suited to targeting individuals who wander around during the several hours between the time the missile is launched and when it lands at its preprogrammed spot.

In all, scores of cruise missiles were launched at the Khost terrorist facility right around nightfall on August 20. The sea-launched Tomahawks had to fly hundreds of miles to reach their targets, including navigating the airspace of Pakistan to get to landlocked Afghanistan. To make sure the Pakistanis didn't think they were under missile attack from India, the vice chairman of the Joint Chiefs of Staff, Gen. Joe Ralston, was dispatched there to alert officials just before the missiles crossed into their airspace that this was a U.S. operation.

We believe that a dozen or more terrorists were killed in the ensuing cruise missile strike, but apparently UBL chose to leave the camp sometime before the missiles arrived, once again dodging a fate he richly deserved. We never were able to determine if his departure was happenstance or if he was somehow tipped off.

Predictably, the plant at al-Shifa was flattened. Later, though, questions arose about how closely it might have been associated with UBL and whether there might be some alternative explanation for the EMPTA trace that had placed the plant on the target list. You can still get a debate within the intelligence community on how good a target al-Shifa was. What's beyond debate is that Bin Ladin's lucky escape only emboldened him for future operations.

Less than two months after the cruise missile attacks, on November 5, 1998, I wrote President Clinton a letter saying that I needed a massive infusion of funds to position the intelligence community where it needed to be in the fight of our lifetime. The signs were everywhere that al-Qa'ida had plans for bigger, more spectacular attacks on U.S. interests. To combat our enemies and to protect American interests, I said, we needed "roughly $2 billion more per year for the intelligence budget above the existing FY-2000–2005 budget." As happened with earlier requests, we received only a small portion of what we asked for. At the same time, I directed Cofer Black, who had become head of the Counterterrorism Center, to put together a new strategy to attack al-Qa'ida. We called it simply "The Plan." But there was nothing simple about it.

The Plan recognized that our first priority was to acquire intelligence about Bin Ladin by penetrating his organization. Without this effort, the United States could not mount a successful covert action program to stop him or his operations. To that end, The Plan laid out a strong, focused effort, using our own sources, our foreign partners, and enhanced technology, to gather the intelligence that would let us track and act against Bin Ladin and his associates in terrorist sanctuaries, including Sudan, Lebanon, Yemen, and, most important, Afghanistan.

To execute The Plan, the Counterterrorist Center developed a program to select and train officers and put them where the terrorists were located. The Center launched a nationwide recruitment program using CIA's Career Training Program resources to identify, vet, and hire qualified personnel for counterterrorist assignments in hostile environments. We sought native fluency in Arabic and other terrorist-associated languages, as well as police and military experience, and appropriate ethnic background. In addition, the Center established an eight-week advanced Counterterrorist Operations Course to teach CIA's hard-won lessons learned and counterterrorism operational methodology.

In reviewing our record against al-Qa'ida, Cofer concluded that our efforts had stopped several planned attacks against U.S. embassies. We had significantly damaged UBL's infrastructure and put some doubt in his mind about the security of his operations. But all this had only set him back. It had not stopped him. Unless we changed our tactics, we would find it harder in the future to achieve operational success against al-Qa'ida. They were learning about us as we were learning about them.

My frustration with the quality and depth of our intelligence regarding al-Qa'ida and Bin Ladin continued to grow. I was tired of relying on one tribal group without much corroborating data to make decisions as to whether we should launch capture operations, or cruise missiles, within narrow windows of time. Our entire intelligence community and our foreign partners needed to be challenged to do better in gathering data from where it mattered most—inside Afghanistan. We needed to get over the threshold of confidence that policy makers needed and wanted. So, on December 3, 1998, I sat at home and furiously drafted in longhand the memo I titled, "We Are at War." In it I told my staff that I wanted no resources or people spared in the effort to go after al-Qa'ida. The 9/11 Commission later said that I declared war but that no one showed up. They were wrong.

While many people were focused exclusively on one man, al-Qa'ida had a leadership structure, with training facilities, all residing in Afghanistan. Our strategic objective was to get more intelligence—human, signals, and imagery—not just to target Usama bin Ladin but also to deal with a movement that was operating in sixty countries. The hub of the enterprise was Afghanistan, and from that hub spoked sanctuaries and, farther afield, other countries where significant operational capability existed.

By the fall of 1999 several things came together. First was CTC's operational plan, and second, the work of forty-year veteran Charlie Allen, the associate deputy director of central intelligence for collection. The most important paragraph in my December 1998 memo was not about holding more meetings and killing more trees, but rather my direction to Charlie Allen to immediately push the rest of the intelligence community to make Bin Ladin and his infrastructure a top priority:

> *I want Charlie Allen to immediately chair a meeting with NSA, NIMA [our imagery agency], CITO [our clandestine information technology operation] and others to ensure we are doing everything we can to meet CTC's requirements.*

Allen wrote me back a week later:

> *Senior collection managers assess that overall the Community's capabilities against UBL and his infrastructure are sharply focused. Collectors have not only taken an extraordinary range of steps since the East African Embassy bombing to enhance the capacities but they continue to develop additional measures where all elements of the community were involved.*

Through 2000, Allen would provide formal detailed updates five more times—we would also have almost daily interaction.

Once Cofer Black had finalized his operational plan in the fall of 1999 to go after al-Qa'ida, Allen created a dedicated al-Qa'ida cell with officers from across the intelligence community. This cell met daily, brought focus to penetrating the Afghan sanctuary, and ensured that collection initiatives were synchronized with operational plans. Allen met with me on a weekly basis to review initiatives under way. His efforts were enabling operations and pursuing longer-range, innovative initiatives around the world against al-Qa'ida. In terrorism, the tactical and strategic blur—operational success on the tactical level yields strategic results, new leads, more data, and better analysis.

You had to destroy terror cells that were trying to kill you, disrupt them, render them to justice, take the data generated, and drive on. The amount of data we collected exploded—CTC's walls were covered with the faces of known terrorists and their connections, their linkages to people on the other side of the world. Cofer understood the imperative. He knew we had disrupted attacks, "that we had damaged UBL's infrastructure, and created doubt inside al-Qa'ida about the security of his operations and operatives." But he intuitively understood something else as well—that we were fighting a worthy opponent and we had no on-the-ground presence in Afghanistan. He knew that without penetrations of Usama bin Ladin's organization, without access to Afghanistan, we were fighting a losing battle.

Allen and Black sat side by side at scores of briefings with me and other senior CIA and FBI officers in the run-up to 9/11. As a result of the intelligence community's efforts, in concert with our foreign partners, by September 11, Afghanistan was covered in human and technical operations.

We were working with eight separate Afghan tribal networks, and by September 11 we had more than one hundred recruited sources inside Afghanistan. Satellites were repositioned. The imagery community had systematically mapped al-Qa'ida camps. We engaged the Special Operations Command and used conventional and innovative collection methods to penetrate al-Qa'ida in Afghanistan and the rest of the world. We expanded our open source coverage (spy-speak for reviewing open media, such as newspapers and radio) of al-Qa'ida. Leadership of the FBI was given full transparency into our efforts.

Some countries allowed their soil to be used to train capture teams and deploy major collection facilities on their borders with Afghanistan. In other sanctuaries and around the world where al-Qa'ida had significant capability, operations and collection initiatives were pursued that allowed us to stop attacks and generate more data. Allen implemented other significant long-term technical enhancements that had nothing to do with day-to-day operations, involving multiple countries and services to target al-Qa'ida leaders and infrastructure. There was nothing tactical or ad hoc about any of this. It was opportunistic and strategic in the same breath.

We identified foreign strategic relationships that would extend our operational reach, services that could infiltrate their own officers into terrorist sanctuaries. Prior to 9/11, we identified nine worldwide hubs where we provided technical assistance, and analytic training—the ability to fuse data essential to rapid operational turnaround. These were places where we knew we would get a huge bang for our buck against al-Qa'ida, strategic investments that would dramatically grow around the world after 9/11.

To scores of other intelligence services, we provided as much assistance as possible, so that when I or my senior colleagues made calls to seek assistance, we had willing partners. In this way we had capital in the bank at the other end when we wanted to make a withdrawal. Amazingly, the 9/11 Commission would later say that my idea of a management strategy for a war on terrorism was simply to rebuild CIA. The commission failed to recognize the sustained comprehensive efforts conducted by the intelligence community prior to 9/11 to penetrate the al-Qa'ida organization. How could a community without a strategic plan tell the president of the United States just four days after 9/11 how to attack the Afghan sanctuary and operate against al-Qa'ida in ninety-two countries around the world?

It was during this same period that I decided that the usual intelligence reporting in the form of Presidential Briefs, finished intelligence reports, National Intelligence Estimates, and the like was insufficient for conveying the seriousness of the threat. So I began sending personal letters to the president and virtually the entire national security community, explicitly laying out why I was concerned about the looming terrorist attacks. I knew that all senior officials had full in-boxes—only something out of the ordinary would get their attention.

Even one such letter would have been an unusual step. During my tenure, I wrote eight of them. My intention was not to cry wolf, and certainly not to scare the recipients out of their wits, although a careful reading of the letters would certainly have accomplished that. I believed the only way to get their attention was to tell them what I knew and what concerned me, and to do so over and over and over again. I am confident that officials in both the Clinton and Bush administrations understood the seriousness of the threat.

In the first letter, dated December 18, 1998, I wrote:

> *I am greatly concerned by recent intelligence reporting indicating that Usama Bin Ladin is planning to conduct another attack against US personnel or facilities soon . . . possibly over the next few days. One of Bin Ladin's deputies has used code words we associated with terrorist operations to order colleagues in East Africa to complete their work.*

In the letter, I noted that Bin Ladin's organization had a presence in more than sixty countries and had forged ties with Sunni extremists around the world. The letter went on to say that UBL was interested in conducting attacks inside the United States or within the territory of allies such as the United Kingdom, France, and Israel.

Ten days later I wrote again, updating the previous letter and quoting a Middle Eastern service as saying that they agreed with our assessment that UBL sought to strike in the near term against at least one U.S. target. I reported that Bin Ladin had purchased ten surface-to-air missiles from Afghan warlords to defend his terrorist camps but noted that the same missiles could be used to attack aircraft on U.S. territory. I wrote again on December 30 and then on January 14, 1999, with additional details picked up from a variety of sources.

My public warnings continued, too. In my annual worldwide threat testimony on February 2, 1999, I told the Senate that "there is not the slightest doubt that Usama Bin Ladin, his worldwide allies, and his sympathizers are planning further attacks against us . . . despite progress against his networks, Bin Ladin's organization has contacts virtually worldwide, including in the United States. . . . He has stated unequivocally that all Americans are targets. . . . I must tell you we are concerned that one or more of Bin Ladin's attacks could occur at any time."

A few days later we received intelligence that told us Bin Ladin was at a hunting camp in southern Afghanistan in the company of a number of sheikhs from the United Arab Emirates. Once again there were those, including some in Alec Station, who were anxious for the United States to obliterate the place in the hopes of getting UBL. If a bunch of Arab princes were killed, too—well, that would be the price they paid for the company they kept. Before a decision could be made as to whether to launch a strike, we got word UBL had moved on.

In hindsight, these on-again, off-again attacks should have been leading policy makers to a serious discussion over the use of force against the al-Qa'ida leader. Instead of considering alternative approaches to the less-than-ideal cruise missile attacks, policy makers seemed to want to have things both ways: they wanted to hit Bin Ladin but without endangering U.S. troops or putting at significant risk our diplomatic relations. As a result, we were constantly ginning up attack plans and making last-minute decisions about whether some snippet of information we had just obtained was good enough to launch missiles and whether UBL might stay put for a few hours so we could get him. I remember one weekend when I was summoned away from my son's lacrosse game to the security vehicle accompanying me so I could take a call. UBL might have been spotted again, and I had to make a recommendation on the spot—do we launch or not? That's no way to do business.

Throughout the fall of 1999, the threat situation was bad. And then it got worse. A steady drumbeat of reports leading up to the millennium told us that al-Qa'ida had entered into the execution phase of numerous planned attacks, although we couldn't say with certainty where or when.

It wasn't just al-Qa'ida and Bin Ladin's millennial ambitions we were worried about. We ran a quiet but effective sweep in East Asia, leading to the arrest or detention of forty-five members of the Hizbollah terrorist network.

We also mounted a disruption campaign against Hezbollah's chief backer, MOIS, the Iranian intelligence service. (The acronym stands for Ministry of Intelligence and Security.) Agency officers approached MOIS officers on the street or wherever we could get close to them and asked them if they would like to come to work for us or sell us information.

In one memorable example, John Brennan, our liaison to the Saudis, handled the local MOIS head himself. John walked up to his car, knocked on the window, and said, "Hello, I'm from the U.S. embassy, and I've got something to tell you." As John tells the story, the guy got out of the car, claimed that Iran was a peace-loving country, then jumped back in the car and sped away. Just being seen with some of our people might cause MOIS officers to fall under suspicion by their own agency. The cold pitches undoubtedly ruined some careers, and maybe even lives, but also occasionally paid off in actual intelligence dividends. It couldn't happen to a nastier bunch of people.

There were scores of operations going on around the world

simultaneously. One of them, the surveillance of a suspicious meeting in Kuala Lumpur, ended up being much more significant than we knew at the time. (That meeting, which involved some future 9/11 hijackers, is described in chapter 11.)

On December 6, 1999, Jordanian authorities arrested a sixteen-man team of terrorists who planned New Year's Eve attacks on pilgrims at John the Baptist's shrine on the Jordan River, and on tourists at the SAS Radisson hotel in Amman. The terrorists planned to use poisons and improvised devices to maximize Jordanian, Israeli, and U.S. casualties. We later learned they intended to disperse hydrogen cyanide in a downtown Amman movie theater. The Jordanian intelligence service, through its able chief, Samih Battikhi, told us that individuals on the team had direct links to Usama bin Ladin.

All the alarm bells were going off at CTC, especially since the millennium period overlapped Ramadan. Jihadists believed the Islamic holy month a propitious time to wage warfare against nonbelievers. In addition, they viewed the millennium as a symbolic deadline for the return of Jerusalem to Muslims. From Cofer Black's perspective, what we saw in Jordan matched Bin Ladin's preference for softer targets, his focus on non-Muslim casualties, and his growing interest in the use of chemical agents. CTC's and Cofer's view was that the next attack would likely be bigger than East Africa. We told President Clinton that Usama bin Ladin was planning between five and fifteen attacks around the world during the millennium and that some of these might be inside the United States. This set off a frenzy of activity. CIA launched operations in fifty-five countries against thirty-eight separate targets. I must have talked to Sandy Berger, Louis Freeh, and Janet Reno three times a day during this period. Foreign Intelligence Surveillance Act (FISA) surveillance warrants were being processed by Fran Townsend at the Department of Justice at a record pace. I made countless phone calls to my counterparts around the world trying to get them to share our anxiety and our efforts.

We alerted our colleagues to the north about the presence of an Algerian terrorist cell in Canada. At about the same time an alert customs official in Port Angeles, Washington, spotted Ahmad Ressam nervously trying to enter the United States. The thirty-two-year-old Algerian panicked and tried to flee but was arrested. A quantity of nitroglycerin and four timing devices were found hidden in his car. He later admitted to being part of a plot to bomb Los Angeles International Airport. In looking back, much more should have been made about the significance of this event. While Ressam's plot was foiled, his arrest signaled that al-Qa'ida was coming here.

The government was exhausted—our northern border vulnerable, the United States did not have a comprehensive and integrated system of homeland security in place. Borders, visas, airline cockpits, watchlists—all were managed haphazardly. We would pay the price in two years, when the lack of a coherent system of protection would be exploited by terrorists.

Dick Clarke, the national coordinator for security and counterterrorism, writes in his memoir that at three o'clock on the morning of January 1, 2000, he walked out on the roof of the White House and popped a bottle of champagne to celebrate the fact that the New Year had arrived on the West Coast without a single terrorist assault on the contiguous United States. In his memoir, Louis Freeh says that when the millennium finally passed that early morning, he was too tired to do anything other than go home and fall in bed. I don't remember the moment arriving or passing, or my celebrating anything. To be sure, the millennium represented a spike in terrorist activity and a serious threat to American interests, but at CIA, the threat was part and parcel of a seamless terrorist onslaught. We had watched this, worried about it, and combated it for years, and we knew we would continue doing so after public attention had waned, the computers had all survived the flip over to a fresh millennium, and the news cameras had deserted Y2K and moved on in search of other stories.

After the millennium, threat reporting mostly settled down to its usual dull roar. Then, in the late summer of 2000, it began to soar once more. Again with the help of liaison services, the fruit of all the bridge building we had been doing over the last several years, we were able to break up terrorist cells planning attacks against civilian targets in the Gulf region. These operations netted anti-aircraft missiles and hundreds of pounds of explosives and brought a Bin Ladin facilitator to justice.

Our technological capacity increased dramatically in 2000 when CIA teams deployed to Central Asia and began operating on an experimental basis a new prototype of the Predator unmanned aerial vehicle. This small, remotely controlled aircraft started flying over Afghanistan and sending back truly remarkable real-time reconnaissance video. Sitting in a command center in Washington, Tampa, or anywhere in the world, you could see with great clarity what was going on in a terrorist compound half a world away.

In the Predator's very first trial run, on September 28, 2000, we observed a tall man in flowing white robes walking around surrounded by a security detail. While the resolution was not sufficient to make out the man's face, I don't know of any analyst who didn't subsequently conclude that we were looking at UBL. Finally, we now had a real-time capability and did not have to rely solely on secondhand information relayed by our tribal assets or picked up in signals intelligence and analyzed days later. What we were looking at, however fuzzy, could have been the shape of evil. Yet, as technologically dazzling as that was, it was frustrating in almost equal measure. Yes, we might have been looking at UBL, but we were not in a position to do anything about it. Later, after much testing and adjustment, the Predator would carry its own weapons load, but for now about the best the military could do was spin up some more cruise missiles and hope that UBL didn't move on.

Then, on October 12, 2000, the undeclared war we were fighting with al-Qa'ida got ratcheted up to a whole new level. Sitting at anchor in port at Aden, in Yemen, the Navy destroyer USS

Cole was attacked by a small explosive-laden suicide boat. The ensuing explosion ripped a huge hole in the side of the *Cole*, rolling it up like the lid of a tin can and killing seventeen American sailors. Only by heroic effort was the crew able to save their ship from sinking.

In the aftermath of the attack, it was clear that known al-Qa'ida operatives were involved, but neither our intelligence nor the FBI's criminal investigation could conclusively prove that Usama bin Ladin and his leadership had had authority, direction, and control over the attack. This is a high threshold to cross. The ultimate question policy makers have to determine is what standard of proof should be used before the United States decides to deploy force? It must always be a standard set by policy makers because ultimately it is they who bear the responsibility for actions taken. What's important from our perspective at CIA is that the FBI investigation had taken primacy in getting to the bottom of the matter.

During the 9/11 Commission's investigation, much was made of the fact that the United States did not immediately retaliate for the attack on the *Cole*. The country was in the middle of the 2000 presidential election, which then turned into a constitutional crisis when no clear winner emerged. Perhaps it would have been difficult to launch new military ventures while the country was fixated on counting chads and Supreme Court votes. Equally important was the fact that we didn't have any inviting targets. By then we didn't need any additional excuses to go after UBL or his organization. But simply firing more cruise missiles into the desert wasn't going to accomplish anything. We needed to get inside the Afghan sanctuary.

On December 18, 2000, with a month left in the administration, I again wrote to the president and representatives of virtually the entire national security bureaucracy:

> *The next several weeks will bring an increased risk of attacks on our country's interests from one or more Middle Eastern terrorist*

groups . . . The volume of credible threat reporting has grown significantly in the past few months, particularly concerning plans by Usama bin Ladin's organization for new attacks in Europe and the Middle East. . . .

Our most credible information on bin Ladin activity suggests his organization is looking at US facilities in the Middle East especially the Arabian peninsula, in Turkey and Western Europe. Bin Ladin's network is global however and capable of attacks in other regions, including the United States.

Iran and Hezbollah also maintain a worldwide terrorism presence and have an extensive array of off-the-shelf contingency plans for terrorist attacks, beyond their recent focus in Israel and the Palestinian areas.

We have the most success where local authorities share our concern—such cooperative efforts often produce valuable information about other terrorist plans as happened after the Millennium plot in London.

Not every government and liaison service shares our concern or is willing to work closely with us, and such resistance often denies us good intelligence we could use to predict attacks or disrupt an operation. As a result, pockets exist where terrorists can establish a foothold, plan attacks and carry them out with little warning.

A new administration would soon arrive, but the old situation awaited it. Al Qa'ida were still coming at us. There was not a meeting held with a foreign partner or leader where either I or our officers did not register al-Qa'ida as our top priority. Many thought we had become obsessed. Others failed to understand fully how terrorists in their countries might be planning for attacks within ours. There is one important moral to the story: you cannot fight terrorism alone. There were clear limitations to what we could do without the help of like-minded governments.

The 9/11 Commission suggested that in the run-up to 9/11

policy makers across two administrations did not fully understand the magnitude of the terrorist threat. This is nonsense.

In authorizing several covert-action authorities, the principal policy makers of the Clinton administration understood fully the nature of the threat we were facing. These documents spelled out in detail why it was necessary to continually ratchet up the pressure against Bin Ladin. These written authorities made clear that Bin Ladin posed a serious, continuing, and imminent threat of violence to U.S. interests throughout the world. They said that CIA considered the threat unprecedented in geographic scope. They took note of the fact that twenty-nine Americans had died during the East African and *Cole* bombings; that Bin Ladin had a presence in at least sixty countries and had forged ties with Sunni extremists worldwide; that the intelligence community had strong indicators that Bin Ladin intended to conduct or sponsor attacks inside the United States. The documents also made clear that Usama bin Ladin's organization was aggressively seeking chemical and biological weapons and that he would use them against American official and civilian targets. I know that the most senior decision makers in the Clinton administration understood the magnitude of what we were facing.

As the new guard arrived, Steve Hadley and Condi Rice also understood the threat as well when they were briefed on the covert authorities they were inheriting as they assumed their jobs.

Terrorism throughout the 1990s fully engaged the highest levels of our government, and while people can argue about what was or was not done, to me, the knowledge and concern of senior officials was indisputable.

Very late in the Clinton administration, Sandy Berger asked me, if I were unconstrained by resources and policies, how I would go after Bin Ladin and al-Qa'ida. I asked Cofer Black and his team in CTC to put together a paper that we might present to the new administration—whoever it turned out to be. We called this the "Blue Sky" paper. It was designed to include our best ideas for how the war on terror might proceed if we were free from resource limitations or past policy decisions that had hampered our progress. We sent the paper to Dick Clarke on December 29. Among other things, it called for a significant effort to disrupt al-Qa'ida in its Afghan sanctuary. The paper also recommended major support for the Northern Alliance so that they could take on the Taliban, and it also sought to provide assistance to neighboring states such as Uzbekistan to help them drive the terrorists out of their backyard. There was "no single silver bullet" available to deal with the problem, we wrote. Instead, a multifaceted strategy was needed to produce change.

To my mind the Blue Sky memo was a compelling blueprint for the future. It was brimming with good ideas—plans and strategies we would roll out less than ten months later, days after 9/11—but the timing of it meant that, for now, most of those good ideas would simply sit in Dick Clarke's safe and await the new administration.

CHAPTER 8

"They're Coming Here"

On December 12, 2000, the U.S. Supreme Court decided in effect, by a vote of 5–4, that George Bush would be the next president of the United States. If you believe some of my critics, I knew the outcome nearly two years earlier, when CIA headquarters was renamed the "George Bush Center for Intelligence," after George W.'s father.

I was pleased to preside at the ceremony on April 26, 1999, honoring the headquarter's new namesake and one of my predecessors, George H. W. Bush. He is a man still fondly remembered for helping the Agency through a very rough patch when he was DCI two decades previously. But I can't claim clairvoyance. An act of Congress directed the name change, not me. At the ceremony, I quoted from President Bush's farewell remarks when he left the Agency in 1977: "I take with me many happy memories," he said then. "I am leaving, but I am not forgetting. I hope I can find some ways in the years ahead to make the American people understand more fully the greatness that is CIA."

Although he served as Director for less than a year, George Bush, with his wife, Barbara, provided Agency employees with a sense of caring and family. They also maintained their connections after his time as DCI ended. As vice president, George Bush chaired a commission looking into the threat of terrorism—and his findings led to the creation of CIA's Counterterrorist Center. As president he was committed to leveraging the power of intelligence to help him handle the burdens of his office, and he insisted on being personally briefed on the latest intelligence six days a week, just as his son would later do.

During that visit to the Agency, he and his wife were greeted like rock stars. They were extraordinarily generous with their time, shaking hands, signing autographs, and reconnecting with a CIA workforce that was genuinely fond of both of them. Barbara Bush spoke at an event hosted by our family advisory board in the Agency's auditorium. The two of them that day left us with a powerful leadership message: Take care of people and they will take care of you. During my tenure as director, 41 (as the first President Bush is known) frequently checked in with me with an encouraging note or phone call. He was always our staunchest public defender.

That spring day in 1999, I was not worrying about who might occupy the Oval Office almost two years hence. At CIA we pay attention to who might win foreign elections, but we have no special insights on U.S. politics. True, whoever the new president might be, it would have a significant impact on my life.

Either candidate was likely to want his own DCI, but if the party in power changed along with the president, the odds of my going were greater still. Intellectually, I accepted that fact, but in my heart I wanted to stay because I felt the job was unfinished. Once the Supreme Court ruled in favor of George W. Bush, I figured the odds of my being gone by January 20 had increased.

David Boren, the former Oklahoma senator and now president of the University of Oklahoma—and one of my closest and most valued mentors—advised me that, if given the opportunity, I should stay on for the first half-year of the new administration, then tender my resignation. That way, he said, I would have worked under presidents of both political persuasions. I also felt that by sticking around I could ease the transition for both the new administration and CIA. Back when he was DCI, the first president Bush offered to stay on at CIA similarly at the start of the Carter administration. Jimmy Carter said, "No, thanks." Had Carter said yes, it is questionable whether George H. W. Bush would ever have reached the presidency.

I was in downtown D.C. in late December, racing to some meeting, when I got a call from Dottie, my invaluable special assistant, the "Miss Moneypenny" of CIA. Dottie said that Rich Haver, who was handling the intelligence transition for Dick Cheney, had just come by my office and was all but measuring the place for new drapes. Donald H. Rumsfeld, Cheney's own esteemed mentor, was going to be the new DCI, Haver gleefully hinted. How soon could I move out? Because the election had been so heavily contested in the courts, the Bush people had gotten a delayed start in filling senior positions. Any day, I expected a call informing me of the name of my successor.

I remember taking time off at the end of the month so that Stephanie, John Michael, and I could spend Christmas with my brother in New York City, and then head off to Boston to celebrate New Year's Eve with our closest personal friends, Steve and Jeryl. Just before we left New York, the media was filled with the Rumsfeld story—the announcement that he was to be the new director was due any hour. Rather than wait around for what amounted to a deathwatch for my tenure, we decided to get an early start to Boston. We were on the interstate—John Michael and I in the lead car, and Stephanie in the follow car—when word came in from the headquarters command post that Rumsfeld had indeed been appointed, but to be secretary of defense, not DCI.

This didn't mean that my job was safe—far from it. At any moment I might get a call that would tell me to start cleaning out my desk. But for the time being, the most frequently rumored candidate to replace me was going elsewhere.

We had started giving George W. Bush intelligence briefings even before he was officially designated president-elect. The administration had authorized us to give him access to the same kinds of data that was being provided to Bill Clinton in his final month in office. Al Gore, of course, continued to be briefed as the sitting vice president.

We sent some of our top analysts down to Austin in late

November to establish contact and start bringing the governor up to speed in case he were about to become commander in chief. The governor scared our briefers one morning when he said after one session, "Well, I assume I will start seeing the good stuff when I become president." We were not sure what his expectations were, but he was already seeing "the good stuff." As a result, though, we redoubled our efforts to upgrade the PDBs. It was clear that if he were certified the winner, this son of a former president and DCI was going to pay very close attention to our business.

A little more than a week before assuming office, the president-elect came to Washington and took up residence at Blair House, across the street from the White House, on Pennsylvania Avenue. On January 13, I went to see him there, to brief him on the state of the world and what we were most worried about. John McLaughlin and the deputy director for operations, Jim Pavitt, were with me. The president was joined by the vice president–elect and Andy Card. We told them that our biggest concerns were terrorism, proliferation, and China. I don't recall Iraq coming up at all. At the end of the briefing the president asked to have a word with me alone. Uh, oh, here it comes, I thought.

"Why don't we just let things go along for a while," I remember him saying, "and we'll see how things work out." I gathered from that I was neither on the team nor off it. I was on probation. As would be expected, there were some adjustments to make.

Under President Clinton, I was a Cabinet member—a legacy of John Deutch's requirement when he took the job as DCI—but my contacts with the president, while always interesting, were sporadic. I could see him as often as I wanted but was not on a regular schedule. Under President Bush, the DCI post lost its Cabinet-level status. But I soon found out that I was to have extraordinary access nonetheless.

The transition team made it clear to us that they wanted the president to receive a regular in-person intelligence briefing six days a week, just as his father had. We selected one of my former executive assistants, Mike Morell, to be the president's personal briefer. I sat in on the first post-inauguration briefing but fully expected to let Morell be our sole daily point of contact. After a couple of briefings without me present, the president pulled Morell aside and asked, "Does George understand that I would like to see him here with you every day?" I hadn't wanted to show up every day for fear that it would look like I was campaigning to keep my job. Making an appearance every now and then would suffice, I figured. But now I got the message loud and clear. My schedule and my life were never the same. That was the downside. My work hours stretched out even longer. My home time shrank again. But the upside was undeniable. Being in regular, direct contact with the president is an incredible boon to a CIA director's ability to do his job.

There were lots of other differences to adjust to. Gore versus Cheney? Both brought very different perspectives to the vice president's office. Gore had served on the House Intelligence Committee many years before. True to his interests, he had a fascination for wonkish issues. He asked lots of questions about the impact on national security of water shortages, disease, and environmental concerns. "Bugs and bunnies," some people called it. But I learned a lot from him on these matters. And he was right. Those kinds of issues can have a profound effect on population flow, migration, civil wars, ethnic strife, and the like. Cheney had a more traditional view and knew a hell of a lot about our business. Both were avid consumers of intelligence and provided considerable assistance to us.

Back in 1999, one of the many times I was scrambling for more resources for CIA, I sent Gore a handwritten note briefly arguing our case and citing what I thought was a necessary supplemental appropriation. "We could use your help here," I concluded. He replied in short order, "You've sold me. Is this enough?" That was music to my ears. Cheney, too, was often extraordinarily helpful. He was always willing to use his personal clout on our behalf—calling world leaders, for example, and leaning on them to give us information or access or whatever we needed. I never failed to get his aid when I asked for it.

The one big difference between the two was that Gore had his national security advisor, Leon Fuerth, represent him at Principals' meetings, while Cheney generally sat in on them himself. That was his privilege, obviously, but having one of the ultimate decision makers actually participating in the debate made it more difficult for Condi Rice, the president's national security advisor, who chaired the meetings. The vice president's presence may also have had an unintended chilling effect on the free flow of views as important policy matters were debated.

For a DCI, the most important relationship with any administration official is generally with the national security advisor—the person who digests everything the intelligence community and State and Defense departments have to say, carries it to the president, and renders counsel. Sandy Berger had performed that job with obvious zeal, although his street-tough manner occasionally rubbed against the more delicate sensibilities in government. His successor, Condoleezza Rice, had served in the Bush 41 NSC under Brent Scowcroft, a man who had twice performed that job and who did it as well as anyone ever had. From the outset, it was obvious that Condi was very disciplined, tough, and smart, but she brought a much different approach to the job than her predecessor. Sandy not only didn't mind rolling up his sleeves and wading into the thick of things; he seemed to relish it. Condi, by contrast, was more remote. She knew the president's mind well but tended to stay out of policy fights that Sandy would have come brawling into.

All of the above falls generally under the category of atmospherics. Administrations change. People are different. You have to get along with a new group, with new ideas. Every new administration wants to evaluate things once they get the offices for which they have been campaigning. And every administration starts out slowly—feeling their way along. The Bush crowd had an especially late start anyway because of the electoral stalemate, and they carried a heavy load of aversion to any policy the Clinton administration had favored. Doing things differently from their predecessors seemed almost an imperative with them.

The slow-motion changeover and the full agenda, domestically and internationally, that the new administration brought with it had the greatest impact, in my estimation, on the war on terror. It wasn't that they didn't care about Usama bin Ladin or al-Qa'ida, or that they got rid of people who did. Below the top level of the new government virtually the entire counterterrorism team stayed in place. But at the top tier, there was a loss of urgency. Unless you have experienced terrorism on your watch—unless you have been on the receiving end of a 4:00 A.M. phone call telling you that one of your embassies or one of your ships has just been attacked, it is hard to fully fathom the impact of such a loss. I know that you should be able to understand intellectually the significance of the threat, but there is nothing like being there when the bomb goes off to get your undivided attention.

The simple fact is that the terrorism challenge was not an easy one to tackle. It wasn't just a matter of going out and getting the bad guys. Policy had to be decided. Diplomacy had to be factored in. These things require time for an administration to wrap its mind around. Take one of the toughest terrorism issues of all—what we thought of as the Pakistan problem.

For years, it had been obvious that without the cooperation of the Pakistanis, it would be almost impossible to root out al-Qa'ida from behind its Taliban protectors. The Pakistanis always knew more than they were telling us, and they had been singularly uncooperative in helping us run these guys down. My own belief, one shared widely within CIA, was that what the Pakistanis really feared was a two-front conflict, with the Indians seeking to reclaim Pakistan and the Taliban mullahs trying to export their radical brand of Islam across the border from Afghanistan. A war with India also posed the grim specter of a nuclear confrontation, but from the ruling generals' point of view, the best way to avoid having their nation Taliban-ized was to keep their enemy close. That meant not cooperating with us in hunting down Bin Ladin and his organization.

The relationship was complicated further by mistrust and resentment. The dominant thinking within the Pakistani officer corps was that the United States had unstated ulterior motives in Afghanistan, specifically the desire to keep the nation unstable and chaotic to discourage construction of oil and gas pipelines through both Afghanistan and Pakistan. The goodwill we had won in Pakistan by helping to drive the Russians out of neighboring Afghanistan had also evaporated over the last dozen years. The Pakistani leadership for the most part felt that the United States had abandoned them, especially when we imposed economic sanctions on both Pakistan and India in the wake of their nuclear tests. Simultaneously, the military-to-military relationship that had once been so strong between our two nations had been allowed to wane over the years. Once, senior Pakistani officers had been trained almost exclusively in the United States. That wasn't true with the younger generation. From an intelligence perspective, we had precious few leverage points on which to build.

Until 9/11, the Bush administration found itself in the same box with regard to Pakistan that had plagued the Clinton years. Even though thousands of terrorists had been trained in al-Qa'ida camps in Afghanistan, policy makers had become consumed with Pakistan's internal stability, the command and control of their nuclear weapons, and the likelihood of a nuclear conflict with India. Obviously, these were legitimate concerns, but terrorism was a serious issue, too. Yet, because of this policy tension, we were never able to get a green light from our government to aid in any serious way Ahmed Shah Masood and his Northern Alliance in their efforts to reclaim Afghanistan from the Taliban.

Even within CIA there was debate over how to proceed with

Pakistan, the Taliban, and al-Qa'ida. If you sat in the Counterterrorism Center with Cofer Black and his team, the choice was clear: immediate action was required to support the Northern Alliance. Policy makers who were fixated on whether we could produce enough actionable intelligence to spin up a missile to take out Bin Ladin and his top lieutenants had totally missed the point. Getting only Bin Ladin was never going to solve the problem. To do that, you had to destroy al-Qa'ida's sanctuary and disrupt the infrastructure that guided and funded operations around the world. That meant action on the ground.

If you sat in Islamabad, however, the world looked very different. For starters, the Northern Alliance had been nurtured for years by Pakistan's mortal enemies, the Indians and the Russians. Aligning ourselves with Masood and his fighters would put us in league with the devil, for potentially little or no gain. Absent significant U.S. military involvement, the Northern Alliance would never defeat the Taliban. If we just made the Alliance a greater threat to the Taliban, we would end up reinforcing the Taliban's need for al-Qa'ida support and thereby strengthen rather than weaken Bin Ladin's position in Afghanistan.

Gen. Mahmood Ahmed, the Pakistani intelligence chief who was in Washington when the 9/11 attacks went down, was emblematic of the problem. I'd met with him over lunch on September 9, 2001, and tried to press him about Mullah Omar, Bin Ladin's most ardent protector within the Taliban regime. Mahmood assured us that Omar was a man who wanted only the best for the Afghan people. Fine, we told him, but he's also harboring a guy who has created a sanctuary for training terrorists who murder American embassy workers and sailors. In fact, his defense of Mullah Omar was typical of Mahmood. As gracious as he could be over the lunch table, the guy was immovable when it came to the Taliban and al-Qa'ida. And bloodless, too. After the USS *Cole* was attacked by Bin Ladin's suicide bombers, Mahmood sent our senior officer in Islamabad a very precisely worded message that managed to convey his condolences for the loss of life without offering a single word of support for our going after al-Qa'ida in its Afghan lair.

What's more, we had to assume that he was an accurate proxy for his boss, Gen. Pervez Musharraf. We knew that Mahmood had been instrumental in rallying critical elements of the Pakistani army to support Musharraf during the 1999 coup against President Nawaz Sharif. In effect, Mahmood had ensured that Musharraf would succeed. Some thought the best we could hope from either of them was that the Pakistani intelligence service might turn a blind eye to whatever actions we undertook in Afghanistan to go after the Arab presence there. Failing that, there was always the chance that the Afghans and perhaps even some Taliban officials might mount a jihad against the predominantly Arab al-Qa'ida, but that, too, seemed a long shot. The Arabs and Bin Ladin had become institutionalized in Afghanistan through their property acquisitions and their largesse to the Taliban leadership. Mahmood's sole suggestion in the first days of his Washington visit was that we try bribing key Taliban officials to get them to turn over Bin Ladin, but even then he made it clear that neither he nor his service would have anything to do with the effort, not even to the extent of advising us whom we might approach.

The events of 9/11 changed that calculus entirely. Until then, the new Bush team had to sort through this incredibly complicated and delicate set of issues, and decide where they stood on the questions and what actions to take and postures to assume. And in truth, for all that they wanted to put daylight between themselves and the Clinton administration, they weren't any more successful at resolving difficult and competing issues in their opening months than their predecessors had been.

At CIA we obviously had a more acute sense of urgency. Lt. Gen. John "Soup" Campbell, the senior active-duty military officer on my staff and one of the finest officers I've ever worked with, was running a series of tabletop exercises regarding Predator operations. Soup wanted to be prepared for the day when the UAV would be able to carry a warhead. Who would operate the aircraft? Who would make the decision as to if and when to fire? How would the U.S. government explain it, if Arab terrorists in Afghanistan suddenly started being blown up? I raised some of these same questions in my first weekly meeting with the new national security advisor, on January 29, 2001, and I kept raising them again and again.

Like me, Dick Clarke had been retained at the start of the administration in his old job and was equally anxious to restore attention to the war on terror. To that end, he took our Blue Sky memo and crafted his own recommendations for jump-starting U.S. efforts against al-Qa'ida. Clarke's memo was called "Strategy for Eliminating the Threat from Jihadist Networks of al Qida: Status and Prospects." He proposed "rolling back" al-Qa'ida over a period of three to five years, talked about using military action to attack al-Qa'ida command-and-control targets and Taliban infrastructure, and even expressed concern that there might be al-Qa'ida operatives in the United States.

I later learned that on January 25, 2001, Clarke sent this memo to Condi Rice saying there was an urgent need for an NSC principals meeting to review his proposed strategy against al-Qa'ida. But this meeting was never held.

One thing was glaringly apparent. If we were going to proceed with anything like what we had in mind—that is, if we were going to switch from a defensive to an offensive posture against the terrorists—we needed new covert-action authorities. Again, let me stress one very important fact: CIA is a policy implementer, not a policy maker. Those entrusted with making policy, beginning with the president, decide what we are allowed to do in pursuit of ends they deem important.

Early in March, I went by to see Stephen J. Hadley, Condi's deputy at the National Security Council, and handed him the list of the expanded authorities we were seeking to go after Bin Ladin.

These authorities would place us much more on the offensive, rather than have us reacting defensively to the terrorist threat. I thought they were critical, but I also knew they required a discussion among policy makers that was long overdue. My hope was that the authorities we were seeking would kick off that discussion.

"I'm giving you this draft now," I told Steve, "but first, you guys need to figure out what your policy is."

The authorities in the draft were very broad and would have explicitly authorized CIA or its partners to plan and carry out operations to kill UBL without first trying to capture him. We believe these authorities were unprecedented in scope.

The next day, Mary McCarthy, a CIA officer then serving as NSC senior director, called John Moseman, my chief of staff, and said basically, "We need you to take back the draft covert-action finding back. If you formally transmit these to the NSC, the clock will be ticking, and we don't want the clock to tick just now."

In other words, the new administration needed more time to figure out what their new policies were, and thus didn't want to be in a position someday to be criticized for not moving quickly enough on a critical intelligence community proposal.

If the new administration had embraced our Blue Sky concept wholeheartedly and granted us all the authorities we sought that day in March, would we have been able to prevent 9/11? I don't know. After all, the plot was already well under way, and the terrorism threat was growing daily.

In my first public testimony during the new administration, in February 2001, I told the Senate that "The threat from terrorism is real, it is immediate, and it is evolving. . . . [A]s we have increased security around government and military facilities, terrorists are seeking out 'softer' targets that provide opportunities for mass casualties. . . . Usama Bin Ladin and his global network of lieutenants and associates remain the most immediate and serious threat. . . . He is capable of planning multiple attacks with little or no warning."

In other testimony later that spring, I told Congress that "We will generally not have specific time and place warning of terrorist attacks. . . . The result . . . is that I consider it likely that over the next year or so that there will be an attempted terrorist attack against U.S. interests." My sense was that something was coming—something big—but to my great frustration we could not determine exactly what, where, when, or how.

We delivered the same message through classified briefings and analysts' reports. A March paper stressed the critical role that Afghanistan played in providing sanctuary for terrorism. A paper the next month talked about the growing belief among jihadists that there was some U.S.-led conspiracy against Islam.

During the spring of 2001, at one of the innumerable Deputies' meetings, John McLaughlin expressed frustration at the lack of action. "I think we should deliver an ultimatum to the Taliban," he said. "They either hand Bin Ladin over or we rain hell on them." An odd silence followed. No one seemed to like the idea. Richard Armitage, the deputy secretary of state, called John after the meeting and offered a friendly word of advice: "You are going to get your suspenders snapped if you keep making policy recommendations. That is not your role."

Throughout my tenure as DCI, under two administrations, I had a weekly private meeting with the national security advisor. Looking back on the notes from those sessions now, I find that in almost every meeting terrorism was high on the agenda but never more so than in the spring and summer of 2001.

For my regularly scheduled meeting with Condi Rice on May 30, I brought along John McLaughlin, Cofer Black, and one of Cofer's top assistants, Rich B. (Rich can't be further identified here.) Joining Condi were Dick Clarke and Mary McCarthy.

Rich ran through the mounting warning signs of a coming attack. They were truly frightening. Among other things, we told Condi that a notorious al-Qa'ida operative named Abu Zubaydah was working on attack plans.

Some intelligence suggested that those plans were ready to

be executed; others suggested they would not be ready for six months. The primary target appeared to be in Israel, but other U.S. assets around the world were at risk.

Condi asked us about taking the offensive against al-Qa'ida. Cofer told her about our efforts to work with other intelligence services, penetrate terrorist organizations, and the like.

"How bad do you think it is?" Condi asked. Cofer told her that during the millennium the terrorist threat situation was an "eight on a ten scale." Right now, he said, we were about at a "seven." Clarke told her that adequate warning notices had been issued to appropriate U.S. entities.

The FAA issued warning notices, embassy security was tightened around the world, military installations in the Middle East went on higher alert levels. We were asked to brief other Cabinet members. We returned to CIA headquarters with the hope that our message had been received.

Information about Zubaydah kept popping up in various bits of intelligence. In June 2001 we were informed by the British that Abu Zubaydah was planning suicide car bomb attacks against U.S. military targets in Saudi Arabia by the end of the month. We learned via the FBI's debriefing of the would-be millennial bomber Ahmad Ressam, for example, that Abu Zubaydah had requested high-quality Canadian passports for smuggling operatives into the United States. As part of his bargaining for a reduced sentence, Ressam told the FBI that Zubaydah was considering attacks in several U.S. cities. Ressam provided no details on specific venues, but he did say that Zubaydah was in it for the long haul—that he was willing to spend a year or more in preparation if that would lead to a successful attack.

(When we captured Zubaydah in Pakistan in March 2002, some media accounts suggested that he was not such an important player. Those accounts are dead wrong. Worse yet, it has been suggested that the Bush administration exaggerated his importance in their comments to the media—again dead wrong.

I believe to this day that Abu Zubaydah was an important player in al-Qa'ida operations.)

Threat information continued to pour in, almost from every nook and cranny of the planet. Some examples of what my top people and I were confronted with on a daily basis throughout the months leading up to 9/11:

• Yemeni terrorists were planning an attack in Jordan.

• A group of Pakistanis was planning to bomb the American community in Jeddah, possibly the U.S. or British schools there.

• The FARC, a terrorist group in Colombia, reportedly was planning to car-bomb several sites in Bogotá, including the U.S. embassy and a mall frequented by embassy employees.

• Hizbollah was readying large-scale terrorist operations in Southeast Asia.

• An extremist group was planning an attack against the U.S. embassy in Sanaa, the capital of Yemen.

• Four Saudi nationals were heading from the United Arab Emirates to Kuwait to attack U.S. interests.

• Three suspects arrested in Malaysia in May for attempted robbery had cased U.S. facilities and U.S. Navy vessels in preparation for an attack.

• An Algerian-based terrorist cell responsible for planning an attack against the U.S. embassy in Rome or the Vatican was broken up by the Italians in July and its members deported.

• Meanwhile, the leading al-Qa'ida operatives involved in the *Cole* bombing were in Afghanistan planning new attacks against the United States.

As for Ayman al-Zawahiri, the former Egyptian Islamic Jihadist leader who had become Bin Ladin's top deputy, it was almost impossible to turn around without finding him entwined in murderous intrigues, planning to renew terrorist operations throughout Europe. Al-Qa'ida was assessing advanced operations for a major attack in Israel against U.S and Israeli targets, to be led by Zawahiri. Zawahiri, we learned, was coordinating terrorists in Saudi Arabia and the Middle East.

Still other intelligence assessments painted a picture of a plot to kidnap Americans in India, Turkey, and Indonesia. That was said to be the work of a renegade Egyptian extremist figure, Rifat Taha Mousa, then living in Damascus. Mousa was so despised throughout most of the Muslim world that he had even been expelled from Iran. Syria had allowed him in after several other Arab countries also handed him his walking papers, then arrested him on a tip we provided. Mousa had put out numerous fatwas against the United States in the several months prior to his arrest. He was also close to the Blind Sheikh, Omar Abdel-Rahman, who was linked to the 1993 bombing at the World Trade Center. In addition, Mousa had shared a podium with Bin Ladin and Zawahiri in Afghanistan during the summer of 2000. We had a photograph of him seated right between the two of them. Talk about a Toxic Trio.

In June we learned that several Arab terrorist camps were closing in Afghanistan. Al Jazeera reported (erroneously, as it turned out) that Bin Ladin was leaving the country, fearing an American strike against him. The Arab satellite channel MBC broadcast an interview with Bin Ladin and his key lieutenants in which he said there will be a "big surprise" in the coming weeks and a "hard hit against U.S. and Israeli interests." MBC also reported that Bin Ladin's forces were in a state of high alert. Other reports told of imminent suicide attacks in the Gulf. Al-Qa'ida operatives were leaving Saudi Arabia to return to Afghanistan, which was a concern to us because, as we learned in the aftermath of the *Cole* attack and East Africa bombings, those responsible had beaten feet just before the attacks occurred. In Afghanistan, Arabs were said to be anticipating as many as eight celebrations. Operatives were being told to await important news within days. Zawahiri was warning colleagues in Yemen to anticipate a crackdown and urging them to flee. To our great frustration, the Saudis, who probably held more keys to unlocking the inner workings of al-Qa'ida than any other liaison service, were slow-rolling us on the feedback we kept requesting. Finally, at our request, Dick Cheney called the Saudi crown prince to break the logjam.

On June 28, 2001—I remember the date exactly and the event vividly—Cofer Black and I sat down for a briefing on the state of the global terrorism threat. Cofer had again brought along Rich B. It was Rich who did most of the talking. We now had more than ten specific pieces of intelligence about impending attacks, he said. The NSA and CTC analysts who had been watching Bin Ladin and al-Qa'ida over the years believed that the intelligence was both unprecedented and virtually 100 percent reliable. Over the last three to five months we had been witness to never-before-seen efforts by Ayman al-Zawahiri to prepare terrorist operations. Abd al-Rahim al-Nashiri, the mastermind of the *Cole* attack, had disappeared. A key Afghan camp commander was reportedly weeping with joy because he believed he could see his trainees in heaven. All around the Muslim world, important operatives were disappearing while others were preparing for martyrdom. Rich's June 28 briefing concluded with a PowerPoint slide saying, "Based on a review of all source reporting, we believe that Usama Bin Ladin will launch a significant terrorist attack against the U.S. and/or Israeli interests in the coming weeks." Five days later, on July 3, we learned as a result of intelligence that Bin Ladin had promised colleagues that an attack was near.

As the threat reporting intensified, so did our efforts overseas. By late June, in cooperation with foreign partners, we had launched disruption efforts in nearly two dozen countries. Almost twenty of our best unilateral extremist terrorist penetrations around the world had been told to gather as much information as possible on the impending attacks. Either leaders of our counterterrorist team or I had been in direct contact with eighteen chiefs of foreign intelligence services, seeking their assistance. We talked about specific demarches to the Pakistanis, to close down the Pakistani-Afghan border, and their border with Iran, the preferred transit choice of al-Qa'ida operatives exiting Afghanistan on their way to the Gulf. A worldwide cable to our stations and bases urged immediate action to run down all extremist leads. In the United States, we were working diligently with the FBI to secure and exploit as many terrorist communications as possible. That meant going through the Foreign Intelligence Surveillance Act Court, which considers government requests to authorize surveillance of suspected foreign agents inside the United States. The FISA Court was tremendously helpful, yet it was becoming increasingly evident by early July of 2001 that further legislative improvements were needed because the existing statutes did not give us the flexibility we needed to get on top of a savvy and increasingly sophisticated terrorist network.

American embassies closed upon our recommendation or beefed up their protection. Navy ships left Middle Eastern ports and headed out to sea. Again, I can't say what *didn't* happen as a result of those warnings and the high level of alert we were broadcasting, but I'm convinced that the summer and fall of 2001 would have been even more catastrophic—and the bloodshed far more widely spread—had we sat on, ignored, or soft-pedaled what we were hearing.

On July 5, several senior CTC officers went to the Justice Department to brief Attorney General John Ashcroft about our concerns. They told him that we believed that a significant terrorist attack was imminent and that preparations for an attack were in the late stages or already completed. We continued to believe, however, that an attack was more likely to be conducted overseas. At the end of the briefing the attorney general turned to some FBI personnel and pointed at CIA officers present. "Why are *they* telling me this?" he asked. "Why am I not hearing this from you?" CIA briefers thought this was an odd reaction.

By July 10, Cofer Black, Rich B., and their counterterrorism team had put this flurry of reporting into a consolidated, strategic assessment. That afternoon, Cofer asked to see me. The briefing he gave me literally made my hair stand on end. When he was through, I picked up the big white secure phone on the left side of my desk—the one with a direct line to Condi Rice—and told her that I needed to see her immediately to provide an update on the al-Qa'ida threat. I can recall no other time in my seven years as DCI that I sought such an urgent meeting at the White House. Condi made the time immediately, and Cofer, Rich, and I made the fifteen-minute ride to the White House.

When we arrived in Condi's office, Dick Clarke and Steve Hadley were waiting for us. Rather than sit on the couch as we usually did for our weekly meetings, I asked if we could arrange ourselves around Condi's conference table so everyone could follow the briefing charts. I thought the more formal setting and stiff-backed chairs were appropriate for what was about to be said. Rich handed out the briefing packages and took it from there. His opening line got everyone's attention, in part because it left no room for misunderstanding: "There will be a significant terrorist attack in the coming weeks or months!"

A specific day was impossible to pick: "We know from past attacks that UBL is not beholden to attacks on particular dates," Rich explained. "Bin Ladin warned of an impending attack in May of 1998, but the attacks against the embassies were not carried out until August. UBL will attack when he believes the attack will be successful." The signs, though, were unmistakable. Key Chechen Islamic terrorist leader Ibn Kattab has promised some "very big news" to his troops, Rich said. A chart displayed seven specific pieces of intelligence gathered over the past twenty-four hours, all of them predicting an imminent attack. Among the items: Islamic extremists were traveling to Afghanistan in greater numbers, and there had been significant departures of extremist families from Yemen. Other signs pointed to new threats against U.S. interests in Lebanon, Morocco, and Mauritania.

Rich's next chart contained what in the business we call a "gisting," a summation of the more chilling statements we had in our possession through intelligence:

- A mid-June statement from UBL to trainees that there will be an attack in the near future.
- Information that talked about moving toward decisive acts.
- Late June information that cited a "big event" that was forthcoming.
- Two separate bits of information collected only a few days before our meeting in which people were predicting a stunning turn of events in the weeks ahead.

The attack will be "spectacular," Rich told Condi and the others, and it will be designed to inflict mass casualties against U.S. facilities and interests. "Attack preparations have been made," he said. "Multiple and simultaneous attacks are possible, and they will occur with little or no warning. Al-Qa'ida is waiting us out and looking for vulnerability."

Rich went on to summarize our efforts to disrupt specific targets tied to Bin Ladin. Our intent, he explained, was not just to startle or stop specific bad guys. We wanted the targets to spread the word that Bin Ladin's plans had been compromised. Our hope was that we might cause him at least to delay the attacks, but that could never be anything more than a stalling action. At the end of this graph, underlined, were these words: "Disruption only delays a terrorist attack. It does not halt a terrorist threat."

As we had arranged, Rich swung from that point into arguing that consideration should be given immediately to moving from a defensive to an offensive posture vis-à-vis al-Qa'ida and Bin Ladin. "We have disrupted or delayed the current attack, but the UBL threat will continue to exist," he said. "UBL's goal is the destruction of the United States. We must consider a proactive instead of a reactive approach to UBL. Attacking him again with cruise missiles after this new terrorist attack will only play to his strategy. We must take the battle to UBL in Afghanistan. We must take advantage of increasing dissatisfaction of some Afghan tribes with the Taliban. We must take advantage of the Afghan armed opposition."

At the end of the briefing, Condi turned to Clarke and said, "Dick, do you agree? Is this true?" Clarke put his elbows on his knees and his head fell into his hands and he gave an exasperated yes.

Condi looked at Cofer and asked, "What should we do?"

Cofer responded, "This country needs to go on a war footing *now*."

"Then what can we do to get on the offensive now?" Condi asked. I can't recall if it was Cofer or I who answered that question. "We need to re-create the authorities that we had previously submitted in March," one of us said. I reminded Condi again that, before the authorities could be okayed, the president needed to align his policy with the new reality, and she assured me that this would happen. It was just the outcome I had expected and hoped for when we left Langley for the White House maybe an hour earlier, but the tragedy is that all this could have been taking place four months earlier, if our initial request for expanded authorities hadn't been so abruptly tabled.

As we were leaving Condi's office, Rich and Cofer congratulated each other. At last, they felt, we had gotten the full attention of the administration.

When press accounts of the July 10, 2001, meeting surfaced in the fall of 2006, some 9/11 Commission officials said that we had never told them about the meeting. Transcripts of my classified testimony in early 2004 showed that I did discuss the meeting with the commission. Why they failed to mention it in their final report is a mystery to me.

Initially some administration officials suggested that the briefing might not have occurred but they later amended their comments to say that while it had taken place, it contained no new or urgent information. Obviously they had not reviewed the briefing slides, especially the one regarding seven pieces of intelligence collected in the previous twenty-four hours that predicted imminent terrorist attacks.

Rich had assured the group gathered in Condi's office that day that the NSA strongly discounted the possibility of disinformation. "Throughout the Arab world," he said, "UBL's threats are known to the public. There will be a loss of face, funds, and popularity if UBL's attacks are not carried out." Everyone, though, still wasn't convinced. Sometime shortly afterward, Steve Cambone, undersecretary of defense for intelligence, came to see me and asked if I had considered the possibility that al-Qa'ida's threats were just a grand deception, a clever ploy to tie up our resources and expend our energies on a phantom enemy that lacked both the power and the will to carry the battle to us.

"No," I said to Steve, "this is not a deception, and, no, I do not need a second opinion. I have been living with this for four years. This is real." I told Steve that it would be a tremendous mistake to dismiss what our experience told us was inevitable. "We are going to get hit," I said. "It's only a matter of time." Steve wasn't alone. Paul Wolfowitz was raising the same question. To Steve's credit, after 9/11 he went out of his way to tell me he had been wrong.

We had hoped that the July 10 meeting would finally get us on track, or at least had pointed us in the right direction. Three days later, a meeting of the Deputies Committee was held to discuss the covert-action authorities we had initially requested back in March. But the bureaucracy moved slowly. The authorities granted on September 17, 2001, were substantially the same as the ones we had requested in March.

More intel kept coming in. On July 13 we received intelligence about Abu Musab al-Zarqawi, who was wanted by the Jordanians for his involvement in the millennial plots (and who would go on to mastermind untold numbers of kidnappings, beheadings, and bombings in Iraq before being killed in a U.S. bombing raid in June 2006). Zarqawi, we learned, wanted to arrange a meeting in Iran for apparent operational planning.

At one of my daily briefings, I found out from the Palestinians about a plan to attack the American embassy in Beirut. Turkish police, I learned, had responded to my calls and begun conducting operations to identify as many Bin Ladin targets in Istanbul as possible. Meanwhile, explosives had been smuggled from Yemen to Saudi Arabia on July 6 for use against U.S. military targets. The Saudis had finally responded to intelligence we had provided them in January, undoubtedly the fruit of the call the vice president had made to Crown Prince Abdullah urging cooperation. In response, we told the Saudis we needed to keep working with them, we needed to keep engaging them, and we needed to keep pushing them toward more timely interaction with us—the same message I would deliver myself to the crown prince two years later, after the al-Qa'ida attacks inside the kingdom.

In mid-July we learned senior al-Qa'ida operatives might be returning to Pakistan contingent on where and when a certain event occurred. Our information told us that some were wondering whether unidentified pressure had halted plans for terrorist attacks. This gave us some hope that our disruption efforts might be having some effect.

The Egyptian service told us that a senior operative from Jemaah Islamiya, a Southeast Asian terrorist organization allied with al-Qa'ida, was planning an attack on U.S. and Israeli interests in order to help win the release of the Blind Sheikh. Four trucks filled with C-4 explosives had been brought to Kampala, in Uganda, and operatives there had begun casing the American embassy. We immediately contacted the Ugandans and also brought in the Tanzanians and Kenyans. Al-Qa'ida had already proved how effective it could be at striking U.S. interests in Africa.

A European intelligence service warned us about a "concrete and serious" threat emanating from a diffuse Mujahideen network in Afghanistan and Pakistan. Al-Qa'ida operatives were traveling to Europe, they said, but the target and timing of the attack were unknown. The next day, that same service provided specific information about the activities of a foreign operative well known to us. That same day, July 17, sources within the Zawahiri network told us of an attack that was to take place inside Saudi Arabia within days. We immediately informed the Saudis. Yemenis arrested a key Bin Ladin passport forger who was involved in a threat against the U.S. embassy in Sanaa, and we provided them with debriefing requirements. A few days later we received six separate reports that an Afghanistan-based narco-trafficker was facilitating the shipment of explosives and bomb-making kits to al-Qa'ida operatives in Yemen, to be used against U.S. and British interests there. Five members of the group had met with Bin Ladin in Khandahar. From Afghanistan came word that the Taliban intelligence chief, Kari Amadullah, was interested in establishing secret contact, outside the country and without Mullah Omar's knowledge, "to save Afghanistan." From the Northern Alliance, Ahmed Shah Masood told us that Bin Ladin was sending twenty-five operatives to Europe for terrorist activities. The operatives, he said, would be traveling through Iran and Bosnia.

The whole world seemed on the edge of eruption.

In a briefing I received on July 24, I learned that Jordan's King Abdullah had sent word that, in his view, Bin Ladin and his command structure in Afghanistan must be dealt with in a decisive and military fashion. To that end, he offered to send two battalions of Jordanian Special Forces to go door to door in Afghanistan, if necessary, to deal with al-Qa'ida. The offer was a wonderful gesture but would have to have been part of a larger overall strategy in order to succeed. To King Abdullah, Bin Ladin was the greatest threat in the world to his nation's security, and he wanted us to know that Jordan was ready to act as the pointy end of the spear. Like father, like son, I thought. That apple had fallen right next to the tree. How could anyone help but respect the king of Jordan and his family after something like that?

A CTC update on the terrorist threat situation brought word from another intelligence source that they had detained an associate of Zarqawi. Interestingly, this person linked Zarqawi with Abu Zubaydah, expanded our knowledge about Zubaydah's network in the Gulf and Europe, and provided leads to other operatives in Sudan, the United Kingdom, and the Balkans. In running down the data, we concluded that Zarqawi's network was larger and better connected than we had anticipated. The operative was moved to Jordan for further questioning.

Also on the agenda from CTC that day: two Egyptian extremists had been identified in Indonesia, where the government was quickly moving to disrupt the pair, arrest them, and send them to a country in which they were wanted. The UAE had arrested Djamel Beghal, who had been planning to bomb the U.S. embassy in Paris.

The operative who was behind the threat to bomb the embassy had arrived in the United Kingdom. We had so informed the Brits and had alerted the Swedes of the operative's onward travel home after he left the UK. The Bolivians had arrested six Pakistanis who were planning an airline hijacking. One of those arrested appeared to be related to Kasi, the man who had killed two CIA officers at the Agency's front gate in 1994. It was likely that the six would be deported to Pakistan, where authorities would question them at our urging.

That same day, we had reporting that Zawahiri was in Yemen and we were pursuing confirmation and a plan to exfiltrate him to the United States. Although we doubted this information, it was our intention to play this hand out. I was also briefed on a major breakthrough in our ongoing effort to technically penetrate al-Qa'ida and Taliban leadership in Afghanistan. Tremendous teamwork with the British service made this possible and was now providing a quantum leap in our coverage of Arabs in Khandahar and of the Taliban leadership.

We were also working on the resumption of a long-stagnant counterterrorist relationship with the Russians. We thought it essential to make the attempt in light of Chechen linkages to al-Qa'ida. To date, the track record of data provision by the Russians had been poor, but we hoped to be able to exploit the unique access we believed they continued to have in Afghanistan.

If you are getting confused, frustrated, or exhausted reading this litany, imagine how we felt at the time living through it. And imagine how I and everyone else in the room reacted during one of my updates in late July when, as we speculated about the kind of attacks we could face, Rich B. suddenly said, with complete conviction, "They're coming here." I'll never forget the silence that followed.

Just about this same time, the National Security Council authorized us to begin deploying the Predator by September 1, in either an armed or unarmed reconnaissance mode. According to the order, we were to work out cost-sharing details with the Defense Department. Our belief was that deploying the Predator in unarmed reconnaissance mode was ill advised and unnecessarily exposed the capability. We preferred that the next time it was over Afghanistan that it be equipped to take immediate action if we spotted UBL. But the testing to date on the Predator's Hellfire warhead had shown mixed results.

I took the NSC action as a positive sign that the policy makers were beginning to engage the difficult issues of the war on terror, but we still needed a Principals' meeting to thrash out once and for all the administration's policy regarding our use of an armed Predator. I wanted to have the meeting as soon as possible, but given the technical difficulties with arming the Predator, the NSC decided to put it off until after Labor Day.

That summer, whenever a PDB contained information about possible al-Qa'ida attacks, the president would ask his PDB briefer, Mike Morell, what information we had that might indicate an attack could come inside the United States. With the president heading off to Crawford for much of August, Mike asked our analysts to prepare a piece that would try to address that question. That was the origin of the now-famous August 6 PDB titled "Bin

Ladin Determined to Strike in the US." Nearly the full text of the item appears in *The 9/11 Commission Report.* The report makes clear that nothing would have pleased UBL more than to attack in our homeland. But although clear about his desire and intent, we did not have and therefore did not convey information about any specific ongoing plot.

A few weeks after the August 6 PDB was delivered, I followed it to Crawford to make sure the president stayed current on events. That was my first visit to the ranch. I remember the president graciously driving me around the spread in his pickup and my trying to make small talk about the flora and fauna, none of which were native to Queens. By then, an eerie quiet had settled over our threat reporting—the lull before the storm. We learned much later that Bin Ladin was waiting for the president and Congress to return to Washington, after Labor Day. He knew our customs and habits well.

In August, I directed a thorough review of our files to identify potential threats. I didn't want to leave any stone unturned, even if that meant replowing old ground. Temporary calm or not, the threat attack was too real for us to sit back and wait. I later learned that CTC officials had begun a similar review even before I asked them to do so. It was during this period that they discovered cables from the year before that suggested that possible al Qa'ida operatives might have entered the United States. The issue involved two men, Khalid al-Mihdhar and Nawaf al-Hazmi, who later boarded American Airlines Flight 77 on the morning of September 11 and helped fly it into the Pentagon. (So much has been written and so much misunderstood about this "watchlisting" issue—and it became such a cornerstone of the 9/11 Commission's critique of the Agency—that I will deal with it in a chapter all its own.) It was also during this time when I first heard the name Zacarias Moussaoui. (This, too, requires a detailed discussion to be handled in a chapter ahead.)

By early September, CIA had a group of assets from a Middle Eastern service working on our behalf. None of the more than twenty individuals knew they were working for us. They were targeted against a range of terrorism issues. One third of them worked against al-Qa'ida. By September 2001, we had two unilateral agents successfully penetrate terrorist training camps in Afghanistan.

On September 4, the principals—Condi, Don Rumsfeld, others, and I—finally reconvened in the White House Situation Room. This was Tuesday, the day after Labor Day. Washington was coming back to life after surviving another sultry August. Under other circumstances, the Principals' meeting might have had the feel of a reunion. This one didn't. The meeting was dominated by the same subject that had been lingering unresolved all summer long: whether the president should approve our request to fly the Predator in a weaponized mode. Unfortunately, the Predator still wasn't ready to do that, although the Hellfire missile system was slowly edging toward being ready for deployment.

We also needed to debate the question of when the armed Predator was functional, who should operate it? There was a legitimate question about whether aircraft firing missiles at enemies of the United States should be the function of the military or CIA. It was an important issue, or so it seemed at the time, and I was skeptical about whether a military weapon should be fired outside of the military chain of command. But that was before 9/11.

Six days later, on September 10, a source we were jointly running with a Middle Eastern country went to see his foreign handler and basically told him that something big was about to go down. The handler dismissed him. Had we known it at the time, however, it would have sounded very much like all the other warnings we received in June, July, August, and early September—frightening but without specificity.

Less than twenty-four hours later, the unthinkable happened. But to us, it wasn't unthinkable at all. We had been thinking about nothing else.

CHAPTER 9

9/11

On the morning of September 11, the day that changed everything, I met former senator David Boren for breakfast at the St. Regis Hotel, at 16th and K Streets in Washington at eight thirty. The president was out of town, traveling in Florida, which meant there was no Presidential Daily Briefing. David had plucked me from obscurity in 1987 to serve as chief of staff of the Senate Select Committee on Intelligence, which he chaired. I looked forward, as I always do, to getting together with him that morning.

We were just starting to catch up when Tim Ward, who was leading my security detail that day, walked over with a worried look on his face. As befits his position, Tim is a calm, unflappable fellow, but his manner was so urgent when he interrupted us that there was no doubt that something important was on his mind. I stepped away from the table, and he told me that a plane had flown into the World Trade Center's South Tower. Most people, I understand, assumed that the first crash was a tragic accident. It took the second plane hitting the second tower to show them that something far worse was going on. That wasn't the case for me. We had been living too intimately with the possibility of a terrorist attack on the United States. I instantly thought that this had to be al-Qa'ida.

I told Senator Boren the news. He recalls my mentioning Bin Ladin and wondering aloud if this is what Moussaoui had been involved with. It was obvious to us both that I had to leave immediately. With Tim Ward, I climbed back into my car and, with lights flashing, began racing back to headquarters.

All the random dots we had been looking at started to fit into a pattern. As I remember it, in those first minutes my head was exploding with connections. I immediately thought about the "Bojinka" plot to blow up twelve U.S. airliners over the Pacific and a subsequent plan to fly a small airplane into CIA headquarters, which was broken up in 1994.

Our safe American world had been turned upside down. The war on terror had come to our shores.

En route, I called my chief of staff, John Moseman, and told him to assemble the senior staff in the conference room next to my office, along with key people from the Counterterrorism Center. With all hell breaking loose, it was hard to get calls through on the secure phone. Essentially, I was in a communications blackout between the St. Regis and Langley, the longest twelve minutes of my life. It wasn't until I arrived at headquarters that I learned that as we were tearing up the George Washington Parkway at something like eighty miles an hour, a second plane had hit the North Tower.

As the first reports came in of the planes hitting the World Trade Center, Lt. Gen. Mahmood Ahmed, head of Pakistan's Inter-Service Intelligence agency (or ISI), and among the people who could have done the most to help us track down Usama bin Ladin pre-9/11, was meeting on Capitol Hill with Congressman Lindsay Graham, Representative Porter Goss, who would eventually replace me as DCI, and others. A half hour later, Mahmood was being chauffeured along Constitution Avenue when someone pointed out a plume of smoke rising from across the Potomac—the first sign that the Pentagon had been struck. Simultaneously, Shafiq bin Ladin, UBL's estranged brother, was attending the annual investor conference of the Carlyle Group at the Ritz-Carlton Hotel, around the corner from me and just blocks from the White House. Three senior CIA officers—Charlie Allen, Don Kerr, and John Russack—were having a long-planned breakfast at the Agency with Navy Commander Kirk Lippold, who had been commanding officer of the USS *Cole* when the ship was attacked in Yemen. Much of the discussion, naturally, focused on terrorism. The Agency participants later told me that Lippold was distressed that the American people still didn't recognize the threat. It will take some "seminal event," he said, to awaken the public. After the breakfast, Lippold went to CTC for some briefings. When the World Trade Center was struck minutes later, Charlie Allen reached the commander and told him, "The seminal event just happened." Amazingly, Lippold rushed back to work, arriving just in time to see American Airlines Flight 77 plow into the Pentagon.

Even now, five years later, I find it hard to describe the mood in the conference room when I finally arrived. The time, I would guess, was about 9:15 A.M. Both World Trade Center towers had been hit, and I don't think there was a person in the room who had the least doubt that we were in the middle of a full-scale assault orchestrated by al-Qa'ida.

CTC head Cofer Black recalls speaking with Dale Watson, the head of counterterrorism for the FBI, in a kind of cryptic code all that day. I think that was probably true of most of us, to a greater or lesser degree. Sentences didn't need to be completed; half-expressed thoughts were fully understood. We had been at this so long, planning for it in so many ways.

But anticipating an attack and having it happen—seeing the collapse of the World Trade Center—are not the same things. The first is intellectual. The second quickly becomes visceral, and the anxiety level in the conference room in that first hour was extraordinary. Only minutes after the South Tower was hit, the Counterterrorism Center received a report that at least one other commercial passenger jet was unaccounted for. At 9:40, John McLaughlin and Cofer Black took part in a secure video conference with Dick Clarke, from the White House. By then, the Pentagon had just been hit, and we knew more planes were loose. On the heels of the Pentagon strike, phone calls started rolling in—not intelligence, just friends and colleagues relaying the rumors that were gripping Washington and expressing hope that we would know what was true and what was false: a bomb had gone off in the West Wing of the White House; the Capitol and the State Department were in flames. The fact was, we had no idea what was real and what wasn't, but everyone was wondering, what next? Reports came in of several airplanes that were not responding to communications from the ground and perhaps heading toward Washington. Several CTC officers reminded us that al-Qa'ida members had once discussed flying an airplane into CIA headquarters, the top floor of which we were presently occupying.

I can remember asking Mike Hohlfelder, the chief of my security detail, what he recommended. "Let's get out of here," he answered. "Let's evacuate." I was reluctant. We didn't want our own workforce or the world to think that we were abandoning ship. But I also didn't want to risk the lives of our own people unnecessarily, and as someone in the conference room pointed out, in case the building had been targeted, we needed to have our leadership intact and able to make decisions.

At about 10:00 A.M. word was sent out for a large number of our multi-thousand-person workforce to go home. They soon joined the horrendous traffic jam that choked Washington's roads. The White House had evacuated fifteen minutes earlier, just after the Pentagon was hit. In New York City, the United Nations complex, nearly twelve thousand employees strong, began clearing out at 10:13. Back in D.C., the State and Justice departments and the World Bank followed suit minutes later.

Initially, our senior leadership team moved from my seventh-floor conference room to one on the first floor—a bit safer, but still too vulnerable if an airplane came crashing into the building. We then left the building altogether, exiting via the southeast corner of the headquarters building and heading diagonally across the campus to the Agency's printing plant, where a makeshift operational capability had been installed.

One group stayed behind in headquarters. Cofer Black felt very strongly that the roughly two hundred employees in his Counterterrorism Center needed to maintain their positions both in the Global Reaction Center on the highly exposed sixth floor, where a shift of eight people routinely worked, and in a safer, windowless facility down low in the building, where the bulk of CTC was located.

"Sir," he said to me after I had issued the evacuation order, "we're going to have to exempt CTC from this [evacuation] because we need to have our people working the computers."

"Well," I responded, "the Global Response Center—they're going to be at risk."

"We're going to have to keep them in place. They have the key function to play in a crisis like this. This is exactly why we have the Global Response Center."

"Well, they could die."

"Well, sir, then they're just going to have to die."

According to Cofer, I paused for a moment, and said, "You're absolutely right."

Now that we were under attack, the Counterterrorism Center, with its vast data banks and sophisticated communications systems, was more vital than ever. Even as we were discussing going or staying, CTC was sending out a global alert to our stations around the world, ordering them to go to their liaison services and agents to collect every shred of information they could lay their hands on. I admired their unwavering courage and dedication. CIA headquarters is pretty much a glass house. If a plane had targeted it, the people in the Global Response Center could have watched their fate flying right at them.

Inside the printing plant, the initial scene was pretty chaotic. We had only rudimentary capabilities for access to all of our data and communications networks. In the aftermath, we all realized that we needed additional backup communications capabilities if and when a similar situation arose again. People were scrambling to get the phones operational and to get in touch with Mike Morell, the president's briefer, who was with George Bush in Florida when the first plane struck. As Mike would later tell the story, he, Karl Rove, and Ari Fleischer, the White House press secretary, were riding in a motorcade van when Ari took a call, then turned to Mike and asked if he knew anything about a small plane hitting the World Trade Center. Mike immediately called our Operations Center and was told that the plane wasn't small. Shortly afterward, waiting for the president to finish meeting with elementary school students and their teachers, Mike saw the second tower struck on TV. Later, aboard Air Force One, the president queried Mike about a Palestinian extremist group, the Popular Front for the Liberation of Palestine, or PFLP, which was taking credit for the attack in the press. Not likely, Mike told him. PFLP simply didn't have the capability for something like this. The president took that in and then told Mike that if we learned anything definitive about the attack, he wanted to be the first to know. Wiry, youthful looking, and extremely bright, Mike speaks in staccato-like bursts that get to the bottom line very quickly. He and George Bush had hit it off almost immediately. In a crisis like this, Mike was the perfect guy for us to have by the commander in chief's side.

Simultaneous with establishing contact with the president and his traveling party, we were trying to reach our office in New York City, to evaluate whether everybody was present and accounted for there, and trying to get as much data as we possibly could for ourselves. As happens in any crisis, anomalies kept surfacing, odd bleeps that in calm times would probably have meant nothing but in these times could have meant almost anything. One example: Airplanes are tracked via transponders. Every one of them emits a unique signal. At least some of the hijackers that morning had known how to turn off the transponders so that their planes would be harder to track. Now a commercial passenger jet on its way to Great Britain was emitting all kinds of squawks, with the transponder going off and on. Had al-Qa'ida launched a two-continent attack? Ultimately, the matter was resolved—there was no nefarious intent; the transponder was simply faulty—but in the interim I called Richard Dearlove, my counterpart at MI-6, to tell him what we were hearing and what we knew.

Although in our collective gut we knew al-Qa'ida was behind the attacks, we needed proof, so CTC requested passenger lists from the planes that had been turned into weapons that morning. Incredibly, I was later told, the initial response from some parts of the bureaucracy (which parts since mercifully forgotten) was that the manifests could not be shared with CIA. There were privacy issues involved. Some gentle reasoning, and a few four-letter words later, the lists were sprung, and an analyst from CTC raced over to the printing plant. "Some of these guys on one of the planes are the ones we've been looking for in the last few weeks." He pointed specifically to two names: Khalid al-Mihdhar and Nawaf al-Hazmi. That was the first time we had absolute proof of what I had been virtually certain of from the moment I heard about the attacks: we were in the middle of an al-Qa'ida plot.

Around this same time, the vice president called to ask if we could anticipate further attacks. By then, a fourth plane, United Flight 93, had gone down in Shanksville, Pennsylvania. There was a lull in the action, and to me that was telling. "No," I told him. "My judgment is that they're done for the day." It was a gut call; I had no data to go on. But the pattern of spectacular multiple attacks within a very tight attack window was consistent with what we knew of al-Qa'ida's modus operandi based on the East African embassy attacks and others. Events happened within a strict timeline, and then they were done.

Like everyone else in America, we were all working through our own personal dramas as the morning progressed. My brother, who happened to be in Washington on business, called early on, anxious to get back to New York City, where his wife and family and our mother live. Was any public transportation running? Was it possible, was it safe, to fly? I told him no, and so he rented a car and headed home. My mother was in a panic, as I knew she would be. I called her and told her I was safe. Stephanie, meanwhile, was phoning other family members, assuring them that the CIA building hadn't been hit.

Our son, John Michael, was then starting ninth grade at Gonzaga, a Jesuit Catholic high school not far from Capitol Hill. A CIA security detail found him there, took him under its wing, and transported him out to our house. Everybody in Washington in a position like mine fears for his kids in this new, terrorist-driven world we live in. Not to John Michael's liking in the least, this was the beginning of a permanent security detail that would follow him just about everywhere while I remained DCI.

Stephanie had the worst of it that morning, by far. Sometime around midday she got a call from Tom Heidenberger. He and his wife, Michele, were old friends. Our sons had gone to elementary school together and now were classmates at Gonzaga. Michele was a flight attendant for American Airlines. Tom wasn't certain, but he thought she had been scheduled to work Flight 77 that morning, the plane that had hit the Pentagon. Although Tom was a pilot himself, for USAir, he couldn't get his own employer or American Airlines to tell him if Michele had been on board. Could Stephanie call me, he asked? She did. I asked to see the manifests that had just come into our possession. The names were in alphabetical order. First the passengers were listed. Below them were the names of the crew. My heart sank as I read the name Michele Heidenberger.

I called Stephanie with the news, and she drove over to the Heidenbergers' home in Chevy Chase to break it personally to Tom. Michele was fifty-seven years old, the mother of two.

Although I didn't immediately notice his name on the list, one of my high school buddies, Bob Speisman, was also a passenger on Flight 77.

My staff and I left the printing plant and returned to headquarters about one o'clock that afternoon. The danger was over for the day, in our estimation, and all of us felt isolated at the printing plant. One of my senior staff later told me that not long before we left the printing plant, he said to a colleague that the attacks were going to be viewed as a huge intelligence failure, and the colleague had looked at him incredulously and replied something like, "Why would this be an intelligence failure? These things happen. This is a war. This is a battle." I don't know what I would have said at that moment if the same suggestion had been made to me. The death count was clearly mounting into the thousands. Finger-pointing of any kind, at us or at someone else, was the remotest thing from my mind. But somewhere, I suppose, the blaming had already begun. Maybe that's inevitable. Maybe it's just the way Washington works.

That afternoon passed mostly in a blur of meetings. The historical record tells me that there was a 3:30 P.M. teleconference, again over a secure line, with the president, who had touched down at Offutt Air Force Base, in Nebraska, while zigzagging his way back to Washington. The president was speaking from the underground headquarters of the U.S. Strategic Command.

I remember him asking me who I thought had done this. I told him the same thing I had told the vice president several hours earlier: al-Qa'ida. The whole operation looked, smelled, and tasted like Bin Ladin, and the passenger manifests had all but confirmed our suspicions. When I told the president particularly about al-Mihdhar and al-Hazmi, he shot Mike Morell one of those "I thought I was supposed to be the first to know" looks. Mike placed an angry call to my executive assistant, Ted Gistaro, who was in his second day on the job, asking to see the talking points I had prepared for the exchange. "Can't do," Ted told him. "They're embargoed." "Embargoed from the president of the United States?" Mike shot back. It was one of those little flaps that happen when everyone is working under great stress. Before long Mike was able to pass what information we had to the president through Andy Card. Also in my talking points that afternoon was a warning we had received from French intelligence that said another group of terrorists was within U.S. borders and was preparing a second wave of attacks.

Throughout the teleconference, the president was focused, in control. That evening's face-to-face meeting with him only served to confirm my first impression.

By the time I arrived there, sometime after nine o'clock, the

White House was an armed fortress. I was too busy reading briefing papers, though, to notice whatever extra protection had been laid on. My car had no sooner pulled to a stop than the Secret Service escorted me through a long, elaborate passageway to the bunker, a place I had never visited before and would never be in again. The president and vice president were both there, along with Dick Clarke, Condi Rice, Colin Powell, Don Rumsfeld, Joint Chiefs chairman Gen. Hugh Shelton, and a few others, including Lynn Cheney and Laura Bush.

"Bunker," I realize, implies sandbag fortifications and artillery shells bursting overhead. This wasn't that. The White House bunker is basically a stripped-down and hardened Situation Room—but there was a definite warlike feel to the room and, that day, more raw emotion in one place than I think I've ever experienced in my life: anger that this could have happened, shock that it had, overwhelming sorrow for the dead, a compelling sense of urgency that we had to respond and do so quickly, and a continuing feeling of dread about what might lie ahead. Al-Qa'ida was through for the day, or so we believed, but plenty of intelligence data suggested that this was intended as the opening act of a multi-day sequence. Even at this early point, too, there was a growing fear—one that would spread in the days ahead as fresh reports came in—that the terrorists had somehow secreted a weapon of mass destruction into the United States and were preparing to detonate it.

At eight thirty that evening, speaking from the Oval Office, the president addressed the nation in terms both stirring and deeply earnest, including the first enunciation of what became known as the Bush Doctrine. "I've directed the full resources of our intelligence and law enforcement communities to find those responsible and to bring them to justice," he told a global audience of some eighty million people. "We will make no distinction between the terrorists who committed these acts and those who harbor them." For us at CIA, the new doctrine meant that the restraints were finally off. We already had on our shelves the game plan for going after both al-Qa'ida and its protectors, the Taliban, in Afghanistan. Now we could begin to implement it. Amid the sorrow of the day, we realized that we were finally going to be given the authorization and the resources to do the job we knew had to be done.

The president followed the Oval Office address with a meeting with the full National Security Council, in this same bunker. Now it was down to what amounted suddenly to a war cabinet. Back only hours earlier from Peru, Colin Powell talked about the problem in diplomatic terms: We had to make it clear to Pakistan as well as to Afghanistan that the time for equivocation was over. I was probably more forward-leaning: Yes, we needed Pakistan's help; it was the country closest to Afghanistan and the one with the most sway over it. But the time for talking with the Taliban had come and gone. To go after Bin Ladin and his shadow army, we had to remove the curtain they hid behind. The president said we had to force countries to choose. The vice president weighed in with several questions about finding targets in Afghanistan worth hitting. But what I remember more than anything else about that meeting was the president's manner, not his words. He was absolutely in charge, determined, and directed. He stressed the urgency of the moment, and he made it clear, by word and example, what his expectations were for us in terms of thinking through how we would respond.

No doubt about it, 9/11 was the galvanizing moment of the Bush presidency. It transformed him in ways I don't think any of us could have fully predicted. His leadership in the months ahead made a huge difference.

My senior staff was waiting for me when I got back to Langley that evening. The official record of my schedule for the day ends at 11:00 P.M., but I think that just means that Dottie Hanson finally went home then. My own recollection is that I left headquarters closer to one o'clock in the morning, for not much more than a long nap, a shower, and a change of clothes. I was due back at the White House early the next morning. A day like 9/11, though, never really ends, except by the clock.

One evening, several days after 9/11, Stephanie and I took some time to visit Tom Heidenberger to see how he was coping with the death of Michele. It was still so hard to believe. Tom wanted to see for himself where she had died, but at the time it was impossible for civilians to get anywhere near the Pentagon, where efforts to recover remains of those killed in the building were continuing. We got in my SUV and were driven to the Pentagon by my security detail. Flashing badges at countless roadblocks, we finally reached an area overlooking the twisted ruins at the Pentagon. Tom brought a bouquet of flowers to leave at the site where his wife and so many others had died. Being there with Tom and knowing that thousands of other American families were enduring similar pain was one of the saddest things I have ever experienced.

John McLaughlin, Jim Pavitt, Cofer Black, and I talked often in the first months after the attacks about the emotional toll the attacks were taking on our employees. Everyone was working overtime; everyone was strained. We kept waiting for and preparing for an emotional response, especially on the part of Cofer's people in the Counterterrorism Center. By and large, though, it never came. Somehow along the way, I missed my own emotional buildup. That came to a head on the day after Thanksgiving.

That Friday was the first day I had taken off in well more than two months, since the weekend before the attacks. I had used up whatever reserve of adrenalin I'd been running on. Sometime during my morning of supposed leisure, I went out in front of our house, sat down in my favorite Adirondack chair, and just lost it. Whatever the trigger was, the whole thing came down on me at that moment. I thought about all the people who had died and what we had been through in the months since. How in God's name had this happened? I remember asking myself. How in hell could I have been on top of all this? What am I doing here? Why me? Why am I living through this? The questions were flying through my head. Stephanie came out about then. I'd been alone up to that moment, except for the security detail watching the house from the street, and thinking who knows what. I recall Stephanie's saying to me, "You're supposed to be here. This is something that you've been working on all your life, and you've got a lot more work to do." And that did it. That snapped me out of it, but it was a black, black time until then.

The one thing that so many people have missed about CIA and 9/11, including the 9/11 Commission so far as I could tell, is that it was personal with us. Fighting terrorism is what we do; it's in our blood. In the months and years leading up to 9/11, we had worked this ground every day. To thwart the terrorists we disrupted attacks, we saved lives. We sacrificed our lives, too, often figuratively and sometimes literally.

If the politicians and press and even the 9/11 Commission often failed to understand this, our global partners in the intelligence business had no doubt. We were still sorting out the details on 9/11 when Avi Dichter, the chief of Shin Bet, called from Israel to express his regrets and say that he and his people were with us, no matter what. This wasn't a bureaucratic call. Avi and I had lived through Arafat together and much more, but there was a connection through that phone call that went far beyond anything that had preceded it. Be strong, Avi told me. Lead your people. He didn't have to say that he had seen hundreds of his own countrymen killed by terrorists, on his watch, and I didn't have to add that I now understood what it was like to be the chief of the service when the same thing happened on my soil. All that was implicit, and stronger because it never had to be spoken. Several years later, though, in taping a farewell message for Avi's retirement ceremony, I put into words what I felt so strongly about 9/11: "We all became Israelis on that day," I told Avi.

Despite the constraints on air travel into the United States, the British came over on September 12: Sir Richard Dearlove, the chief of MI-6; Eliza Manningham-Buller, the deputy chief of

MI-5; and David Manning, Prime Minister Blair's foreign policy advisor. I still don't know how they got flight clearance into the country, but they came on a private plane, just for the night, to express their condolences and to be with us. We had dinner that night at Langley, an affirmation of the special relationship between our two nations and as touching an event as I experienced during my seven years as DCI.

Signs of support kept pouring in. King Abdullah and Queen Rania of Jordan called to express their condolences. Gen. Mohammed Mediene, the Algerian intelligence chief, was in Washington when al-Qa'ida struck. Like Avi Dichter, he knew up close the pain and challenge of terrorism, and he, too, could not have conducted himself in a more dignified manner or been more sympathetic to our suffering.

All of these people knew how much 9/11 had struck at the core of each of us at CIA. They'd been there; they'd shared our same fears; they knew that each of the thousands of dead was a personal defeat for us. And I'm sure they would have understood as well as anyone outside CIA the reaction so many of us—at the leadership level and in the ranks—had in the hours and days immediately after the attack. We're going to run these bastards down no matter where they are, we told ourselves. We're going to lead, and everybody else is going to follow. And that's what we set out to do.

CHAPTER 10

"We're at War"

On September 12, the president chaired an NSC meeting and stressed in stronger terms what he had said on television the evening before: he wanted not just to punish those behind the previous day's attacks but to go after terrorists and those around the globe who harbored them.

The next day, in the White House Situation Room, I briefed the president and War Cabinet for the first time on our war plan. "We're prepared to launch in short order an aggressive covert-action program that will carry the fight to the enemy, particularly al-Qa'ida and its Taliban protectors," I said. "To do that, we will deploy a CIA paramilitary team inside Afghanistan to work with opposition forces, most notably the Northern Alliance, and to prepare the way for the introduction of U.S. Special Forces." There were challenges, I told the Cabinet. Ahmed Masood's assassination on September 9 had left the Northern Alliance without a powerful and widely respected central figure, but we had technology on our side and an extensive network of sources already in country, and we would succeed.

Cofer Black followed me with a PowerPoint presentation that detailed our covert action capability, projected deployments, and the like. As I had, Cofer made it clear that we would be taking on not just al-Qa'ida but the Taliban as well. The two were inseparable unless the Taliban chose to make the separation itself, and that seemed unlikely, despite our best efforts to drive a wedge between them. We would be undertaking war, in short, not just a search-and-destroy mission for Bin Ladin and his lieutenants—war against an enemy that for the most part would rather blow itself up than be captured. That meant casualties on their side and on ours. Cofer made no effort to predict how many Americans might be killed, but he did make certain the president understood that the mission wouldn't be bloodless. Bush assured him that he did.

"How quickly could we deploy the CIA teams?" the president asked.

"In short order," Cofer answered.

"How quickly, then, could we defeat the Taliban and al-Qa'ida?"

"A matter of weeks," Cofer told him.

I didn't think that was possible; and in fact it wasn't. The president had been disappointed to learn that the Pentagon had no contingency plan in place for going after al-Qa'ida and the Taliban. George Bush was going a hundred miles an hour by then, completely engaged. If you couldn't keep up, he wasn't interested in you.

The point Cofer and I both wanted to make was that this war would be driven by intelligence, not the pure projection of power. The challenge wasn't to defeat the enemy militarily. The challenge was to find the enemy. Once that was done, defeating him would be easy.

On Friday, September 14, we refined our plan further so that Afghanistan was only the opening act of a comprehensive strategy for combating international terrorism. Then we did a dry run in preparation for my presenting the plan the next day at Camp David. That evening, the NSC sent us stacks of papers to review before we arrived at Camp David, input from what must have been every stakeholder in the intelligence and military sectors of government. I remember thinking as I waded through them that hundreds of trees had been killed for no good reason. The papers were irrelevant, as near as I could tell, to anything I was going to say, and by then I was so confident in the rightness of our approach that I had little use for the half measures and unformed strategies that other agencies were beginning to trot out.

Saturday, September 15, accompanied by John McLaughlin and Cofer Black, I briefed the War Cabinet at Camp David. The president was sitting directly opposite me across the big square table in the rustic Camp David conference room, with the vice president and Colin Powell on either side of him. Others present included Don Rumsfeld and Paul Wolfowitz, sitting side by side, Condi Rice, Steve Hadley, Rich Armitage, Attorney General John Ashcroft, and the new FBI director, Robert Mueller.

The title of the briefing was "Destroying International Terrorism." The heading on the first page read: "The 'Initial Hook': Destroying al-Qa'ida and Closing the Safe Haven." Cofer Black and I launched into the distinct pieces of the plan.

We had to close off Afghanistan by providing immediate assistance to the Northern Alliance and their remaining leaders, and accelerate our contacts with southern Pashtun leaders, including six senior Taliban military commanders, who appeared willing to remove Mullah Omar from power. This built on work we had begun in early 2001 to engineer a split between the Taliban leadership and Bin Ladin and his Arab fighters. We had to seal off Afghanistan's borders by directly engaging the Iranians, Turks, Tajiks, Uzbeks, and Pakistanis.

We told the president that our only real ally on the Afghan border thus far had been Uzbekistan, where we had established important intelligence-collection capabilities and had trained a special team to launch operations inside Afghanistan. We knew that Uzbekistan would be our most important jumping-off point in aiding the Northern Alliance.

We raised the importance of being able to detain unilaterally al-Qa'ida operatives around the world. We understood that to succeed both inside and outside Afghanistan we would have to use the large infusion of money coming our way to take the activities of our foreign partners to new levels in operating against al-Qa'ida.

Some of our most important regional allies could create a cadre of officers who could blend seamlessly into environments where it would be difficult for us to operate on our own. We told the president that we would be relentless in maximizing the number of human agents reporting on terrorist organizations. We also proposed immediate engagement with the Libyans and Syrians to target Islamic extremists.

We suggested using armed Predator UAVs to kill Bin Ladin's key lieutenants, and using our contacts around the world to pursue al-Qa'ida's sources of funding, through identifying nongovernmental organizations (NGOs) and individuals who funded terrorist operations.

We were going to strangle their safe haven in Afghanistan, seal the borders, go after the leadership, shut off their money, and pursue al-Qa'ida terrorists in ninety-two countries around the world. *We were ready to carry out all these actions immediately, because we had been preparing for this moment for years.* We were ready because our plan allowed us to be. With the right authorities, policy determination, and great officers, we were confident we could get it done. Others may have seen it as a roll of the dice. But we were ready, and the president was going to take the chance.

Sure, it was a risky proposition when you looked at it from a policy maker's point of view. We were asking for and we would be given as many authorities as CIA had ever had. Things could blow up. People, me among them, could end up spending some of the worst days of our lives justifying before congressional overseers our new freedom to act. But everything we asked for that day at Camp David and in subsequent days was based on the solid knowledge of what we needed. Nobody knew this target like we knew it. Others hadn't been paying attention to this for years as we had been doing. And nobody else had a coordinated plan for expanding out of Afghanistan to combat terrorism across the globe. Operationally, as far as we were concerned, the risk was acceptable. That didn't mean we weren't going to lose people—Cofer had made that crystal clear—but this was the right way to go, and we were the right people to do it.

The morning session at Camp David was freewheeling, all over the place. Sometime around noon, the president suggested we take a break. When we reassembled that afternoon, the discussion was much more directed, and the president was in full agreement with just about everything we had said during the day. "That's great," he said about our war plan. The whole mood was one of growing optimism.

The next day, September 16, I fired off a memo titled "We're at War" to top officials at my own shop and throughout the intelligence community, which said in part:

> *There can be no bureaucratic impediments to success. All the rules have changed. There must be an absolute and full sharing of information, ideas, and capabilities. We do not have time to hold meetings to fix problems—fix them—quickly and smartly. Each person must assume an unprecedented degree of personal responsibility.*

Four days later, on September 20, in an address to the nation before a joint session of Congress, the president said, "Our war on terror begins with al-Qa'ida, but it does not end there. It will not end until every terrorist group of global reach has been found, stopped, and defeated." By then, as I remember, the president had already granted us the broad operational authority I had asked for.

Now that we had been thrown on to a war footing, issues that had seemed intractable just days earlier suddenly seemed far less set in concrete. The Pakistan problem is one such example. On September 13, Rich Armitage invited Pakistani ambassador Maleeha Lodhi and Mahmood Ahmed, the Pakistan intelligence chief, who was still in Washington, over to the State Department and dropped the hammer on them. The time for fence-sitting was over. There would be no more games. George Bush had said in his 9/11 address to the nation that the United States would make no distinction between terrorists and the nations that protected them.

Pakistan was either with us or against us. Specifically, Armitage demanded that Pakistan begin stopping al-Qa'ida agents at its border, grant the United States blanket overflight and landing rights for all necessary military and intelligence operations, provide territorial access to American and allied intelligence agencies, and cut off all fuel shipments to the Taliban. Armitage is a bull of a man. Mahmood must have felt like he had been run over by a stampede by the time he left Rich's office. I seriously doubt, however, that Rich actually threatened to "bomb Pakistan back to the stone age," as General Mahmood reportedly later told President Musharraf. Meanwhile, I was playing the good cop—or at least a better one—in my meetings with Mahmood. Couldn't he at least meet with Mullah Omar and make it crystal clear to him that the Taliban was going to pay a terrible price if it insisted on continuing to protect al-Qa'ida and Bin Ladin?

The president, too, became engaged in the matter in a way he had never been before the attacks. At the September 13 morning briefing, he asked me for a country-by-country review of the fight against Islamic extremism and Bin Ladin. What had their liaison services done in the past year to help us? What more could we ask of them? Would a call from the president or some other senior government official be useful? As always, Pakistan was at or near the top of the list.

All those factors played a role in edging Mahmood toward our position, but the simple fact that he was in Washington when the attacks occurred probably had the greatest influence. He saw the plume of smoke rising from the Pentagon. He watched the reaction all around him, and he understood as he never could have if he had been following events from Islamabad how deep and viscerally Americans felt the attacks. "It was like a wounded animal," is how he put it to us. That didn't stop him from continuing to throw up lots of cautions—even after the attacks, Mahmood was still trying to save the Taliban—but now he knew that if we did not get satisfaction, we were still coming after al-Qa'ida no matter who objected or who tried to stand in the way.

That, I'm sure, is why Mahmood finally did agree to meet with Mullah Omar after he returned home. As a result, Omar called a two-day ulama—a kind of national religious council—to decide what to do about al-Qa'ida and our demand that the Taliban stop sheltering terrorists. Ultimately, of course, that availed us nothing, despite some initial optimism on our part. Bin Ladin wasn't handed over, which assured that the full might of the U.S. military would come crashing down on the Taliban's head. But across the border in Pakistan, Pervez Musharraf clearly got the message we were sending him and, I can only assume, the message Mahmood sent back to Pakistan immediately after the attacks. Within hours of Armitage's delivering his ultimatums, and despite some violent internal opposition, Musharraf agreed to them. In this period, Pakistan had done a complete about-face and become one of our most valuable allies in the war on terrorism. On October 8, as a final measure of his determination to aid America in rooting out al-Qa'ida, Musharraf replaced Mahmood Ahmed as head of the ISI, even though he had been instrumental in Musharraf's rise to power. Like us, Musharraf must have concluded that in the new global reality, his intel chief was just too close to the enemy. Whatever the reason, I've always considered Musharraf's reversal to be the most important post-9/11 strategic development after the takedown of the Afghan sanctuary itself.

Hard on the heels of 9/11, we also ramped up our own intelligence collection procedures. In normal times, principal agents gather information via runners who have penetrated into or near the heart of an organization of interest. Episodically, runners and the agents who control them meet, information gets exchanged, and whatever qualifies even marginally as "intelligence" is passed up the chain, either directly to the analysts back at Langley or via the remote chain of command that the principal agents report to. Like all bureaucratic models, this one has its drawbacks, principally of time—working even fast channels creates enough friction to sometimes turn fresh news stale—but it does provide maximum security for all involved.

If 9/11 had taught us anything, however, it was that we couldn't let the people who were dedicated to our destruction sit comfortably in their safe havens while we followed the usual routines and employed the normal safeguards. We needed real-time reporting from the field, and to get it we threw out the book.

We were beefing up our contingent in Pakistan by the hour. Carpenters hammered and sawed through the middle of the night to create new offices, including one room where we had phones lined up to receive calls, each one marked with an index card so the duty officer would know who was checking in and what language—Farsi, Dari, whatever it was—would be needed to take the message.

We made our own pass at coopting the Taliban. As Mahmood was preparing for his meeting with Mullah Omar, Bob Grenier, a senior CIA officer in the region, traveled to a hotel in the mountains of Baluchistan, in Pakistan, to meet with Mullah Osmani, the commander of the Taliban's Khandahar Corps, a man then widely acknowledged to be the second-most powerful figure in the movement, next to Mullah Omar. The general and his small entourage had traveled overland from Khandahar. Surrounded by the luxuries of a five-star hotel, and with one of the general's aides taking painstaking notes so that the proceedings could be carried back to Omar, Grenier first explained the obvious: al-Qa'ida was going to pay dearly for what had been done to the United States, and if the Taliban stood in the way, it would suffer equally. Then he proposed multiple solutions. The Taliban could turn Bin Ladin over to the United States for prosecution. If that violated their religious obligation to be good hosts, they could administer justice themselves, in a way that clearly took him off the table. Or if they wanted to save face altogether, they could stand aside and let the Americans find Bin Ladin and extricate him on their own. That night, Bob slept fitfully in a hotel room directly across the hall from Osmani—"a stone-cold killer," as he describes him—and the next morning he departed and filed a report that reads like a chapter from a spy novel.

When I carried it to the White House, President Bush read the report with rapt attention.

Not surprisingly, Omar spurned our suggestions, so in a subsequent October 2 meeting with Osmani at a villa in Baluchistan, Grenier proposed an alternative solution: overthrowing Omar. Osmani could secure Khandahar with his corps, seize the radio station there, and put out a message that the al-Qa'ida Arabs were no friends of the Afghans and had brought nothing but harm to the country and that Bin Ladin must be seized and turned over immediately. That, too, came to nothing, but just to make the proposal to a killer such as Osmani took considerable guts on Grenier's part.

While we were accelerating intelligence-gathering and doing our best to turn the screws on al-Qa'ida and the Taliban, we were also loosening constraints on our own people and their imaginations. In less than a century, warfare had evolved from massed armies and trench-to-trench battles to guerrilla confrontations and mutually assured destruction to the jihadist-terrorist model that dominates our own time. To keep up, we had to toss out old systems and shake loose from outdated stereotypes.

We had worked hard prior to 9/11 to break down the old protocols, to make ourselves less of a top-down organization. CIA has one of the deepest and most varied pools of talent in the world; our field officers have done things that you will not read about in spy novels. To me, it made no sense to bring a deputy director or associate director to a meeting with, say, the president, just because rank seemed to demand it. I wanted to take the person closest to the action, the one with hands-on experience, to tell the commander in chief what was really happening. Sometimes I had to drag them along, especially if they had just flown in from some hot spot on the far side of the world and wanted a good shower and a day to sleep, but for the most part, I think, they took it as a sign of respect for what they had done and sacrificed, and for the knowledge they had gained as a result.

Post-9/11, we redoubled that effort. I'd show up at the White

House or at Camp David with people with dirt under their nails and in rumpled clothes, their having just gotten off an airplane returning from the war zone. No government bureaucracy can ever be entirely flat, but those of us in the top positions at CIA worked hard to make our bureaucracy as horizontal as it could be.

We did essentially the same thing with our officers in the field—we gave them the go-ahead to make calls on their own at the point of contact with the enemy. Flattening the authority pyramid gave us real-time decision making. In part, we had no choice. Terrorism wasn't just al-Qa'ida. If there was to be war—and that seemed inevitable—it wouldn't be fought only in Afghanistan. We were facing a worldwide threat matrix, and we had to respond globally with a labor pool that was already stretched perilously thin.

As the fall of 2001 went on, we would meet daily at headquarters to review the threat reporting—what we'd heard about over the last day, whether we'd notified those who were threatened, what we were doing about the threats. It was amazing how often we would pick up a lead in, say, South America about someone in Yemen we wanted to take off the street. Terrorists are as interconnected as the rest of us in the borderless cyber world. If the operation was high risk, John McLaughlin or I would have to make the call to go ahead. Far more often than not, though, the call would be made at a lower level or out in the field. We gave our people plenty of running room because they needed it, because we made sure they were fully briefed about what the Agency was trying to achieve and because they were, in the overwhelming majority, incredibly competent. The war in Afghanistan only accelerated that trend. If we had tried to micromanage that roll across the desert from the seventh floor of headquarters, we would still be on the road to Kabul today.

Around midnight on September 12, after a late dinner with the British intelligence chiefs who'd flown over to express their condolences, I was sitting in my office kicking ideas around with Jami Miscik, our second-most senior analyst at the time. I told her that I wanted to create a group within CIA whose sole purpose in life would be to think contrarian thoughts. The cliché in Washington is to "think outside the box," but I didn't want us to get just beyond the edge of the ordinary. I wanted people so far out of the box they would be in a different zip code. Jami loved the idea, and within fifteen minutes or so, we had dubbed the group the "Red Cell."

We picked out participants as we sat there, called them that night despite the late hour, and told them to be in Jami's office at eight the next morning. One of the leaders was Paul Frandano, a Harvard-trained senior analyst with a goatee and a liking for colorful bow ties. Not your typical academic, Paul has a mischievous sense of humor and delights in contrarian thinking. Our goal was to free some of our best people from purely objective considerations. These were men and women steeped in analysis. Their intellectual foundation was built solidly on fact, or as close to "fact" as intelligence work often gets. Now we asked them to take an imaginative leap from that, to try to get inside the mind and imagination of our enemy. Over the months ahead, we gave them a variety of specific topics to write about. Among them: "How Usama Might Try to Sink the U.S. Economy," "Deconstructing the Plots—An Approach to Stopping the Next Attack," and everyone's favorite, "The View from Usama's Cave." The latter—issued on October 27 and number twenty-two in the series—gave Red Cell participants a chance to speculate on what was going through Usama bin Ladin's mind and what he might be saying to his key lieutenants three weeks into the U.S. attacks on Afghanistan. Among the quotes it imagined for UBL were these: "I see no need to rush out with new strikes against America" and "I will give more operational scope to my lieutenants. I will instruct them to hold to my standards, but they will make their own decisions about when to strike."

Every Red Cell report was accompanied by a statement on the left-hand side of the front page: "In response to the events of 11 September, the Director of Central Intelligence commissioned

CIA's Deputy Director for Intelligence to create a 'red cell' that would think unconventionally about the full range of relevant analytic issues. The DCI Red Cell is thus charged with taking a pronounced 'out-of-the-box' approach and will periodically produce memoranda and reports intended to provoke thought rather than to provide authoritative assessment." For all I know the other government agencies who received the reports thought we'd gone round the bend, but I believe the reports worked extraordinarily well, in terms of both their imaginative content and the insight they offered into the real world. The events of September 11 weren't business as usual; we couldn't begin to shape our response in the usual way. To my mind, at least, that spirit had a domino effect throughout CIA in the days and weeks after 9/11.

Our December 2000 Blue Sky memo was the template for the war plan against al-Qa'ida that we would set out to follow within hours of the first plane hitting the World Trade Center. Ever since that template had first been laid out, a group of specialists from our Counterterrorism Center had been massaging and refining the plan, and by 9/11 they had it as right as anything can be in an undefined and constantly changing war theater. I'll never forget what one of our top Afghan strategists, a much-decorated veteran of the Agency, told me after the war there had been fought and won, because it encapsulates everything I feel about the campaign and the great pride I take in having the opportunity to serve with such people: "What I thought was really remarkable about the Bin Ladin program," he said, "wasn't just the hard work, the people going around the clock, but their intellectual development. They were able to coordinate all these different pieces and work with liaisons and send teams out. It was remarkably complex, and I think they paved the way for the successes we're having today. No one else in the U.S. government had ever done that—this is really the beginning of the evolving global battlefield—and a little team down in CTC basically figured this out and set the course for how we wage counter CIA-centric focus terrorism war on the global battlefield."

I couldn't agree more. Maybe it's my own obsession, but I can't stress this enough. We—CIA, the intelligence community, investigative bodies, the government at large—missed the exact "when and where" of 9/11. We didn't have enough dots to connect, and we'll always have to live with that. But at CIA we knew al-Qa'ida was coming, and afterward we took the fight to them in a way that I feel certain Usama bin Ladin and his lieutenants and protectors never expected in their worst-case scenarios.

On September 27, sixteen days after the World Trade Center and the Pentagon had been hit, we inserted our first covert teams into Afghanistan. Less than two and a half months later, a core group of ninety CIA paramilitary officers, along with a small number of Special Forces units, in combination with Afghan militias and supported by a massive aerial bombardment by the U.S. military, had defeated the Taliban and killed or captured one quarter of Usama bin Ladin's top lieutenants, including his military commander, Mohammed Atef, a key player in the 9/11 attacks. Kabul had been liberated, and Hamid Karzai named president by a national council. Afghanistan would be CIA's finest hour.

For years I had been trying to convince two administrations that the terrorist threat was seamless—that what had happened overseas to our East African embassies and the USS *Cole* could happen here. Now the seamlessness could no longer be ignored. "There" and "here" had become the same place. The world was one single war theater.

John McLaughlin remembers my calling him from the White House sometime shortly after the attacks and saying, "We have to put down on paper what we think al-Qa'ida's targets are. I know we don't know—but place your bets." We got all our top people around the table, ran through all the possibilities, and came up with a potential hit list. High on it were symbols of

American culture such as movie studios, amusement parks, and sports stadiums, and transportation hubs such as airports, harbors, and bridges. Corporate headquarters and other elements of the economic system were also listed along with military sites; the energy infrastructure, especially targets that would make a visible statement about energy dependence; icons of our national identity (the Washington Monument, the Statue of Liberty, even Mount Rushmore); and the nodes of the global telecommunications central nervous system, including the Internet and electronic bank transactions. We also noted that Bin Ladin often took years to plan his attacks and liked to return to the same targets, as witnessed by the World Trade Center. It would be reckless to provide more details—the last thing I want is to do the terrorists' work for them—but the effect of seeing so many prime targets in one four- or five-page report was galvanizing.

Based on our assessment, I called Jack Valenti, then head of the Motion Picture Association of America, and told him to make sure his industry was buttoned down. I also met with people such as Michael Eisner from Disney; Gary Bettman, the commissioner of the National Hockey League; and National Basketball Association commissioner David Stern; to urge them to step up security at their venues.

Our stark assessment, I believe, played a large part in the president's conclusion that somebody needed to be paying attention full time to protecting Americans inside our own borders, and in the subsequent decision to establish a Department of Homeland Security. For years, we at CIA had been playing offense against the terrorists overseas, but no one had been playing defense against them at home. It's an old axiom among football coaches: offense alone never wins.

The president asked John McLaughlin in late September, "Why do you think nothing else has happened?" To me, there's no mystery. We'd done what the president had asked: we all were up on our toes. It's hard to prove a proposition by the absence, in this case, of follow-up attacks on American soil, but I can't help but think that somewhere along the way in those first weeks after 9/11, someone who was supposed to do something crucial—buy forged passports, say, for a second team of terrorists, or sneak some kind of weapon or explosive over the border—was discouraged or disrupted or otherwise thwarted by what we and the FBI and the border patrol and city police forces and lots of other newly alert Americans were doing. In the battle against terrorism, I truly believe that heroes are everywhere.

CHAPTER 11

Missed Opportunities

Could anything have prevented 9/11? Despite a vast amount of fact-finding by the 9/11 Commission, journalists, authors, and many others, that question continues to haunt all of us involved in U.S. counterterrorism. Both the 9/11 Commission and the Congressional Joint Inquiry said that stopping the attacks would have been unlikely, but that doesn't prevent all of us from asking—what if? I certainly don't pretend to offer definitive answers here, but I will try to strip away some of the confusion and bluster surrounding two complex and frequently misunderstood missed opportunities: the oddly intersecting matters of "watchlisting" (placing suspected terrorists on lists to prevent their entry into the United States) and the arrest of Zacarias Moussaoui.

These two issues illustrate how Washington operates under its own laws of physics. One rule inside the Beltway is that for every action there is an unequal and opposite overreaction. Here is an example. The cover of the June 3, 2002, edition of *Time* magazine read "The Bombshell Memo." Inside was an article titled "How the FBI Blew the Case." The lengthy piece recounted how an unknown FBI agent, Coleen Rowley, had just sent a thirteen-page letter to FBI director Bob Mueller, copying members of the Senate Intelligence Committee. In the letter Rowley criticized the Bureau for failing to act on requests from her Minneapolis field office for permission to obtain a warrant to search the belongings of Zacarias Moussaoui, a French-born al-Qa'ida operative who had been arrested on August 17, 2001. The article also tied in complaints from FBI special agents in Phoenix who had sent a memo to their headquarters on July 10, 2001, trying and failing to draw attention to potential Islamic terrorists attending flight schools in the United States.

As news magazine stories go, this one was pretty devastating. A proud organization such as the FBI never likes to hear that it has blown any case, much less the biggest terrorism assault in our history. No organization, though, is better at defending itself than the FBI, and it had no intention of taking this rap lying down. The Bureau knows that when you get slugged in *Time*, you punch back in *Newsweek*, and that's just what it did.

The very next week the cover of *Newsweek* screamed, "The 9/11 Terrorists the CIA Should Have Caught." The story inside, titled "The Hijackers We Let Escape," described how CIA picked up the trail of two men, later to become 9/11 hijackers, when they attended a meeting in Kuala Lumpur, Malaysia, in January 2000. The article said, somewhat incorrectly, that CIA "tracked one of the terrorists, Nawaf al-Hazmi, as he flew from the meeting to Los Angeles." *Newsweek* went on to say that "astonishingly, the CIA did nothing with this information," and that CIA did not notify the FBI, "which could have covertly tracked [the terrorists] to find out their mission." An unnamed FBI official was quoted as saying that CIA's not sharing the information about the two men was "unforgivable." Bureau sources told the news magazine that if they had known of the two men, they could have connected them to all the other hijackers—an argument *Newsweek* found "compelling." The article set off a firestorm and became a pillar of the conventional wisdom that CIA had intentionally withheld information from the Bureau.

A few days later, on June 8, *Newsweek* senior writer Evan Thomas was discussing the article on *Inside Washington*, a syndicated talk show, when host Gordon Peterson asked, "How is *Newsweek*'s relationship with the FBI these days?" Thomas answered, "Well, it was pretty good since we did their bidding." Thomas, who is a very knowledgeable reporter steeped in the intricacies of national security and intelligence reporting, later called CIA's press office to claim that he had misspoken and didn't really know what he was talking about in this instance. Whether he did or not, a very complex story had been reduced to a bumper sticker—"CIA Intentionally Withheld Information"—and despite our best efforts, the 9/11 Commission, the Congressional Joint Inquiry, and the mass media largely bought into it.

To me, what's important to realize is that the watchlisting problem was not, as is so often claimed, an example of CIA and FBI not working with each other. Throughout this pre-9/11 period both agencies were coordinating closely. Louis Freeh and I worked very hard to overcome historical animosities and misunderstandings and to get both organizations to recognize that they were on the same team. Through two administrations, I had no closer relationships in Washington than with Louis Freeh, Bob Mueller, and their senior officers. While our cultures and missions may have been different, there was no difference in the heartfelt way CIA officers and FBI special agents tried to protect the country. We frequently held high-level coordination meetings, committed to assigning some of our best people to each others' headquarters (jokingly referred to as the "hostage exchange program"), and tried to help each other in every way possible.

Six FBI officers were assigned to CIA headquarters at the time of 9/11; their role was to ensure that the Bureau's interests were always considered and that information valuable to the Bureau was passed back to the home office through official and unofficial channels. A similar group of CIA officers worked out of the FBI offices to help translate CIA's needs and capabilities to our law enforcement partners. Of course, there were coordination problems—agencies are bound to have different perspectives over their equally important missions. (The post-9/11 Patriot Act went a long way toward fixing some of these issues.) What's critical—and what the 9/11 Commission and others missed—is that the so-called wall preventing a free flow of intelligence to FBI criminal investigators was not really the heart of the matter. The main problems were old-fashioned ones: too few people on both sides working on too many issues. We needed more people, better communications, and, particularly on the FBI side, better information technology support. After 9/11, Bob Mueller and I sought even more ways to drive our organizations closer together. In the aftermath of such a tragedy it was perhaps inevitable that people would try to drive wedges between us.

The watchlisting story begins as part of the investigation into the August 1998 bombing of the two U.S. embassies in Africa. FBI agents pursuing that case came up with a telephone number of a suspected terrorist facility in the Middle East believed to be associated with al-Qa'ida or Egyptian Islamic Jihad terrorists. That suspicious phone number was shared with CIA, NSA, DIA, the State and Treasury departments, and others. About a year later, in December 1999, intelligence collected from that phone indicated that several men would be traveling to Kuala Lumpur for a meeting to be held in Malaysia early the next month. The information about the meeting was distributed to a number of agencies, including the FBI, at the same time.

As is often the case, the intercepted communications did not include the full names of any of the participants. We had only first names to go on. Nonetheless, CIA launched a major effort to see if we could identify who the attendees were and what they were up to. With the help of a local intelligence agency, on January 4, 2000, one person whom we initially knew only as "Khalid" was identified as he passed through a third country en route to Malaysia. The local intelligence service copied the man's passport, which identified him as Khalid al-Mihdhar. The passport also carried a stamp indicating that al-Mihdhar held a valid entry visa for the United States. That information was sent back to Washington electronically.

We did not know who al-Mihdhar was at first. At the time of the Malaysia meeting, we were in the midst of the largest counterterrorist operation in history, dealing with the Millennium threat. We wanted to be sure that meeting participants were not headed to Southeast Asia to launch an attack. Based on the first name, Khalid, and a phone number, a CIA desk officer initiated surveillance of the individual during his overnight layover on the way to Malaysia.

In a cable dated January 4, 2000, CIA's officers at the intermediate stop reported both to CIA headquarters and to our officers in Kuala Lumpur that a Khalid al-Mihdhar had been identified by local authorities and a copy of his passport had been obtained.

The next day, January 5, CIA officers in Saudi Arabia e-mailed headquarters stating that al-Mihdhar's visa application from the previous year had been reviewed and he had listed his destination as New York and his intended travel date as May 2, 1999. The cable also stated that the information on the visa application form matched the information in the visa, indicating that the visa was still valid.

Once this e-mail came to CIA, it was opened by CIA officers and three FBI officers detailed to the Counterterrorism Center. A senior CIA officer on the scene recently said to me, "Once Mihdhar's picture and visa information were received, everyone agreed that the information should immediately be sent to the FBI. Instructions were given to do so. There was a contemporaneous e-mail in CIA staff traffic, which CIA and FBI employees had access to, indicating that the data had in fact been sent to the FBI. Everyone believed it had been done. The parts of our operation that got the most criticism were the parts where CIA and the FBI were working most closely together."

What never happened was a formal transmission to the FBI, in a report called a CIR (Central Intelligence Report), documenting what everyone believed had already occurred, the sending of al-Mihdhar's photo and visa data. An FBI officer assigned to CIA, known as a "detailee," in fact initiated the drafting of the formal report, but it was never cleared for transmission. The same senior officer said to me, "The CIR was a separate process, providing retroactive documentation of the fact the stuff had already been passed, not to convey new information."

No excuses. However, overworked men and women who, by their actions, were saving lives around the world all believed the information had been shared with the FBI.

Meanwhile, on the ground in Malaysia, we learned that the meeting was being hosted in a condo owned by someone named Yazid Sufaat. We could tell that those in attendance were acting suspiciously, but at the time, we were unable to learn what was being discussed.

On January 6, in an e-mail to a colleague back at Langley, a CIA officer serving at FBI headquarters stated that he had shown an FBI special agent an NSA report on some of the Malaysia meeting's participants, but that the FBI agent was already aware of the meeting. The CIA officer described in extensive detail surveillance efforts against the group in Malaysia and shared this information with several FBI officers. Twice while the surveillance operation was ongoing, then FBI director Louis Freeh was briefed on the effort by his own staff.

Once we had learned the names of several of the individuals who were attending the Malaysia meeting, CIA should have placed them on a watchlist that might have prevented their entering the United States. A half a dozen other agencies, including the FBI, also had the names and could have done so as well, but did not. That does not absolve CIA from blame. We later discovered that there was inadequate staff training on how to handle watchlist submissions. Officers in the field, where primary responsibility for watchlisting resided, thought headquarters would do it, and vice versa. Clearly, a communication breakdown occurred, and we worked hard to rectify the shortcoming once we were aware of it after 9/11.

While we were able to get the names of some of the participants, we were never able to determine what went on at the meeting in Malaysia. When the session in Kuala Lumpur broke up, the participants dispersed. Two, al-Mihdhar and Nawaf al-Hazmi, flew to Bangkok (not directly to Los Angeles, as *Newsweek* contended in al-Hazmi's case). We asked the local intelligence service to keep an eye on them. Almost two months after the fact, on March 5, 2000, the Thais passed on information that said that Nawaf al-Hazmi had arrived in Bangkok in early January and departed for Los Angeles about a week later, arriving on January 15 on United Airlines Flight 2. The information made no reference to al-Mihdhar, although we learned much later that he, too, was on the same United Airlines flight.

CIA officers in the field sent this information back to headquarters but included it at the end of a cable that contained routine information. The cable was marked as being for "information" rather than "action." Unfortunately, no one—not the CIA officers nor their FBI colleagues detailed to CTC—connected the name Nawaf al-Hazmi with the meeting of eight weeks before.

What would later prove a raw point between CIA and FBI involved an al-Qa'ida operative we at first knew only as "Khallad." FBI had developed sketchy intelligence about Khallad before the October 2000 attack on the USS *Cole.* After the attack, we discovered further intelligence linking Khallad to the phone number in Yemen that had been associated with the Kuala Lumpur meeting. In a meeting in November, a senior FBI official, John O'Neill, received Khallad's full name and a copy of his photo. (John would later retire from the FBI and take a job as chief of security at the World Trade Center, and tragically die there in what was his third week on the job.) By the end of November 2000, CIA and FBI both knew Khallad's full name, Khallad bin Attash, had his picture, and knew he was a senior security official for Bin Ladin. Both organizations knew he had supported the *Cole* attack.

By December 2000, investigators began wondering whether Khallad bin Attash and Khalid al-Mihdhar (who was at the Malaysia meeting the previous January) might be one and the same. It turned out that both were at the meeting, but they were two different individuals. That month a CIA officer and his FBI colleague based in Islamabad showed the photo O'Neill had obtained to a jointly run intelligence source who had insights into al-Qa'ida.

They conducted what is known in the intelligence business as a "rolling car meeting," or "RCM." To avoid compromising the source, they picked him up at nighttime on a busy street and conducted their business while driving around. A second armed female CIA case officer was in the backseat for security. The asset was shown the several photos and correctly picked out the one of Khallad by flashlight.

At a follow-up meeting in January, this time at the U.S. embassy in Islamabad, the source was shown surveillance photos taken in Malaysia. With the FBI assistant legal attaché and two CIA case officers present, he identified someone who he said was Khallad. (He had the wrong person, but we would not know that until after 9/11.) Two weeks later, according to CIA message traffic, a group of FBI analysts from the New York field office were sent on temporary duty to Pakistan in part to debrief this same asset.

On June 11, 2001, an analyst from FBI headquarters, another FBI analyst assigned to CIA's CTC, and a lone CIA analyst traveled to the Bureau's New York field office to brainstorm the *Cole* investigation. The FBI analyst carried with her the surveillance photos taken in Malaysia. The photos were discussed with the local special agents, who reportedly had requested copies. The FBI analyst told them that she would try to get the photos "over the wall." After 9/11, several FBI officials would allege that CIA had refused to share these photos with the Bureau. On the day of 9/11 itself, CIA and FBI officers from CTC were on the way to brief Director Mueller on the case investigations, with photos in hand. They never got there.

By July 2001, indications were everywhere that a major terrorist attack was about to occur. As I later told the 9/11 Commission, "the system was blinking red." I instructed the people in CTC to review everything in their files to search for any clue that might suggest what was coming. The request, though, was redundant. Everyone in CTC felt as strongly as I did that something catastrophic was about to happen, and they had already begun such a review.

In mid-August analysts reviewing the Kuala Lumpur meeting came across the cable that said that Nawaf al-Hazmi had come to the United States in January 2000. Contact with the U.S. Immigration and Naturalization Service showed no record of al-Hazmi's ever having left the country. The analysts then checked on the other named individuals believed to have attended the Malaysian meeting and found that Khalid al-Mihdhar had arrived in the United States along with al-Hazmi, departed on June 10, 2001, and then returned July 4, 2001.

This alarmed us sufficiently that on August 23 an immediate message went out alerting the State Department, FBI, INS, Customs, and others about the pair and asking that they be barred from entering the country if they were outside the United States, and tracked down if they were still here. Even though they were watchlisted, that act alone did not ensure that they would be automatically placed on a no-fly list preventing them from boarding an airplane. In fact, this did not occur, and even though they were watchlisted nineteen days before 9/11, they were not found. Obviously, if we had watchlisted the two a year and a half earlier, when they first came across our radar screen, we would have had a far better chance of preventing them from subsequently entering the United States. That was essentially what happened to Ramzi bin al-Shibh, who, for other reasons, was several times denied entry into the United States. Al Qa'ida simply replaced him among the plotters, and I feel certain the same would have happened with al-Hazmi and al-Mihdhar.

CIA had multiple opportunities to notice the significant information in our holdings and watchlist al-Hazmi and al-Mihdhar. Unfortunately, until August, we missed them all. What if we had noticed our mistake after al-Mihdhar and al-Hazmi entered the United States, but months rather than weeks before the plot unfolded? Most likely the two men would have been deported. In theory, the FBI might have secretly followed them, which might have led to our learning of some of their collaborators in this country, but that may have run counter to Bureau practice at the time. Deportation might have delayed but probably would not have stopped 9/11. In the final analysis, al-Mihdhar and al-Hazmi were soldiers, not generals—replaceable parts in a determined killing machine.

In my view, another opportunity may have been lost by the inability of the FBI lawyers to figure out a way to search the luggage of Zacarias Moussaoui. The first time I heard of him was on August 23, 2001, when CTC provided me with a terrorist threat update covering a large number of topics. Included in the twelve items on the agenda was information regarding the arrest of an associate of Abu Musab al-Zarqawi; al-Qa'ida kidnapping threats in Turkey, India, and Indonesia; a discussion of the pending deportation from the UAE to France of Djamel Beghal, who intended to blow up the U.S. embassy in Paris; the arrest of six Pakistanis in La Paz, Bolivia, who were intending to hijack an aircraft; and other items. The last item was about Moussaoui. The briefing chart was entitled "Islamic Extremist Learns to Fly."

A French national, Moussaoui was arrested on August 16, 2001, by the FBI on the grounds that he had overstayed his U.S. visa, but it wasn't the visa problem that brought him to the FBI's attention. Moussaoui had enrolled in flight school in Minnesota and paid for his training in cash. He was interested in learning to fly 747s, but not in taking off or landing. He was interested to learn that 747 doors do not open in flight. He wanted training on London–JFK flights. Moussaoui's flight instructors did not like what they were seeing with this obviously unqualified student, and they alerted the FBI.

We immediately went to work on the case with the Bureau.

As alarming as the information on Moussaoui was, I was comforted by the fact that FBI had its hands on the guy. My assumption was that the Bureau would, as standard practice, brief Dick Clarke's Counterterrorism Security Group at the NSC, and the case would be well covered.

During the 9/11 Commission hearings, I was stunned to hear Tom Pickard, who was acting FBI director in August 2001, suggest that *I* had somehow failed to notify *him* about Moussaoui. Failed to tell him? Hell, it was the FBI's case, their arrest. I had no idea that the Bureau wasn't aware of what its own people were doing.

More than four and a half years later, in the spring of 2006, I was subpoenaed as a possible witness for Moussaoui at his trial, held in the U.S. District Court in Alexandria, Virginia. In the end I was never called to testify. Moussaoui was duly found guilty of conspiracy to kill Americans and sentenced to life in prison. But in preparation for my possible testimony, and with the help of CIA's General Counsel's Office, I set out to learn everything I could about what the Agency had been able to put together after Moussaoui's arrest. The following account relies heavily on that information.

Let me stress that most of this is *not* information I knew in 2001.

On August 15, 2001, CIA officers in the field were told by FBI's Minneapolis office that Moussaoui would be arrested the next day. The CIA officers, in turn, informed the CTC of the impending arrest, and the CTC did a "trace" on Moussaoui, looking for anything that we might have on him in our files. That search came up negative. Before August 15, we had never heard of Zacarias Moussaoui, at least under his real name. Later, in January 2002, one of our sources told us that in Baku in 1997 he had met someone whom he now knew to be Moussaoui. At the time, Moussaoui was using the nom de guerre of Abu Khalid al Francia. The source reported on him to us in April 2001, using only the "al Francia" name. By August 18, Minneapolis special agent Harry Samit was in direct contact with Chuck Frahm, an FBI special agent assigned to CIA who was then the deputy group chief for al-Qa'ida operations. Samit provided everything Minneapolis had on Moussaoui, which Frahm passed on to CIA officers.

Even though Moussaoui was taken into custody on August 16, lawyers at the FBI believed that they did not have sufficient cause to obtain authority to search his belongings, but, at least from our perspective, that would soon change. On August 24, 2001, CIA learned that Moussaoui was a known quantity to the French internal service, the very capable Direction de la Surveillance du Territoire, or DST. They said that Moussaoui had recruited a friend of his into Ibn Khattab's Chechnyan Mujahideen. Khattab's group had been accused of, among other things, attacks on a Red Cross hospital in Chechnya in 1996 and blowing up an apartment building in Moscow in 1999. The French investigated Moussaoui's extremist connections and assessed him as highly intelligent, extremely cynical, cold, stubborn, full of hatred and intolerance, and completely devoted to the Saudi-based extremist Wahabi cause.

On August 24, Harry Samit again e-mailed Chuck Frahm specifically asking Frahm to ask CIA's lead analyst, "Is there anything you have that establishes Ibn Khattab's connection to UBL/Al-Qa'ida other than their past association? We are trying to close the wiggle room for FBI headquarters to claim that there is no connection to a foreign power. Since al-Qa'ida is a designated group, anything that you have which indicates an al-Qa'ida connection to Moussaoui via Ibn Khattab would help." Frahm asked the CIA analyst to jump on his computer to respond to Samit. She wrote: "Am not sure why the French info is not enough to firmly link Moussaoui to a terrorist group. Ibn al Khattab is well known to be the leader of the Chechen Mujahidin movement and to be a close buddy with bin Ladin from their earlier fighting days. From a read of the DST info, Moussaoui is a recruiter for Khattab." That same day, a CTC officer passed the Khattab connection via e-mail to the CIA representative at the FBI. "No one in the FBI seems to have latched on to this. Perhaps you can educate them on Moussaoui. This may be all they need to open a FISA on Moussaoui." The "FISA" would have authorized the necessary search.

For us, the Khattab tie-in was sufficient evidence to show that Moussaoui was a terrorist, and thus we sent out a worldwide query through our own channels to the French, British, and other countries. Despite the FBI Minneapolis field office's view that Moussaoui might be engaging in flight training for the purpose of conspiring to use an airplane in the commission of a terrorist act, lawyers and others at FBI headquarters did not believe that the French information was enough to get a court-authorized search warrant. They felt that the information did not meet the threshold of the FISA statute making Moussaoui an "agent of a foreign power."

On August 30, the CIA officer again contacted a fellow CIA officer on assignment at the FBI. "Please excuse my obvious frustration in this case. I am highly concerned that this is not paid the amount of attention it deserves. I do not want to be responsible when they [*sic*] surface again as members [*sic*] of a suicide terrorist op." The officer wasn't through. "I want an answer from a named FBI group chief for the record on these questions . . . several of which I have been asking since a week and a half ago. It is critical that a paper trail be established and clear. If this guy is let go, two years from now he will be talking to a control tower while aiming a 747 at the White House." This comment was particularly prescient because we later learned after 9/11 that Moussaoui had in fact asked Usama bin Ladin for permission to be able to attack the White House. FBI and CIA officers worked the legal obstacles from both ends. The Minneapolis field office was in touch with CTC; FBI and CIA officers at both respective headquarters tried to influence the outcome of the legal debate. When legal hurdles could not be overcome, they came up with a plan.

By August 31, with no FISA warrant in sight to allow access to Moussaoui's belongings, we began working up a scheme with the FBI that would have had Moussaoui deported to France. Our plan was to load Moussaoui's belongings separately, then turn his laptop and luggage over to French authorities for exploitation once he arrived in Paris. (The French did not require the same high level of probable cause that the FBI thought it needed in order to conduct a search.)

Ultimately, we learned that the key lay not in Moussaoui's computer but in his luggage. On September 18, 2001, a week after the attacks on the World Trade Center and the Pentagon, we were informed that a trunk belonging to Moussaoui contained letters indicating that he was the U.S. marketing consultant for a Malaysian company called In Focus Tech. The next day, our officers told us that the general manager of In Focus Tech was Yazid Sufaat, and with that the circle closed and things started to come together in a hurry. Recall that this was the same Yazid Sufaat whose condo in Kuala Lumpur had been the venue for what turned out to be the first operational meeting in the planning for 9/11—the meeting, as noted earlier, that was also attended by al-Mihdhar and al-Hazmi.

If we'd had those letters in Moussaoui's luggage connecting him to Sufaat and—through Sufaat, back to al-Mihdhar and al-Hazmi, who had just been placed on our watchlist—is it possible that enough bells and whistles might have gone off to allow us to make all the necessary connections? While all of us involved lie awake at night asking ourselves this question, I do not believe there was a silver bullet available to us to stop the tragedy of 9/11.

CIA did not watchlist al-Hazmi and al-Mihdhar until August 23, 2001. FBI did not get into Moussaoui's luggage. The famous Phoenix memo, outlining concerns about terrorists being trained at flight schools, was not shared. The FBI's effort to find al-Hazmi and al-Mihdhar was pursued with too few resources. Simply using commercially available software to track their credit card usage might have been decisive, but no such effort was made.

These missed opportunities obscured the hundreds of successful operations conducted by CIA and FBI together and stood out in high relief when discovered. They pointed out larger systemic shortcomings, in resources, people, and technology. They also highlighted something equally important: The al-Qa'ida operatives who killed three thousand people on September 11 understood that the United States had never thought about how to protect itself within its borders. Policies had never been put in place to address just how disconnected our airline security, watch-listing, border control, and visa policies were at the time. There was no comprehensive, layered system of domestic protection in place to compensate for the internal weaknesses that later came into full view. Yes, people made mistakes; every human interaction was far from where it needed to be. We, the entire government, owed the families of 9/11 better than they got from us. *All* of us.

CHAPTER 12

Into the Sanctuary

"We need to go in fast, hard and light," we told the president. "Everyone, including al-Qa'ida and the Taliban, are expecting us to invade Afghanistan the same way the Soviets did in the 1980s. Bin Ladin and his followers expect a massive invasion. They believe we will withdraw in the face of casualties and never engage them in hand-to-hand combat. They are going to get the surprise of their lives." Ours was a strategy unlike any other in recent American history. The plan CIA laid out for the president on September 13 and expanded at Camp David two days later stressed one thing: we would be the insurgents. Working closely with military Special Forces, CIA teams would be the ones using speed and agility to dislodge an emplaced foe. Our plan was to build on relationships that had been carefully forged with regional factions over recent years to give us allies who might help oust the Taliban. This war would never be "Americans against Afghans," we told the president. Rather, it would always be about helping Afghans rid their own country of a foreign menace, al-Qa'ida, and of the Taliban, who had allowed terrorists to hijack their country.

Five times in the two years *prior* to 9/11, CIA teams deployed to the Panjshir Valley of northern Afghanistan to meet with various tribal warlords, and particularly with Ahmed Shah Masood, the head of the Northern Alliance—a loose network of competitive tribal forces made up largely of ethnic Tajiks, Uzbeks, and others who fought against the Taliban rulers of Afghanistan. We bolstered Masood's intelligence capability against Bin Ladin and al-Qa'ida. Masood's brutal murder by al-Qa'ida on the eve of the

9/11 attacks might have undone our plan before it got under way if we hadn't maintained contact with other warlords in the north. And we also had long-standing, if much weaker, relationships with Pashtun tribes in the south. We knew who the players and who the pretenders were. By September 10, 2001, CIA had more than one hundred sources and subsources, and relationships with eight tribal networks spread across Afghanistan. Although these sources proved insufficient to steal the secret that would have predicted and prevented the attacks of 9/11, we were confident that, with the right authorities, we could get those responsible for the tragedy.

The president approved our recommendations on Monday, September 17, and provided us broad authorities to engage al-Qa'ida. As Cofer Black later told Congress, "the gloves came off" that day.

At the White House meeting that same day, the president declared, "I want the CIA to be first on the ground." I sent a memorandum to CIA senior officers stressing that "There can be no bureaucratic impediments to success. All the rules have changed. There must be an absolute and full sharing of information, ideas, and capabilities. We do not have time to hold meetings to fix problems—fix them quickly and smartly. Each person must assume an unprecedented degree of personal responsibility."

There has been a lot written about how Don Rumsfeld was supposedly unhappy that CIA was playing such a prominent role at the time. I never had that sense. We had a good plan. I was seeing my boss, the president of the United States, every day, and he was telling us "Go, go, go." It never occurred to me that we should do anything else.

Speed was everything. We needed to get a team into northern Afghanistan as soon as possible, to engage the various anti-Taliban leaders there and to measure the effect that the assassination of Masood had had on the Northern Alliance. Our bench of Afghan experts was strong but not deep, so we moved quickly to enhance it. To lead the mission, we found the perfect person, attending a pre-retirement seminar. Gary Schroen, deeply knowledgeable about the region, was friendly with many of the senior Afghan warlords and fluent in the local languages of Dari and Farsi. Instead of leaving government service as he had been planning before 9/11, Gary arrived in northern Afghanistan within two weeks of the attacks, at the head of a small team that would be the forerunner of Agency operations there for the next several years.

Sending a senior officer like Gary illustrates the way the Agency operates. Gary was equivalent to a three-star general in rank, and he was first in with a squad of eight men who averaged forty-five years of age and twenty-five years of professional experience. Empowered to speak on behalf of the Agency, Gary was able to enter into agreements, make demands, and, not inconsequentially, dole out some of the millions of dollars in cash that he flew in with.

The CIA Northern Alliance Liaison Team, led by Gary Schroen, traveled to Afghanistan on an old Russian helicopter that we had purchased a year before 9/11 to facilitate our movements in the region. The NALT, as the team was known, set up shop in the village of Barak, at an elevation of 6,700 feet and surrounded by mountains as high as 9,000 feet. Living conditions in Barak were spartan to say the least. The NALT reported that sanitation conditions were "circa mid 12th century" but that the team was "healthy, motivated, and working hard." To remind themselves why they were there, they repainted the tail number on their MI-17 helicopter shortly after arriving, giving it the designation "091101."

Gary quickly established contact with Fahim Khan, one of the Northern Alliance leaders who figured prominently after the assassination of Masood, while also reaching out to other tribal leaders to learn who was with us and who was against us. Simultaneously, NALT team members sent back intelligence that would form the basis of targeting decisions in the military air campaign that was to follow.

Some of the contacts with tribal leaders were face to face. Others were conducted by radio and satellite telephones. Tribal leaders were asked, "Can we count on you to help drive al-Qa'ida and their Taliban protectors out of Afghanistan?" If the answer was yes, food, medical supplies, military equipment, and weapons would soon be air-dropped to them. Between mid-October and mid-December 2001, U.S. aircraft delivered 1.69 million pounds of goods in 108 airdrops to 41 locations throughout Afghanistan. Each drop was tailored to the specific requests and needs of the teams on the ground. One ethnic Uzbek leader told us that his most critical need was horse feed. Others needed saddles. These were shipped along with arms, portable hospitals, and food. Some of our officers slept on millions of dollars in cash, which was used to capitalize on the Afghan tradition of switching sides. A tribal leader who sided with the United States would, within hours, see the answer to his clan's prayers drop from the sky. It gave those warlords tremendous clout within their organizations. But if a tribal leader refused to work with us, essentially declaring himself and his clan our enemies, his clan might find themselves on the receiving end of a different kind of airdrop—a two-thousand-pound bomb courtesy of the U.S. military. Subtle, it wasn't, but neither were the terrorist attacks on Washington and New York that had brought us to Afghanistan.

In addition to working with various warlords, Agency officers in Afghanistan also secretly contacted Taliban officials to try to get them to turn over Bin Ladin. In one case, an agency team traveled to a virtual no-man's-land outside of Kabul for what they hoped would be a meeting with a very senior Taliban intelligence official. CIA headquarters gave the team wide latitude on deciding how to handle the matter. The Taliban official failed to show up, however, but did send his deputy. The stand-in made it clear that they had no intention of being helpful to us. That was a mistake. The CIA team literally rolled him up—in a carpet—threw him in the back of a truck in broad daylight, and spirited him back to U.S.-controlled territory, where he could be questioned. Scores of al-Qa'ida and Taliban were killed in U.S. airstrikes based on what we learned from that Taliban deputy.

On September 26, President Bush paid a visit to CIA headquarters. In a speech in the Agency lobby, in front of a wall of honor memorializing CIA officers who had died in the line of duty, he told our workforce how much confidence he had in them. He also reminded them that the American people expected "a 100-percent effort, a full-time, no-stop effort on not only securing our homeland but bringing to justice terrorists, no matter where they live, no matter where they hide." That, he noted, was "exactly what we're going to do." After the president's remarks, we briefed him on the first reports coming in from the NALT, which, unbeknownst to most of the world, had landed in Afghanistan that same day.

CIA was built to gather intelligence, not conduct wars. When it became clear that we were going to be asked to play a leading role in ousting al-Qa'ida, we added a new branch to our Counterterrorism Center—CTC Special Operations, or CTC/SO. To head up this new branch, we tapped Hank Crumpton, a slow-talking, quick-witted CIA officer who had recently completed a three-year tour of duty in Washington, including two years in CTC and one working with the FBI. Hank was the perfect man for the mission. He had spent ten years in sub-Saharan Africa working around insurgent groups; had extensive interagency experience, including a recent tour of duty with the FBI; and had led the CIA team that went to Yemen to investigate the USS *Cole* bombing. Hank and his family had just arrived in an attractive overseas capital for what was supposed to be a three-year posting. A day or so later he got a call from headquarters: Stop unpacking. We need you back in Washington. To no one's surprise, Hank didn't hesitate for a moment. He knew that the decision to come back would be tough on his three kids. They'd just made the adjustment to a new home, the family belongings had arrived, they were ensconced in new schools, and the family dog had just gotten out of quarantine. "I know you are unhappy," Hank told them, "but think about the families of three thousand people who have just lost their lives. You've got it good. I need you to suck it up and help your mother repack. Let's go home."

Upon his return from overseas, Hank headed directly to Langley from the airport. There, he met with Cofer Black, who outlined his expectations. "Your mission is to find al-Qa'ida, engage it, and destroy it."

Like Gary Schroen, John M. (who remains undercover and cannot be fully identified), a Naval Academy graduate with twenty-six years of government service, was on his way out of the Agency on 9/11. In fact, he was in the second day of a preretirement program at an outlying CIA facility in northern Virginia when the terrorists struck. John jumped in his car and found himself drawn to CIA headquarters. Having no specific assignment, he spent the first day pitching in where he could, delivering messages and helping make sense of a chaotic situation. He told senior officials in the Directorate of Operations that if we had a job for him, he would withdraw his retirement papers. In the meantime, he traveled to New York City and volunteered to help dig through the rubble near the World Trade Center. When Hank heard about John's determination and availability, he quickly tapped him to be one of his deputies.

Another key player in the effort was Frank A., a hulking longtime veteran of CIA's clandestine service who planned and implemented the psychological operations of the Afghan campaign. Throughout the war he became one of our most valued strategic thinkers. Frank had enlisted in the Marine Corps as a young man and later joined the Agency, where he served with distinction on three continents. He is a no-nonsense, can-do kind of guy. I remember one Saturday morning, shortly after 9/11, I was in CTC getting a briefing on operations. It turned out that someone had decided that that was a good day to test the headquarters fire alarms. The briefing kept getting interrupted. We could barely hear ourselves think. Frank calmly got up and ripped the wires out of the alarm in the room we were in. The briefing proceeded.

One of the biggest problems we faced in Afghanistan at the outset was how to foster cooperation with the mostly Tajik tribes in the Northern Alliance without alienating the country's Pashtuns, largely in the south, many of whom had once been supportive of the Taliban. The last thing we wanted on our hands was a civil war.

CIA was split into its own factions on the matter. Some officers, particularly those serving in Pakistan, argued that we should not align ourselves too closely with the Northern Alliance. In general, CTC/SO, our officers in Uzbekistan and Tajikistan, and the NALT unit in northern Afghanistan disagreed. In their view, we couldn't wait for opposition forces to rise up in the south. Instead, we had to take advantage of the Northern Alliance's willingness to engage the enemy right away. I appreciated both arguments, but I agreed with Gary and Hank that momentum was critical.

The original NALT unit was followed by the deployment of six additional CIA teams in the first two months of the war. Like the first, each new team averaged eight members and included experienced officers with Farsi/Dari, Uzbek, Russian, and Arabic language capabilities. These officers were assigned to work with tribal warlords across a broad expanse of northern and western Afghanistan.

The Northern Alliance controlled the mountainous northeastern corner of Afghanistan, including the Panjshir Valley, which led to the Shomali Plains, north of the capital of Kabul, along with some small patches in the central portion of the country. As yet, we had no allies in control of territory in the south. All we could do was hope that the south's participation would fall in place as events progressed.

The war plan was for Northern Alliance forces, with the aid of U.S. airpower and targeting provided by CIA and Special Forces teams, to drive toward north-central Afghanistan and take the town of Mazar-i-Sharif. From there they could establish a land bridge to Uzbekistan, from which supplies could flow. At the same time other Northern Alliance forces would attack the town of Konduz, in the north, while still others would try to take Bamiyan, in central Afghanistan. Then Northern Alliance troops, assisted by the NALT, would head south through the Shomali Plains, toward Kabul.

The key to our strategy was in the way our Afghan allies could be motivated. Based on years of experience in the region, CIA officers knew that the way to galvanize the local units was to appeal to their sense of prestige and honor as it was defined in tribal terms. This required a cultural understanding based on trust and confidence.

At the outset of the war in Afghanistan, CIA's senior officer dealing with Pakistan recommended a limited air campaign in the south, focusing on Taliban air defenses, facilities physically and symbolically associated with Mullah Omar and UBL, and al-Qa'ida–associated training camps. The plan was intended not to alienate the country's large Pashtun ethnic group, which formed the basis of the Taliban's support. A heavy bombing campaign against the Taliban and al-Qa'ida in the north might be seen as the U.S. siding with the mostly Tajik Northern Alliance to the detriment of the Pashtuns in the south. The idea was that such a limited campaign might create fissures within the Taliban, and induce Taliban officials to turn over Bin Ladin. None of this happened. The Pashtuns sat on their hands.

The Northern Alliance warlords had the impression that the American bombing effort was tepid at best. CIA officers in the north argued forcefully that the only way to get the Northern Alliance fully into the fight was to show them that we were serious, with a more aggressive bombing campaign. They said that Afghan military resistance and public support for the Taliban would both collapse under increased U.S. military pressure. The Pashtuns would switch sides, as long as they did not face an imminent threat from the Northern Alliance.

During the first week of the bombing campaign, Gen. Tommy Franks followed our recommendation regarding the gradual application of force but began to feel the heat for being so closely aligned with CIA. The new chairman of the Joint Chiefs, Air Force general Dick Myers, felt the bombing campaign wasn't working, and that CIA's plan was flawed. Tommy and I were both frustrated, and he certainly understood that CIA did not want to micromanage the campaign. But he and I were close enough to be able to talk candidly. It would soon be winter in Afghanistan, and we both knew it was time to act.

On October 17, U.S. Special Forces arrived on the ground. By late October, with CIA officers providing targeting intelligence, military Special Forces troops courageously closing in on Taliban and al-Qa'ida units to provide laser target designation, and fixed-wing aircraft dropping precision weapons, the pace of the air war soon stepped up, and made the critical difference in overwhelming the foe.

There was a lot of bureaucratic tension. In early October, I was taking part in a secure teleconference with the vice president, the secretary of defense, and others when Don Rumsfeld questioned who was in charge on the ground in Afghanistan. CIA and the Defense Department operated under different authorities. I understood Don's sense of order and desire for clarity of command, but this was a different kind of war. It was opportunistic, and required flexibility. CIA and Special Forces personnel on the ground melded together immediately. They did not worry about who was in charge. It was essential to give teams on the ground the tactical autonomy they needed. Our job in Washington was to provide support and guidance, but basically to get the hell out of the way. We understood that, in the end, CIA would support Tommy Franks's efforts and take his lead. But in the beginning, CIA's knowledge of tribal relationships had primacy. I remember not saying much, and Rumsfeld not letting go of the issue until the vice president intervened by saying, "Don, just let the CIA do their job."

He did, for the moment, but that wasn't the last we would hear of the matter. A few weeks later Franks paid me a visit at CIA headquarters.

"I want you to subordinate your officers in Afghanistan to me," he said. That's military talk for "you guys need to work for me."

"It ain't gonna happen, Tommy," I told him.

I have tremendous respect for the military and for Franks in particular, but in this case I knew that if we fell under Pentagon control, the big bureaucracy would stifle our initiative and prevent us from doing the job we were best equipped to do. Tommy was just carrying water for the folks at the Pentagon. He and his staff had long had a great working relationship with the Agency, and we weren't about to screw that up. He and I agreed that CIA would enter into some sort of "Memorandum of Understanding" with CENTCOM on relations between our two organizations. I gave the task of writing the memorandum to Lt. Gen. John "Soup" Campbell. I made it clear that the memo should be written in a manner that did not compromise CIA's prerogatives. Soup had taught me a few things, most notably a great military expression for when you really do not want to get sucked into something: "Go dumb early." And that is exactly what we did with the MOU: drafted it, coordinated it with CENTCOM, and put it on the shelf.

With the Northern Alliance yet to be fully unleashed and bombing in the north still to take its toll on Taliban front lines, some pessimism began to creep in as to whether our strategy would succeed before the onset of winter. On October 25, Rumsfeld sent around a paper that had been produced for him by the Defense Intelligence Agency. He passed out copies of the document at a meeting in the Situation Room. I read it quickly and shot a look at Hank Crumpton, who was sitting behind me. Among DIA's key points was the bold assertion that "Northern Alliance forces are incapable of overcoming Taliban resistance in northern Afghanistan, particularly the strategic city of Mazar-I Sharif, given current conditions." The paper also flatly stated that "The Northern Alliance will not capture the capital of Kabul before winter arrives, nor does it possess sufficient forces to encircle and isolate the city." DIA was equally glum about prospects in the south, saying that "No viable Pashtun alternative exists to [the] Taliban." In its summary, the DIA said, "Barring widespread defections, the Northern Alliance will not secure any major gains before winter."

Pessimism wasn't limited to official sources. On October 31, *New York Times* correspondent R. W. "Johnny" Apple wrote that, "Like an unwelcome specter from an unhappy past, the ominous word 'quagmire' has begun to haunt conversations among government officials and students of foreign policy, both here and abroad. Could Afghanistan become another Vietnam?"

Contrary to what the Pentagon and Johnny Apple were saying, we were closing in on our objectives, but we still had a hard time convincing our own national security team that the plan was working. But we had honed our plan down to four main objectives: capturing Mazar-i-Sharif in the north, pushing south to Khandahar (Mullah Omar's headquarters), unifying the east and west areas of Northern Alliance control, and finally taking Kabul. Throughout it all, the president never wavered.

On the morning of Friday, November 9, Pentagon officials again briefed the White House that things were not going well in Mazar-i-Sharif. Hank Crumpton, whom I had brought along to the session, disagreed. "Mazar will fall in the next twenty-four to forty-eight hours," he boldly stated. Not everyone in the room agreed with Hank's analysis.

Hank proved right; Mazar fell the next day, and Taliban resistance quickly began to dissipate elsewhere in the country. Suddenly, the concern in Washington shifted from things moving too slowly to things moving too fast. The worry now was that the Northern Alliance was getting ahead of the nascent resistance in southern Afghanistan, and that if they took the capital of Kabul too quickly, intertribal fighting and score-settling would break out and chaos would reign.

Granted, that danger existed, but I told Condi Rice and other

NSC officials that it would be impossible to tell the Northern Alliance, after years of resistance to the Taliban, that they should stand down and not retake their country's capital when it lay before them. What's more, I said, we had teams inserted with all the major warlords and could monitor events closely; and indeed, when the Northern Alliance did roll into Kabul on November 14, they demonstrated remarkable restraint in their actions.

As successful as the northern campaign was, the southern one limped along in search of tribal support and, most important, a charismatic Afghan to rally the tribes there against the Taliban. As always, we were getting lots of advice, sometimes from odd precincts. Former national security advisor Bud McFarlane and two wealthy Chicago brothers all weighed in, urging us to support someone by the name of Abdul Haq. Haq had gained prominence and lost a leg fighting the Soviets in Afghanistan in the late eighties.

We dutifully sent officers to meet with him in Pakistan to assess his capabilities. It turned out they were minimal. Haq had only a handful of supporters. CIA officials urged him not to enter Afghanistan until he could muster more forces. We offered him a satellite phone with which he could communicate with us, but he turned it down, apparently, as we later learned, because he feared we would use the phone to track his whereabouts. Tragically, Haq ignored our advice and entered Afghanistan on the back of a mule. Reportedly, by then he had with him nineteen men sharing four rifles. Beforc long we were receiving frantic calls from Haq's American admirers, telling us that he was besieged by the Taliban and demanding that we save him. Unfortunately, there were no American assets anywhere in the vicinity of his uncoordinated entry. CIA did have an armed Predator UAV close by, and we sent it looking for Haq. When we found him surrounded, Agency officers remotely fired the Predator's Hellfire missile, hoping to divert Haq's attackers, but a single missile was insufficient to the task. Haq was captured and executed on October 25.

(Later, in March of 2002, our Predator went to the rescue of U.S. Rangers in a downed helicopter on Roberts Ridge in Shaikot. We were able to alert the Rangers about enemy forces surrounding them. The Predator marked enemy forces for a successful French Mirage attack, and circled overhead until the Rangers were safely extracted.)

Happily, other Afghan leaders in the south showed greater promise. Chief among them was Hamid Karzai, the leader of the Popalzai tribe, which was traditionally based in the Tarin Kowt region of Afghanistan. Although Karzai's following was small, it was loyal, and he was widely respected among the various Afghan factions. He also had incentive: his father had been assassinated by the Taliban in 1999.

On October 9, Karzai entered Afghanistan from Pakistan, where he had been in exile, on the back of a motorcycle and joined up with about 350 of his supporters. Four days later, they seized the town of Tarin Kowt, the dusty capital of Oruzgan province and the area from which Karzai's tribe originated. Taliban forces came down from Khandahar and counterattacked Karzai's lightly armed troops. Unlike Abdul Haq, however, Karzai had accepted our offer of a satellite phone and used it to tell us he was in trouble and to request a resupply of arms and ammo.

We couldn't comply right away—CIA officers in the south had to compete with other urgent requests for matériel support to Afghan units in the north—but finally, on October 30, Karzai received his much-needed airdrop. Still, the situation around Tarin Kowt was desperate. On November 3, Karzai called his CIA contact, someone I can identify only as "Greg V.," and asked to be extracted by helicopter. Greg quickly contacted CIA headquarters and made the case that Karzai represented the only credible opposition leader identified in the south. His survival, Greg said, was critical to maintaining the momentum for the southern uprising.

Greg got the go-ahead to fly in to Tarin Kowt along with a

U.S. Special Forces unit to airlift Karzai and seven of his senior tribal leaders to safety in Pakistan on the night of November 4–5. Karzai made it clear to us that his withdrawal was just a temporary one and that he planned to reenter Afghanistan within days. He hoped that news of his tactical retreat would not be disclosed for fear that it might demoralize some of his supporters. Unfortunately, Don Rumsfeld happened to be in Pakistan at the time and told a press contingent about the evacuation before we could get word to him of Karzai's desire for secrecy.

Karzai's plan was to return to Afghanistan as soon as possible. We agreed, but we also wanted to send a small, joint CIA-DOD team back in with him. On November 14, Karzai and his tribal elders, accompanied by a six-man CIA team, a twelve-man Special Forces unit, and a three-man Joint Special Operations Command (JSOC) unit, made a dangerous nighttime insertion into the Tarin Kowt area. By the next day, Taliban forces had fled Tarin Kowt and about two thousand Pashtun tribal fighters loyal to Karzai awaited his arrival in the town. For the next several days Karzai went from village to village rallying support against the Taliban. As his support grew, U.S. airdrops of machine guns, recoilless rifles, mortars, and communications gear increased as well. Unfortunately, he also attracted the enemy's attention.

On November 16, we received reports of a large force of Taliban fighters moving toward the area. The next day a major battle erupted, and some of Karzai's newly recruited supporters turned and ran. Greg V. took command of the situation, sprinting from one defensive position to another, telling the Afghans that this was their chance to prove their worth and make history. "If necessary, die like men!" he shouted. Backbones stiffened; Karzai's forces repulsed the Taliban attack. For the Afghan war, it was a seminal moment. Had Karzai's position been overrun, as appeared likely for much of November 17, the entire future of the Pashtun rebellion in the south could have ended.

Dramatic events were happening all over Afghanistan. CIA's

NALT Team Delta accompanied tribal warlord Abdul Karim Khalili on a tour of the recently liberated town of Bamiyan, his ancestral home. The town is famous for two huge statues of Buddha carved into an overlooking mountainside. The Taliban had blasted these third-century relics with dynamite and artillery fire in March 2001, saying that proper Muslims should not look upon idols. Khalili sadly noted that "Bamiyan is not Bamiyan without the statues of Buddha." Together he and Delta team drove around the town square, which sits atop the several-hundred-foot plateau where the statues were carved. As daylight faded, they looked out on the snowcapped peaks in the distance. Khalili asked our officer to pass along his heartfelt thanks to the CIA and the U.S. government for allowing him the bittersweet opportunity to see Bamiyan at sunset again.

As the situation in the south solidified around Hamid Karzai, conditions in large parts of the north remained fluid and chaotic. After the city of Konduz fell on November 24, Northern Alliance forces incarcerated many hundreds of prisoners in a nineteenth-century fortress called Qala-i-Jangi, on the outskirts of Mazar-i-Sharif. Many of the Taliban POWs were foreigners, including at least fifty Arabs from Saudi Arabia, Qatar, Iraq, and elsewhere. Also in the mix were Russians, Chinese, and a few Africans. More than just Taliban supporters, many of these people were hardcore al Qa'ida members. We later learned that the prisoners also included one American, John Walker Lindh.

It was a volatile combination in a volatile place, and the explosion wasn't long in coming. I vividly remember receiving an operational cable that described the incident in detail. On Sunday, November 25, two CIA officers from Team Alpha—Johnny Micheal Spann and another man I'll call "Dave"—were dispatched to the fortress to gather intelligence from the prisoners. They set about questioning the detainees in an open prison yard guarded by a few Northern Alliance soldiers. As we later learned, the guards were not only too few in number; they had also done a very poor job of searching the detainees to ensure they had no weapons.

About two hours into the interviewing, Dave heard several explosions and automatic gunfire. He looked over and saw Mike Spann being tackled by several prisoners. Running toward him, Dave drew his nine-millimeter pistol and shot four, including one who was attempting to grab Spann's AK-47 rifle. At least three of the prisoners fell on top of Mike as Dave wrested the rifle away from the fourth.

Looking up, Dave saw another prisoner running toward him and firing a pistol from fewer than ten yards away. Dave shot him and then saw a large group, many still bound with rope, rushing toward him. Dave opened up with Spann's AK-47 while backpedaling. He later estimated that he shot at least fifteen before running out of ammunition and having to replace the empty magazine.

While running for cover, Dave stumbled over the bodies of several dead and wounded Uzbek guards. Eventually, he was able to reach temporary shelter in one of the buildings on the perimeter of the compound. There he ran into five foreign journalists, who asked his assistance in getting out of Qala-i-Jangi. Using one of the journalist's satellite phones, Dave called in reinforcements and air support. The small group holed up in various locations in the building for over five hours while a battle raged outside. During this period Dave was unsure about the status of his partner. One of the journalists said that he had seen Mike escape. As it started to get dark, Dave, the journalists, and several others managed to descend the north wall of the fortress and eventually reach safety.

It was a Sunday afternoon when I got word that we potentially had an officer down. I came into headquarters immediately to monitor developments. Shortly after 9/11, Cofer Black had told me that CIA might lose thirty to forty officers in carrying out our attack strategy. For a relatively small force such as ours, that was a stunning number. But even with such grim expectations—expectations that thankfully never were met—hearing that the first CIA officer was down struck us hard. I went to Hank Crumpton's small office in the CIA headquarters, where we waited in agony for hours, desperately trying to get information from the scene.

Despite the journalist's optimistic account of Mike Spann's escape, we feared the worst for him. Two painful days would pass before U.S. and Afghan allied forces could put down the rebellion, get inside the fortress, and determine for certain that Mike was dead. Word of the riot and the possible death of a U.S. official did not wait for confirmation. Reports of the clash were soon airing around the world, and Pentagon spokesmen were quick to tell the media that no U.S. military personnel were unaccounted for. That led reporters to leap quickly and accurately to the conclusion that a CIA officer was the victim.

Mike Spann was a thirty-two-year-old former Marine who had been with CIA for only a short period. His wife, Shannon, was also a member of the CIA's clandestine service and was on the West Coast with her infant son, visiting family, at the time of the attack. Shannon was out driving when she heard a radio report about the possibility of a CIA officer missing. Immediately, she pulled her car over to the side of the road and called headquarters to find out what she could. I dispatched some officers to California to be with her, and others to Alabama to assist Mike's parents, even before we were able to verify his status.

Once Mike's body was recovered and his family informed, we made the decision to confirm his death to the media. Such confirmation is routine for the military but not always so for CIA. In this case, however, the fact of Mike's Agency background had already leaked. His family wanted to acknowledge who he was and express their pride in his service. There was no way to keep the Agency connection secret and little reason to try. Yet we were quickly criticized by pundits, who accused us of seeking publicity over the first American to die in combat in Afghanistan.

As it turned out, I had to take a trip to Pakistan shortly after Mike was killed to meet with President Musharraf over urgent intelligence we had received regarding possible follow-up al-Qa'ida attacks against the United States. On the way back to the United States, I had my plane divert to Germany, where Mike's body had been taken. On December 2, we brought him on his final trip home. I've never made a more somber journey.

Eight days later, Mike Spann was buried with full honors at Arlington National Cemetery. Shannon impressed us all with her grace, dignity, and strength. The family asked me to make remarks at the graveside, and I was honored to do so. In going to Afghanistan, to "that place of danger and terror, he sought to bring justice and freedom," I said. I told his family, friends, colleagues, and the nation that Mike Spann was a "patriot who knew that information saves lives, and that its collection is a risk worth taking."

The bureaucracy had initially balked at burying Mike in Arlington, since he had been neither retired military nor on active duty at the time of his death. John McLaughlin called Paul Wolfowitz, who quickly said he would support Mike's being given the honor of an Arlington interment. John then called Andy Card, who, based on McLaughlin and Wolfowitz's recommendation, cut through the red tape and made it happen.

Mike's is one of many remarkable stories of heroism by CIA officers in the opening months of the Afghan campaign. Although they are accustomed to working without much support or infrastructure, Afghanistan took that to new heights. Agency officers participated in cavalry charges and called in air strikes while on horseback. One CIA medic attempted to save an Afghan's life by performing an emergency amputation of the soldier's leg using the only device available to him—a large Leatherman pocketknife.

The definitive moment of the CIA's entire campaign may have been saving the life of the country's future leader. By the very early days of December, Hamid Karzai had proved himself not only a fearless fighter but also the indispensable man in the Afghan equation. As a result, CIA and U.S. Special Forces units began to worry about not just supporting him but also ensuring his survival. That, though, became increasingly difficult.

On December 5, Karzai was leading his troops in an assault on Khandahar, one of the last Taliban strongholds. U.S. military personnel were calling in air strikes in support of the assault using Global Positioning System devices. As they were doing so, one soldier replaced the batteries in his GPS unit, forgetting that doing so caused the unit to erase previously entered data and to reset itself at its own location. As a result, an air strike from a B-52 was called in on the soldier's own position. Three Americans and five Afghans died in the mishap. Karzai might have, too, if Greg V. hadn't thrown himself on him, knocking him to the ground just as the bombs struck. It turned out to be an eventful Wednesday for Karzai. That same day, he was selected to be the interim prime minister of Afghanistan.

The routing of the Taliban and al-Qa'ida from Afghanistan in a matter of weeks was accomplished by 110 CIA officers, 316 Special Forces personnel, and scores of Joint Special Operations Command raiders creating havoc behind enemy lines—a band of brothers with the support of U.S. airpower, following a CIA plan, that has to rank as one of the great successes in Agency history.

As we forced out al-Qa'ida and the Taliban leaders from the sanctuary, we continued our focus on capturing or killing Usama bin Ladin. We believed he was in the mountains of southern Nangarhar province, only miles from the Pakistani border. This area had long been an al-Qa'ida stronghold, particularly south of Jalalabad in the Tora Bora Mountains.

By early November, our intelligence reporting was indicating that UBL had fled to the Tora Bora region. When Kabul fell, on November 14, we figured that Bin Ladin and his cohorts would be even more likely to try to flee Afghanistan, perhaps for the ungoverned regions of Pakistan. CIA rushed to set up counterterrorist pursuit teams, made up of Northern Alliance fighters with U.S. advisors, but the vast reaches of the territory made this a difficult mission. Bin Ladin had chosen a good place to hide. The rugged hills of Tora Bora contain dozens of tunnels and caves. As one CIA officer put it, "He had mountains to his back, clear fields of fire in front of him, and a local population unwilling to confront or eject him."

Agency and military officers tried to motivate Afghan forces with the usual combination of exhortations and a liberal allocation of cash to press the attack against suspected al-Qa'ida strongholds. A joint CIA/JSOC team of five men infiltrated into the heart of enemy territory and, for more than seventy-two hours, directed air strikes. At one point, the team requested B-52s to deliver bombs to within twelve hundred yards of their position. In all, about seven hundred thousand pounds of ordnance was dropped between the fourth and seventh of December alone. Hundreds of al-Qa'ida operatives were killed. But CIA officers on the scene began to doubt whether they could rely on the Afghan ground force for this critical push of the campaign. Worse, there were concerns that some of the Afghan units might be actively cooperating with al-Qa'ida elements, helping them to escape.

We had sensitive intelligence that strongly suggested Bin Ladin was in the Tora Bora area and likely was plotting a quick escape through soon-to-be-completed tunnels. U.S. air power was brought to bear on this very difficult terrain.

Aerial bombardment, though, can do only so much. Truly confronting an enemy entrenched in a network of caves requires getting into the caves yourself, and the Afghan troops we were working with were distinctly reluctant to undertake that risk. It was also the Islamic holy month of Ramadan, and Afghan troops were not much interested in conducting attacks. Agency officers in the field and at headquarters started lobbying hard for the insertion of U.S. troops to try to complete the job. Hank Crumpton called Tommy Franks to discuss the situation. Tommy said that if he were to deploy a large contingent of U.S. military to the region it would take weeks to get them in place during that period, and UBL might slip away. He made the call that it was better to press ahead with the units in place at the moment than to wait for reinforcements. We urged the Pakistanis to do the best they could to place troops along the Pakistani-Afghan border. We plotted all available escape routes that Bin Ladin might choose.

I remember the president asking Hank one morning if the Pakistanis could seal the border. "No sir," he said. "No one has enough troops to prevent any possibility of escape in a region like that." The Pakistani military did manage to capture hundreds of al-Qa'ida members slipping across the border, but not the one we wanted most.

CHAPTER 13

Threat Matrix

The attacks of 9/11 were not the end of anything. They were the beginning. That was the message I was getting from my Counterterrorism Center. As far as al-Qa'ida was concerned, 9/11 was just the opening shot.

As traumatic as the attacks were, however, we knew what actions we could take. We knew what needed to be done, and there was a tremendous sense of urgency about it. Over the next several years we were able to achieve remarkable success against the terrorist threat for three strategic reasons.

First was the loss of al-Qa'ida's safe haven in Afghanistan. Because we were able to get into the sanctuary, we suddenly had access to people and documents that laid bare the future plans and intentions of al-Qa'ida. The key to success was rapidly to collect, fuse, and analyze the data in real time and to use it to drive operations.

The second strategic reason for success was Pakistani president Musharraf's decision to join the fight on our side. Pakistan switched sides—from aiding the Taliban to fighting al-Qa'ida. Pakistani intelligence chief Ehsan Ulhaq became a pivotal figure. With the arrest of well over five hundred al-Qa'ida operatives, Pakistan, in concert with U.S. intelligence, denied al-Qa'ida the luxury of a safe haven within the country's settled areas. (For his efforts, al-Qa'ida twice tried to assassinate President Musharraf.)

The third reason was the decisive action on the part of the Saudi leadership following the Riyadh bombings in May 2003. Saudi authorities have detained or killed many of the top known al-Qa'ida cell leaders in the kingdom and hundreds of foot soldiers. They have captured thousands of pounds of explosives. They have also reduced the financial resources at al-Qa'ida's disposal.

Afghanistan, Pakistan, and Saudi Arabia were just part of the puzzle. With the new authorities, money, and confidence that the U.S. president gave to us, we were able to leverage the rest of the world's counterterrorism efforts.

There were a few countries that "got it" long before 9/11. The Jordanians, Egyptians, Uzbeks, Moroccans, and Algerians always understood what we were talking about. It was ironic that, pre-9/11, we had more success in getting help within the Islamic world than elsewhere. The British and French were also always helpful. Both had lived through their own terrorist threats. But until September 11, it was hard to convince most of the world of the legitimacy of our concerns.

In addition to the strategic reasons for our success, there were several tactical steps that were important. One of the most significant keys to our accomplishments against the terrorists came from something that sounds quite mundane: a daily meeting. This meeting would be repeated at 5:00 P.M. every weekday for the three years after 9/11. At these sessions we would try to get a handle on the flood of information about terrorism pouring in from around the world. Virtually every day you would hear something about a possible impending threat that would scare you to death. But you would also hear about opportunities to work with allies, new and old, against this threat. These sessions grew out of biweekly terrorism update meetings I started when I was deputy DCI in 1996. In 1998, after the embassy bombings, the meetings became weekly. Initially we called it "the small group." That title quickly became a joke, because the number of participants expanded until they packed the large wood-paneled conference room down the hall from my office.

The point of the meeting was to pull together in one place everyone who needed to take action in the next twenty-four hours in both our war in Afghanistan and the broader war on terrorism. My intent was to cut short the time it took for information to flow from the people in the field to me and to slash the time between orders being issued in Washington and executed half a world away.

This wasn't CIA talking to itself; we had FBI, NSA, and military officers there as well. The windowless room features a long, highly polished wooden conference table with about twenty chairs around it. The conference room needed its long table because briefers would occasionally roll out charts the size of bedsheets showing analysis that connected terrorists around the world through family, phone, and/or financial contacts. Just before the session started, any maps, charts, or documents to be used in the presentations would be passed out, and at the end they would be just as efficiently collected to keep control of the information. Always there was a palpable fear in the room that the United States was about to be hit again—either here or our interests abroad. No one present thought there was a minute to waste.

Five or six Agency components would lead off the meeting every afternoon. The first briefer was usually from the Office of Terrorism Analysis, initially Pattie Kindsvater, Phil Mudd, and other analysts. Later it was Mark Rossini from the FBI, whom we affectionately called "The Voice," because his deep baritone imparted a special sense of urgency. These briefers would run down the latest threat information. The terrorist acts of 9/11 unleashed a torrent of information from around the world. Suddenly friend and foe alike started reporting information that a day or two earlier they might have withheld or ignored. Some of it would later prove to be questionable, but at the time, we could not afford to dismiss any potential threat—and there were thousands of them.

To help senior administration officials visualize the range of possible plots we were tracking, we developed, in coordination with the FBI, what we called the "threat matrix." A multipage document, the matrix was given to the president each morning as part of his PDB session. Copies of it were also provided to other top officials. In it were the newest threats that had emerged over the past twenty-four hours.

The matrix soon became an important part of the five o'clock meeting. At each session, we went over the next day's matrix, recognizing that many, perhaps most, of the threats contained in it were bogus. We just didn't know which ones. In a typical matrix you might see tales of impending doom picked up from people walking into U.S. embassies overseas, cryptic comments gathered through intercepted foreign communications, anonymous correspondence received by major media outlets, and leads given to us by human assets.

We recognized that the matrix was a blunt instrument. You could drive yourself crazy believing all or even half of what was in it. It was exceptionally useful, however, and an unprecedented mechanism for systematically organizing, tracking, validating, cross-checking, and debunking the voluminous amount of threat data flowing into the intelligence community. The very massiveness of it prompted officials to think through vulnerabilities. Have we done enough to secure major landmarks, theme parks, or water supplies? Are our watchlists tight enough? Sometimes the threats mentioned would strike you as absurd, and then al-Qa'ida would do something to convince you that nothing was out of the range of possibility. Who, for example, would have thought that exploding footwear could be a major air travel problem—until, that is, December 21, 2001, when Richard Reid was subdued on an American Airlines flight from Paris to Miami trying to light explosives hidden in his shoes?

After the discussion of the threat matrix, Hank Crumpton, chief of the CTC Special Operations Group, would come next. He'd be followed by the chief of Alec Station's Bin Ladin Unit, initially Hendrik V., and later Marty M.; then Rolf Mowatt-Larssen, head of CTC's WMD branch, would brief. On occasion we would hear from Phil R., who was in charge of CTC's efforts involving international financial operations. Charlie Allen would carefully listen to our operational requirements and translate them into information requirements, which our intelligence communities, both foreign and domestic, would have to pursue. This was both to meet imminent operational needs and to position us to stay one step ahead of the terrorists.

Also at my side at the five o'clock meetings were John McLaughlin; the heads of the Directorates of Operations, Intelligence, and Science and Technology; the senior leadership of CTC; and others whose goal was to help clear obstacles for those who were on the front lines. Attendance at the five o'clock meetings became a critical part of each person's day. If, for some reason, you missed a meeting, you'd have to struggle the next day to follow the plot lines—so much interconnected information flowed each time.

November 6, 2001, was a typical five o'clock session. On that day I was briefed on a wide variety of freshly collected intelligence: A report had been collected about an Arab, of Persian Gulf origin, who reportedly knew of a planned second strike against the United States that was imminent and who claimed that the operatives were already in place. Additionally, he claimed to know of a third and final attack after which he would be free to come home. Similarly there was information on someone apparently in Jordan who had posted on a website a prediction that another attack on the United States was imminent. You might ask, so what? Until you learned that this same person had posted a note saying they were close to "zero hour" on September 10, 2001.

Another snippet of intelligence that day told us that a known al-Qa'ida associate who had been in the United States from 1999 to the fall of 2001 was aware of big events expected on November 5 and 6. We also learned that an Egyptian who worked for the embassy in Saudi Arabia had suddenly, without explanation, faxed in his resignation. Subsequent investigation showed that the man had ties to al-Qa'ida's partner, Egyptian Islamic Jihad, and was wanted by authorities in his home country. Could his disappearance presage some new attack? We had to try to find him fast.

That same evening, I heard about intelligence gleaned from a senior UBL operative that provided the name of an al-Qa'ida associate determined to conduct a suicide operation. We had the name, biographical data, but no idea where the man was.

Nearly two months after the attacks of 9/11 there was still great skepticism in Saudi Arabia that any of their countrymen had been involved. My staff came to me that night with a proposal that we share the chilling cockpit audio recordings made from United Airlines Flight 93 before it crashed in Pennsylvania. The Saudi-accented voices heard on the tape might remove any doubts.

We had intelligence of three al-Qa'ida–associated people, possibly connected to Abu Musab al-Zarqawi, traveling for unknown reasons; we passed along the intelligence to three countries, all mentioned as possible transit points.

We heard from Russian intelligence about increased concerns over terrorist actions in Chechnya.

A Middle Eastern country captured a terrorist wanted in a third country. Could we help get him there? We could.

The FBI had conducted a polygraph on a source of the U.S. Customs Service who said he knew of a possible nuclear threat to the United States; that source flunked the test, which showed "deception indicated."

The intelligence we heard that night, and every night, were just tiny threads. They had to be woven into a tapestry before we could make sense of what we were seeing. And this was just one day; it is difficult to put in words the number of reports, and the intensity of those reports, that came in every day. As one officer said to me, "I never want to live that again. The pace was furious. The constant refrain was: It must be done tonight, it must be done tomorrow. We have to have that for the president tomorrow. That pace wasn't kept up for days or weeks; it was years."

The five o'clock meetings were decision-making sessions, not briefings. If someone told me he was having trouble getting needed information out of an allied government, I'd often grab the phone right after leaving the meeting, call the head of the intelligence service involved, and light a fire under him. Other times I would order up talking points to be in my hands by six the next morning.

Other governments weren't the only concern. Sometimes we would hear of potential threats that weren't being internalized quickly enough within our *own* government. Countless times someone in the room was directed to get up that second, find a phone, and call the Pentagon, the FBI, the State Department, or some other entity, to make absolutely sure that the right people knew everything we knew and that they were going to get on top of that particular threat. The key was imparting information and context quickly; we had no time for more briefings.

On many occasions, I would be briefed on matters that were, as they say in Washington, "outside my lane." When that happened, I would tell, say, the FBI representative to call Director Bob Mueller and bring him up to speed on a domestic issue, because we intended to mention it in the next day's PDB session in the Oval Office. Without doubt, the president was going to turn to Bob and ask what he was doing about this; it was in everyone's interest that he had a good answer.

Our morning sessions with the president were also intense. He quickly became steeped in our strategy, with regard to activities not only in Afghanistan but also in the rest of the world. He was focused on results yet at the same time did not seek to micromanage our operations. He spent time with the substantive experts we brought to daily meetings and to longer sessions at Camp David on Saturdays. The president never became the action officer, but there was no doubt the leader was in the trenches with us. If you told him about an imminent operation on Monday, you could be certain after a few days he would ask about it, if we had not provided the necessary follow-up.

A PDB session would lead to a broader meeting with Bob Mueller, Tom Ridge, later Fran Townsend, and their staffs, to review the threat matrix, the actions that were being taken, the gaps in our knowledge, and the interventions the president or vice president could undertake to help. Over time, at Andy Card's insistence, we modified the items in the matrix the president would see, to ensure that only those with the necessary weight and quality consumed his attention. When you have been accused of failing to connect the dots, your initial reaction is to ensure that all the dots are briefed. Until our knowledge became more refined, our inclination was to overbrief.

At the core of our effort was the Counterterrorism Center. It was the hub around which all of our efforts revolved. From there CIA stations worldwide were tapped to work both unilaterally and with host government intelligence services to improve the information sharing we relied upon. The long-standing relationships that Agency officers had with counterparts around the world became essential to our success. Even former adversaries seemed more willing to work with us.

As we made progress overseas, we found ourselves struggling domestically. It was stunning how little reliable information was immediately available inside our own borders. There was no good data on how many foreigners had overstayed their visas and no tracking system to see if young men who came into this country to attend university had actually shown up for classes—or if they had changed their major from music to nuclear physics. Nor was there any way for a police department in one part of the country to share suspicious activity data with counterparts across the state or the nation. There was no seamless way to communicate from Beirut to Seattle; there was no communications backbone. And while there were mountains of data within the United States, no one knew how to access it all, and little had been done to train people to put it together and report it, much less analyze it. In the early days, what we did not know about what was going on in the United States haunted us. We had to make judgments based on instinct.

Few understand the palpable sense of uncertainty and even fear that gripped those in the storm's center in the immediate aftermath of 9/11. One particular concern was the fact that, although there wasn't any tracking system in place, there were thousands of foreigners in the United States whose visas had expired. The most important thing we needed to do was to prove the negative: that there were not more al-Qa'ida cells within the country poised to conduct a second wave of attacks. At the time, I remember reflecting on testimony Gen. Mike Hayden, then the director of NSA, had given to a public hearing of the House Intelligence Committee in 2000. Mike created quite a stir when he said that if Usama bin Ladin had crossed the bridge from Niagara Falls, Ontario, to Niagara Falls, New York, there were provisions of U.S. law that would offer him protections with regard to how NSA could cover him. Mike would later say that he was using this as a stark hypothetical. On September 12, 2001, it became real.

After the 9/11 attacks, using his existing authorities, Hayden implemented a program to monitor communications to and from Afghanistan, where the 9/11 attacks were planned. With regard to NSA's policy of minimization, balancing U.S. privacy and inherent intelligence value, Mike moved from a peacetime to a wartime standard. He briefed me on this, and I approved. By early October 2001, Hayden had briefed the full House Intelligence Committee and the leadership of the Senate Intelligence Committee.

Soon thereafter, the vice president asked me if NSA could do more. Our ability to monitor al-Qa'ida's planning was limited because of constraints we had imposed on ourselves through the passing of certain U.S. laws in the late 1970s. I called Mike to relay the vice president's inquiry. Mike made it clear that he could do no more within the existing authorities. We went to see the vice president together. Mike laid out what could be done that would be feasible, prudent, and effective.

Within a week new authorities were granted to allow NSA to pursue what is now known as the "terrorist surveillance program." The rules required that at least one side of the phone call being surveilled be outside the United States and that there be probable cause to believe that at least one end of the communication was with someone associated with al-Qa'ida. Elaborate protocols were set up to ensure that the program was carried out in accordance with these regulations. Within weeks of the program's inception, senior congressional leaders were called to the White House and briefed on it. Prior to its disclosure, twelve such briefings were hosted by the vice president for the leaders of the House and Senate Intelligence Committees. The briefings were thorough and disciplined. From my perspective, Mike gave the members full insight into how the program was being managed, the care that was being taken to ensure that it lived up to its intent, and offered the best analysis he could provide with regard to its results. The program was reauthorized by the president about every forty-five days prior to its disclosure. Each reauthorization was accompanied by an intelligence review, each of which I signed prior to my retirement. This included a comprehensive assessment of the value of continuing the program.

At one point in 2004 there was even a discussion with the congressional leadership in the White House Situation Room with regard to whether new legislation should be introduced to amend the FISA statute, to put the program on a broader legal foundation. The view that day on the part of members of Congress was that this could not be done without jeopardizing the program.

Mike Hayden has persuasively argued that the FISA statute enacted in 1978 could not have contemplated the technology available for terrorist use today, nor provided for the speed needed to deter today's terrorist acts. A bipartisan effort to amend the statute would be wise, so long as it is done in a manner that does not jeopardize critical operational equities. The trauma of 9/11 led, in the words of Mike Hayden, to a program to protect our liberty by making us all feel safer. It was never about violating the privacy of our citizens.

Had this program existed prior to 9/11, Mike Hayden has said that, in his professional judgment, we would have detected some of the al-Qa'ida operatives in the United States and we would have identified them as such. I agree.

As we were coming up with the new terrorist surveillance program, our working assumption had always been that the attacks of 9/11 were simply the first wave. Al-Qa'ida had declared its intention to destroy our country. Why then would it be satisfied with just three thousand deaths? It was inconceivable to us that Bin Ladin had not already positioned people to conduct second, and possibly third and fourth waves of attacks inside the United States. Getting people into this country—legally or illegally—was no challenge before 9/11. Al-Qa'ida had to have known that things would tighten up after the attacks, so logic suggested that they would have acted in advance to prepare for that inevitability. We considered the possibility that in addition to carrying out the September 11 attacks, the nineteen hijackers might also have done casing and provided surveillance for whatever attack would come next. Nothing that I learned in the ensuing three years ever led me to believe that our initial working assumption that al-Qa'ida had cells here was wrong.

Increasingly, we began to concentrate on the possible connections between the domestic front and the data we were collecting overseas. We would identify al-Qa'ida members and other terrorists overseas and often discover that they had relatives, acquaintances, or business ties with people in the United States. Each rock overturned abroad led to ants scurrying every which way, including many toward the United States. These concerns, in part, led to the establishment of the NSA program wrongly described by the media as "domestic spying." The program grew out of concrete evidence that foreign terrorists planning new attacks on America were in communication with colleagues in this country. Oddly, the farther terrorists were from our shores, the more vulnerable they were to our intelligence-collection efforts. In some ways, the safest place for an al-Qa'ida member to hide was inside the United States.

As much as our government would have liked to capture or

kill Usama bin Ladin and Ayman al-Zawahiri, we recognized that the key to crippling al-Qa'ida would be to take down the next tier of leadership, the facilitators, planners, financiers, document forgers, and the like. These were the people who would have the actual links to the terrorist operatives. If we could disrupt or destroy the efforts of these individuals, we might prevent the follow-on attack that we feared so much. Our strategy was clear: to weaken al-Qa'ida's ability to plan and execute attacks, by forcing them to move less capable individuals into positions of leadership. In particular, our focus was on the individuals in charge of planning operations against the United States. Once Khalid Sheikh Mohammed was captured, Abu Faraj al-Libi took over. He was captured in Pakistan in May 2005 and replaced by Hamza Rabi'a, who was reportedly killed in the North Waziristan province of Pakistan seven months later.

One of the first dominoes to fall was Abu Zubaydah. Before 9/11, his name had been all over our threat reporting. After the attacks, he gained an even more prominent role in al-Qa'ida, especially once the United States killed the group's number three man, Mohammed Atef, in a November 2001 air strike in Afghanistan. Time and again in our five o'clock meeting we discussed how to run Abu Zubaydah to the ground.

By March 2002 we had identified a large number of sites in Pakistan that appeared to be al-Qa'ida safe houses. We got the increasingly helpful Pakistani authorities to raid thirteen of them simultaneously; they captured more than two dozen al-Qa'ida members. We were hopeful that a big fish like Abu Zubaydah would be in one of the safe houses, and we were not disappointed. In Pakistan's third largest city, Faisalabad, a gunfight broke out when Pakistani security officials stormed a second-floor apartment. Abu Zubaydah, who was inside, was shot three times and critically wounded.

Ironically, we found ourselves suddenly concerned with trying to save a terrorist's life. Not that we had any sympathy for Zubaydah; we just didn't want him dying before we could learn what he might have to tell us about plans for future attacks. Fortunately, Buzzy Krongard, our executive director, was also on the board of directors of Johns Hopkins Medical Center. Using his contacts there, he arranged for a world-class medical expert to jump aboard an aircraft we had chartered so he could be flown to Pakistan and save a killer's life. Once Abu Zubaydah was stabilized, the Pakistanis turned him over to CIA custody. It was at this point that we got into holding and interrogating high-value detainees—"HVDs," as we called them—in a serious way.

Detainees, in general, had become a critical issue. By this time, many Taliban and al-Qa'ida prisoners were in military custody. Yet the quantity and quality of intelligence produced from their interrogation was disappointing. The detainees were either too low ranking to know much or too disciplined to reveal useful information.

Abu Zubaydah's capture altered that equation. Now that we had an undoubted resource in our hands—the highest-ranking al-Qa'ida official captured to date—we opened discussions within the National Security Council as to how to handle him, since holding and interrogating large numbers of al-Qa'ida operatives had never been part of our plan. But Zubaydah and a small number of other extremely highly placed terrorists potentially had information that might save thousands of lives. We wondered what we could legitimately do to get that information. Despite what Hollywood might have you believe, in situations like this you don't call in the tough guys; you call in the lawyers. It took until August to get clear guidance on what Agency officers could legally do. Without such legal determinations from the Department of Justice, our officers would have been at risk for future second-guessing. We knew that, like almost everything else in Washington, the program would eventually be leaked and our Agency and its people would be inaccurately portrayed in the worst possible light. Out of those conversations came a decision that CIA would hold and interrogate a small number of HVDs.

CIA officers came up with a series of interrogation techniques that would be carefully monitored at all times to ensure the safety of the prisoner. The administration and the Department of Justice were fully briefed and approved the use of these tactics. After we received written Department of Justice guidance on the interrogation issue, we briefed the chairmen and ranking members of our oversight committees. While they were not asked to formally approve the program, as it was conducted under the president's unilateral authorities, I can recall no objections being raised.

The most aggressive interrogation techniques conducted by CIA personnel were applied to only a handful of the worst terrorists on the planet, including people who had planned the 9/11 attacks and who, among other things, were responsible for journalist Daniel Pearl's death. The interrogation of these few individuals was conducted in a precisely monitored, measured way intended to try to prevent what we believed to be an imminent follow-on attack. Information from these interrogations helped disrupt plots aimed at locations in the United States, the United Kingdom, the Middle East, South Asia, and Central Asia.

The president confirmed the existence of the interrogation program on September 6, 2006, when he announced that fourteen HVDs who had been held under CIA control would be transferred to Guantánamo Bay.

Like many of the al-Qa'ida detainees, Abu Zubaydah originally thought that he could outsmart his questioners. He would offer up bits and pieces of information that he thought would give the impression of his providing useful material, without really compromising operational security.

But Abu Zubaydah ultimately provided a motherlode of information, and not just from his interrogation. We were able to exploit data found on his cell phone, computer, and documents in his possession that greatly added to our understanding of his contacts and involvement in terrorism plotting.

Interrogating Abu Zubaydah led us to Ramzi bin al-Shibh. A Yemeni by birth, Bin al-Shibh had studied in Germany with three of the eventual 9/11 hijackers. He had intended to be one of them and was deterred only after four attempts to obtain a U.S. visa failed. Instead, he served as the primary communication link between the hijackers and al-Qa'ida central, meeting with the plot's ringleader, Mohammed Atta, in Germany and Spain, and staying in touch with the terrorists via phone and e-mail. With Zubaydah's unintentional help, Bin al-Shibh was captured by Pakistani authorities on the first anniversary of the 9/11 attacks, after a gun battle in Karachi.

But no success story lasts long in Washington before someone tries to minimize it. A published report in 2006 contended that Abu Zubaydah was mentally unstable and that the administration had overstated his importance. Baloney. Abu Zubaydah had been at the crossroads of many al-Qa'ida operations and was in position to—and did—share critical information with his interrogators. Apparently, the source of the rumor that Abu Zubaydah was unbalanced was his personal diary, in which he adopted various personas. From that shaky perch, some junior Freudians leapt to the conclusion that Zubaydah had multiple personalities. In fact, Agency psychiatrists eventually determined that in his diary he was using a sophisticated literary device to express himself. And, boy, did he express himself.

Abu Zubaydah's diary was hundreds of pages long. Agency linguists translated enough of it to determine there was nothing of operational use in it, yet some Pentagon officials, including Paul Wolfowitz, seemed fascinated with the subject and kept bugging us to translate the whole document. We kept resisting. One day Wolfowitz hounded his CIA briefer. "Why wouldn't we devote the resources to convert the book to English?" he demanded. "We know enough about the diary," the briefer explained, "to know that it simply contains a young man's thoughts about life—and especially about what he wanted to do with women." "Well, what have you learned from that?" Wolfowitz asked. Without missing a beat, the briefer responded, "That men are pigs!" Wolfowitz's military assistant laughed so hard he fell off his chair.

But in Afghanistan there was no time for laughter. As we

achieved success in driving al-Qa'ida out of Afghanistan, they began to search for other sanctuaries for their leadership. The organization sought places where they could plan future attacks against the United States with impunity from law enforcement, intelligence, and military operations. First, al-Qa'ida established itself in the settled areas of Pakistan. Later they moved into the ungoverned tribal areas of South Waziristan. Later still, Pakistani military operations drove them farther north, to areas where I believe their senior leaders continue to operate.

In mid-2002 we learned that portions of al-Qa'ida's leadership structure had relocated to Iran. This became much more problematic, leading to overtures to Iran and eventually face-to-face discussions with Iranian officials in December 2002 and early 2003. Ultimately, the al-Qa'ida leaders in Iran were placed under some form of house arrest, although the Iranians refused to deport them to their countries of origin, as we had requested.

In the spring of 2002, computers, phone records, and other data from al-Qa'ida takedowns in Pakistan, Afghanistan, and elsewhere started to suggest troubling connections to individuals in the United States, particularly in the Buffalo, New York, area. As with so much else in those hectic days, I first learned about all this at one of our five o'clock meetings. I told the lead analyst on the matter to share her concerns immediately with the FBI. We had her take all her data to the regional FBI office, where, initially, she got a skeptical reception. Even in the aftermath of 9/11, there was a reluctance to believe that sleeper cells could be operating in the United States, particularly cells made up of American citizens. But as the FBI dug into the matter, the Bureau became believers. Six Yemeni Americans, all of whom had received training at an al-Qa'ida camp in Afghanistan prior to 9/11, were arrested in September 2002. The group, which became known as the Lackawanna Six, later pled guilty to terrorism-related charges and received prison terms ranging from eight to ten years each.

The five o'clock meetings did more than coordinate the takedown of individual terrorists and unravel future plots. We also used them to track the ebb and flow of overall threat concerns. Throughout the three years after 9/11 there was a lot more "flow" than there was "ebb."

These times of heightened concern would often translate into increasing the terrorist threat warning levels from yellow to orange. We did so on four occasions. In each instance there was a credible intelligence basis for doing so. Initially, there was no choice but to burden the entire country. Over time, we became more sophisticated and surgical in focusing on specific geographical locations and sectors of the economy. In developing the system of protection, the initial option was imprecise. Some pundits alleged that the administration was only elevating the threat level for political purposes, but I can assure you that in each case we believed that the threat was real and imminent and that we had no other reasonable option.

While we raised the threat level on four occasions during my tenure, one period stands out in my mind: the spring and summer of 2004. There were several streams of concern. First, we came into the possession of casing and surveillance reports focused on financial institutions in New York, New Jersey, and Washington. What was noteworthy about the reports was their specificity and attention to detail regarding the buildings themselves, perceived structural deficiencies, the location of security, and the types of alarms in specific locations within the buildings. The reports were written as though produced by an engineering consulting firm and were of a quality consistent with what a sophisticated intelligence service might produce. Only one dot to connect, perhaps, but there were more.

The strategic context for concern was compelling. We were approaching national political conventions and an election. Al-Qa'ida had paid attention to the fact that the March 11 attack in Madrid had brought down the Aznar government in Spain. We believed that Bin Ladin himself had assessed that a logical time to attack the United States was just before the U.S. election, when he perceived the uncertainty created by a potential transition of government would make a response more difficult.

There was the fear that the arrests of operatives in Canada, Pakistan, and New York suspected of planning attacks in London might force al-Qa'ida to accelerate the timing of attacks inside the United States. Because of military operations conducted by Pakistan in the southern tribal areas of Waziristan, al-Qa'ida was under enormous pressure, stimulating the need for a high-stakes showdown with the United States. The plotting against Musharraf's life continued.

The intelligence that we received was more frightening. By July 2004 we believed that the major elements of the plot were in place and moving toward execution and that the plot had been sanctioned by the al-Qa'ida leadership. We believed that al-Qa'ida facilitators were already inside the United States, in an organized group—which to the best of my knowledge has never been found—and that they had selected non-Arab operatives to carry out the attacks.

A separate stream of reporting told us of al-Qa'ida plans to smuggle operatives through Mexico to conduct suicide operations inside the United States. This was linked directly back to direction being provided by al-Qa'ida's leaders. All of this was consistent with the intelligence dating back to 2001 of either the presence of, or attempts to infiltrate, operatives inside the United States.

There was strategic warning, further arrests, and disruption activities overseas and in the United States by CIA, our foreign partners, and the FBI. NSA was operating at a fever pitch attempting to determine linkages from dirty numbers overseas to numbers inside the United States. Detainees were questioned and financial data mined for operational activity, all in real time. We posited likely targets and methods of attack. It was a period of furious activity.

The attacks—based on very credible reporting—didn't happen.

Why? Had the effectiveness of law enforcement and intelligence disrupted the planning? Quite possibly. Was it a conscious decision on the part of al-Qa'ida to delay for its own reasons, out of concern for its weaknesses and the rally-round-the-flag impact an attack would have in the United States? Equally plausible. It was yet another period of high threat that had not come to much, other than exhaustion. I do not know why attacks didn't occur. But I do know one thing in my gut: al-Qa'ida is here and waiting.

The threat was not just within the United States. Often information I heard at the five o'clock meeting would cause me to schedule abrupt overseas trips to key Middle East capitals. At one such meeting, I learned of intelligence that al-Qa'ida operatives were planning to assassinate members of the Saudi royal family and overthrow the Saudi government. I quickly scheduled a meeting with the Crown Prince.

Then–Crown Prince Abdullah is an incredibly impressive man, a billionaire like many Saudi princes, yet one who has never allowed himself to forget his roots. Alone among the top royals, he'll go off and live in the desert for weeks on end to reconnect with the Saud family's past. As cooperative as he could be in our pursuit of intelligence on terrorists, from our perspective, Saudi cooperation against al-Qa'ida could be slow and frustrating.

The Saudis were equally frustrated with us for not sharing enough information, but the speed with which we needed Saudi action came only after the kingdom itself was attacked in May of 2003. Thirty-five people, including ten Americans and seven Saudis, died, and more than two hundred were injured in the al-Qa'ida attack on a Western housing compound in Riyadh. That brought the message home to the royal family in a way nothing else had.

When I first heard about the Riyadh attacks, I knew I had to go see the Crown Prince, to offer condolences and to make a point while the wound was still fresh. I cleared the trip with the president and the national security advisor and gave them a rough idea of what I was going to say. But I wrote out my own talking points for use with the Crown Prince, and I didn't clear them with anyone. There was no reason to do so. I knew what had to be said. I doubt if I've ever had a more direct conversation with anyone in my life.

First, I started with an intelligence briefing on what had just occurred:

• The debate within al-Qa'ida over conducting attacks in Saudi Arabia dates back to the fall of 2002. It was never about whether to strike, but about when and how.

• The loss of sanctuary in Afghanistan, the settled areas of Pakistan, and northeastern Iraq raised an important question: Could the group afford to lose its position in the kingdom and, with it, its chief source of funds?

• Bin Ladin, who prior to 9/11 had imposed a ban on attacks in Saudi Arabia, made his position clear when he urged a key Saudi-based operative, Abu Hazim al-Sha'ir, to move forward with the attacks at any price.

• Khalid Sheikh Mohammed told us later that Bin Ladin's highest priority is to spur a revolution in Saudi Arabia and overthrow the government and that al-Qa'ida operatives in the kingdom had blanket autonomy to conduct attacks on their own.

"Your Royal Highness," I said, "your family and the end of its rule is the objective now. Al-Qa'ida operatives are prepared to assassinate members of the royal family and to attack key economic targets."

I told the Crown Prince that a Saudi-based contact of Saad al-Faqih, a London-based dissident, responded to Faqih's call for the overthrow of the Saudi royal family in February by saying, "The assassination phase has already begun."

I said, "We know that senior al-Qa'ida operatives inside the kingdom are planning attacks against American interests, both in the United States and in Europe. Your Royal Highness, we are exactly where we were before September 11, but with some important differences. We have great specificity with regard to the planning. It's directed against your family and religious leadership. It is directed from within the kingdom against the United States with the same apocalyptic language I saw before the attacks on September 11. Our relationship cannot sustain another attack. So what do we do about this? We either declare war, and act like we mean it, or we accept the catastrophic consequences."

It was a long meeting and an emotional one. Prince Bandar, the longtime Saudi ambassador to the United States, who had ridden with me to the palace, had encouraged me to lay everything on the line, and I did, chapter and verse.

I have rarely been more direct in my life. By the time I was through with my presentation, the room was energized—by my words and by the attacks of a few days earlier—and virtually that very day, the Crown Prince began to implement a plan we'd helped create.

The world is still not a safe place, but it is a safer place now because of the aggressive steps that the Saudis began to take. They arrested, captured, or killed many (if not all) of the senior al-Qa'ida operatives involved in the plotting. One major capture involved Abu Bakr al-Azdi, who confirmed that indeed plotting against the United States was occurring from within the kingdom. They began to clamp down on al-Qa'ida's finances, and engaged with their clerical establishment to overturn fatwas urging mass violence as a tactic. Al-Qa'ida made an important strategic miscalculation, never counting on the Crown Prince's reaction. The anger of this honest man at what had happened to his country was palpable that day. As frustrating as the U.S.-Saudi relationship had been over the years, our patience had paid off.

Particularly important at that time, and from then on, were the efforts of Prince Mohammad bin Naif, interior minister Prince Naif's son, who worked for his father as deputy interior minister for security affairs. MBN, as we called him, became my most important interlocutor. A relatively young man, he is someone in whom we developed a great deal of trust and respect. Many of the successes in rolling up al-Qa'ida in the kingdom are a result of his courageous efforts.

Let's be clear: the Saudis acted out of self-interest. At stake were not only plots against the United States but the stability of Saudi Arabia as well. While sustained Saudi action had been a long time coming, the Crown Prince's sense of urgency was matched by our determination to deny al-Qa'ida the key elements of their political strategy. Al-Qai'da wanted the destruction of the House of Saud and the creation of a Bin Ladin–inspired caliphate, with the economic muscle that oil would confer. The accommodation that the House of Saud had made with the Wahabi branch of Islam had turned the kingdom into a ready source of finance, recruitment, and inspiration for al-Qa'ida. We now had the beginning of a sustained counterterrorism partnership that has carried on since. It has been vital to eliminating an al-Qa'ida safe haven that had operated within Saudi Arabia.

As important as our relationship with the Saudis was, we depended on foreign partners all over the world. Of all the terrorist takedowns, none was more important or memorable than the capture in Pakistan of Khalid Sheikh Mohammed, whom everyone in our business referred to simply as KSM. No person, other than perhaps Usama bin Ladin, was more responsible for the attacks of 9/11 than KSM, and none, other than UBL, more deserved to be brought to justice.

Although KSM grew up in Kuwait, his family comes from the Baluchistan region, which straddles the Iran-Pakistan border. During the mid-1980s, he attended college in North Carolina.

The future Most Wanted list all-star first came to the attention of U.S. intelligence about the time it was learned that his nephew, Ramzi Yousef, had been involved in planning the 1993 World Trade Center attack. Yousef was arrested in Islamabad, Pakistan, in 1995 and later tried and convicted in U.S. courts for his part in planning "Operation Bojinka," which envisioned simultaneously blowing up twelve airliners over the Pacific. Yousef had also been involved in plots to assassinate Pope John Paul II during an official visit to the Philippines and in a plan to have a suicide pilot fly a small plane loaded with explosives into CIA headquarters. Clearly, he and KSM came from the same gene pool.

During the mid-1990s, CIA chased KSM around three continents. We attempted to bring him to justice in Qatar, the Philippines, and even Brazil. He eluded us and ended up in Afghanistan, where he first met Usama bin Ladin. Through the late 1990s, we knew that KSM was taking on an increasingly important role with al-Qa'ida. It was only after the capture of Abu Zubaydah that we learned how significant that role had become. From our interrogations of Abu Zubaydah and later KSM himself, we would learn that it was KSM who first proposed the idea of flying aircraft into the World Trade Center. Initially he suggested stealing small private aircraft and filling them with explosives. Usama bin Ladin reportedly asked, "Why do you use an axe when you can use a bulldozer?" and altered the plan to use commercial airliners full of passengers.

By early 2002, we believed that KSM, like much of the al-Qa'ida leadership, was in hiding in the teeming cities of Pakistan. To find him, CIA ran elaborate human intelligence operations.

I vividly remember Marty M., the then chief of the Sunni Extremist Group of CTC, asking me at the end of one of our Friday five o'clock meetings, "Boss, where are you going to be this weekend? Stay in touch. I just might get some good news."

Later that evening, Pakistani security officials surrounded a house in Rawalpindi where they suspected KSM was hiding. The Pakistanis stormed the residence and were wrestling KSM to the ground when he grabbed for a rifle. In the melee, the weapon went off, shooting one of the Pakistanis in the foot, before KSM was subdued for good.

Marty woke me with the good news. "Boss," he said. "We got KSM." You don't take down a major terrorist in the middle of a large city and have it go unnoticed. Before sunrise, Pakistani media were reporting that KSM had been taken into custody.

By the next morning, Sunday, March 2, U.S. media outlets were carrying news of the capture as well. Some of the stories described the worldly KSM as an al-Qa'ida James Bond. To illustrate the point, they showed photos of him with a full dark beard wearing what were supposedly his traditional robes. It didn't take long for Marty to phone me and relay his disgust at some of the coverage. A native of Louisiana, Marty speaks with a Cajun patois that is sometimes hard to decipher. We used to joke that he speaks "level 5" (fluent) Arabic but only "level 2" English.

"Boss," he said, "this ain't right. The media are making this bum look like a hero. That ain't right. You should see the way this bird looked when we took him down. I want to show the world what terrorists look like!"

Turns out, our officers on the scene in Rawalpindi had snapped and sent back some digital photos of KSM just after his capture, so I suggested that Marty call the Agency spokesman, Bill Harlow, and work something out. Within an hour, Harlow was in CTC looking over a selection of photos that made KSM look nothing like James Bond. Together they picked out the most evocative photo. Then Harlow, armed with a digital copy, called up a reporter at the Associated Press and told him, "I'm about to make your day." Asking only that the AP not reveal where they got the picture, he released the image of a stunned, disheveled, scroungy KSM wearing a ratty T-shirt. The photo became one of the iconic images of the war on terrorism. If we could have copyrighted it, we might have funded CTC for a year on the profits. Foreign intelligence services later told us that the single best thing we ever did was release that picture. It sent a message more eloquently than ten thousand words ever could that the life of a terrorist on the run is anything but glamorous.

Just after KSM's capture, I left on a trip to a half-dozen Middle Eastern countries. Among my stops was Islamabad. I wanted to personally thank the courageous Pakistani security officials who had captured KSM, and indeed I gave several of them CIA medals. I particularly remember the man who had been shot in the foot during the takedown painfully limping forward to receive his medal. From their side, the Pakistanis presented me with the rifle they had seized from KSM.

There have been published reports that CIA paid millions of dollars in "prize money" for capturing al-Qa'ida figures. That is absolutely right. It seemed to us entirely appropriate to tell countries around the world that there is both a price to pay if they cooperate with terrorists, and an appropriate reward to be earned for bringing them to justice. While we could, and sometimes did, simply present a check to the intelligence service responsible for helping us capture a major terrorist, we would occasionally opt for a more dramatic approach. We would show up in someone's office, offer our thanks, and we would leave behind a briefcase full of crisp one-hundred-dollar bills, sometimes totaling more than a million in a single transaction. Post–September 11, the influx of cash in our hands made a huge difference. We were able to fund training, support technology upgrades of our key partners, and generally reward good performance.

I also had the opportunity at one of our stops to meet the foreign agent who had led us to KSM. The man bought his first suit to wear to our meeting. I thanked him for his courage and expressed our gratitude for what he had done. He embraced me, looked me in the eye, and asked just one question: "Do you think President Bush knows of my role in this capture?" I smiled. "Yes, he does," I said, "because I told him." The fellow beamed with pride. "Does he know my name?" he asked. "No. Because that is a secret that he doesn't need to know," I replied. I asked the man why he had agreed to help us and to place his life at risk. His answer goes to the heart of the struggle we're involved in against terrorists worldwide: "I want my children free of these madmen who distort our religion and kill innocent people," he told me.

The benefits of capturing someone like KSM went far beyond simply taking a killer off the street. Through hard work, each success cascaded into others. It was amazing to watch. For example, the same day that KSM was captured, a senior al-Qa'ida financial operator by the name of Majid Khan was also taken into custody.

In interrogation, KSM told us that Majid Khan had recently provided fifty thousand dollars to operatives working for a major al-Qa'ida figure in Southeast Asia known as "Hambali." When confronted with this allegation, Khan confirmed it and said he gave the money to someone named Zubair, and he provided the man's phone number. Before long, Zubair was in custody and provided fragmentary information that led us to capture another senior Hambali associate named Bashir bin Lap, aka "Lilie." That person provided information that led to the capture of Hambali, in Thailand.

The importance of Hambali's capture cannot be overestimated. He was the leader of the Jemaah Islamiya, a Sunni extremist organization that has established an operational infrastructure in Southeast Asia. Hambali swore allegiance to Bin Ladin in the late 1990s, offering him a critical operational advantage: a non-Arab face to attack the United States and our allies. While moderate Islam thrives in Southeast Asia, its geographic expanse offers the opportunity to create dispersed sanctuaries throughout the continent.

What Hambali's arrest demonstrated is that our campaign was targeted not just against al-Qa'ida but also against Sunni extremism around the world. What we are fighting today is bigger than the al-Qa'ida central management structure and more diverse than Arab males between the ages of eighteen and forty. What we have to contend with has an Arab, Asian, European, African, and perhaps even a homegrown American face.

After Hambali was arrested, we went back to KSM and asked him to speculate on who might fill Hambali's shoes. KSM suggested that the likely candidate would be Hambali's brother, Rusman "Gun Gun" Gunawan. So we went back to Hambali, and while being debriefed, he inadvertently provided information that led to the detention of his brother, in Karachi, in September 2003.

In custody, "Gun Gun" identified a cell of Jemaah Islamiya members hidden in Karachi that his brother planned to use for future al-Qa'ida operations. Hambali confirmed that the non-Arab men were being groomed for future attacks in the United States, at the behest of KSM, and were probably intended to conduct a future airborne attack on America's West Coast.

I believe none of these successes would have happened if we had had to treat KSM like a white-collar criminal—read him his Miranda rights and get him a lawyer who surely would have insisted that his client simply shut up. In his initial interrogation by CIA officers, KSM was defiant. "I'll talk to you guys," he said, "after I get to New York and see my lawyer." Apparently he thought he would be immediately shipped to the United States and indicted in the Southern District of New York. Had that happened, I am confident that we would have obtained none of the information he had in his head about imminent threats against the American people.

From our interrogation of KSM and other senior al-Qa'ida members, and our examination of documents found on them, we learned many things—not just tactical information leading to the next capture. For example, more than twenty plots had been put in motion by al-Qa'ida against U.S. infrastructure targets, including communications nodes, nuclear power plants, dams, bridges, and tunnels. All these plots were in various stages of planning when we captured or killed the pre-9/11 al-Qa'ida leaders behind them.

In my view, it wasn't one single thing that hindered a major follow-on attack, but rather a combination of three things. We were successful with information gained from NSA's terrorist surveillance program, CIA's interrogation of a handful of high-value detainees, and leads provided by another highly classified program that tracked terrorist financial transactions. Each of these programs informed and enabled the others. And each was carefully monitored to ensure that it was appropriately conducted.

As much as some things change, many things remain the same. Al-Qa'ida's fixation on the use of airplanes as weapons did not end on 9/11. In the ensuing years, plots to use airliners as weapons were broken in Europe, Asia, and the Middle East. What started in 1995 as the Manila air conspiracy was taken forward to London in April 2006, when British intelligence broke the back of a plot to use liquid explosives on aircraft transiting the Atlantic in the same way that was attempted in 1995. In the years in between, airline plots were directed against Heathrow airport, and there were four separate operations to target both coasts of the United States.

During the Millennium threat, actions in Amman by the Jordanians uncovered the intent to use hydrogen cyanide in a movie theater. Today al-Qa'ida disseminates instructions on how to acquire simple materials that can be purchased in hardware stores to disperse lethal gasses in enclosed facilities, using a simple but effective device they called the "mobtaker." What this tells you about al-Qa'ida is that history matters. They will return to plots previously attempted whether they succeeded or failed.

What the detainees gave us was insight into people, strategy, thinking, individuals, and how they would all be used against us. What they gave us was worth more than CIA, NSA, the FBI, and our military operations had achieved collectively. We were able to corroborate what they told us with other data we had collected. What we now have is an exhaustive menu and knowledge about how al-Qa'ida thinks, operates, and trains its members to conduct operations against us. What we have in our possession is a road map to put in place a systematic program of protection, to deny al-Qa'ida the operational latitude it once enjoyed. The questions are: How effective will we be in relentlessly closing the seams of our vulnerability? How urgently will we pursue the sacrifices required to avert the next attack?

One thing is certain: the United States remains the crown jewel in al-Qa'ida's planning. Its desire to pull off multiple spectacular attacks in the United States that inflict economic and psychological damage is undiminished.

We have learned that al-Qa'ida is a very adaptive organization. Prior to 9/11 they understood the security weaknesses of the United States. They understood our laws, our banking regulations, and the large gaps in our domestic security preparations. They also recognize that we are prone to "fighting the last war." So after the 9/11 attacks, while the United States and our allies have focused on a threat posed by certain young Arab males, al-Qa'ida has shifted its recruitment to bring in jihadists with different backgrounds. I am convinced the next major attack against the United States may well be conducted by people with Asian or African faces, not the ones that many Americans are alert to.

It would be easy for al-Qa'ida or another terrorist group to send suicide bombers to cause chaos in a half-dozen American shopping malls on any given day. Why haven't they? The real answer is that we do not know. (It would be easy to do and would spread the kind of fear and economic damage they desire.) I believe it is because they have set for themselves a bigger goal. They want to hurt us in a measure commensurate with our status as a superpower. To date, the techniques the terrorists gladly employ in places like Iraq and Israel have not been used in the United States.

Our successes against al-Qa'ida have not come without a price. As time passes since 9/11, I fear that Americans will once again begin to think of terrorism as something that happens "over there." That is exactly the mind-set our enemy wants us to have. The lessons of the past and the attacks in England, Spain, Morocco, Bali, Turkey, and elsewhere tell us how they are going to attack, the targets they are interested in attacking, and, most important, that they are intent on coming here again. We will rarely know the "when," but there is no longer any excuse for not understanding the "how" and not doing our best to protect against it. History matters.

CHAPTER 14

They Want to Change History

Acquiring weapons for the defense of Muslims is a religious duty. If I have indeed acquired these weapons, then I thank God for enabling me to do so.

—Usama bin Ladin quoted in *Time* magazine, December 24, 1998, when asked if al-Qa'ida had nuclear and chemical weapons

There was not a shred of doubt that Bin Ladin meant what he said, nor any doubt that he would go to any length to fulfill his "religious duty." Long before 9/11, in public testimony and in secret counsel to two administrations, I raised the alarm about al-Qa'ida. Now, in the aftermath of the attacks on the World Trade Center and Pentagon, I asked my staff, "What's next?"

Although we had his own statements to give us great concern, the consensus inside and outside our own government could be boiled down to this: "Guys in caves can't get WMD." But this was an issue about which we could not afford to be wrong. So soon after 9/11, I directed CIA's CTC to establish a new capability to focus exclusively on terrorist WMD. Even the people I put in charge of that effort were skeptical, hopeful that they would simply be proving a negative. We began to review the historical record. We combed our files and sent teams around the world to share our leads and ask foreign intelligence services about information in their possession. We interrogated al-Qa'ida prisoners and pored over documents found in safe houses and on computers captured in Afghanistan. What we discovered stunned us all.

The threats were real. Our intelligence confirmed that the most senior leaders of al-Qa'ida are still singularly focused on acquiring

WMD. Bin Ladin may have provided the spiritual guidance to develop WMD, but the program was personally managed at the top by his deputy, Ayman al-Zawahiri. Moreover, we established beyond any reasonable doubt that al-Qa'ida had clear intent to acquire chemical, biological, and radiological/nuclear (CBRN) weapons, to possess not as a deterrent but to cause mass casualties in the United States. The assessment prior to 9/11 that terrorists were not working to develop strategic weapons of mass destruction was simply wrong. They were determined to have, and to use, these weapons.

Over time, we were able to link the top echelon of al-Qa'ida's leadership to the group's highly compartmentalized chemical, biological, and nuclear networks. This group included al-Qa'ida's operational chief, Sayf al-Adl; the group's logistics chief, Abu Hafs; Jemaah Islamiya chief Ruidin Isomuddin (Hambali); 9/11 planners Khalid Sheikh Mohammed and Ramzi bin al-Shibh; Egyptian CBRN expert Abu Khabab al-Masri; self-described "CEO of anthrax," Yazid Sufaat; and explosives expert and "nuclear CEO," Abdel al-Aziz al-Masri.

As we researched the information we were slowly gathering from myriad sources, we unlocked a disturbing secret: the group's interest in WMD was not new. They had been searching for these weapons long before we had been looking for *them*. As far as we know, al-Qa'ida's fascination with chemical weapons goes back to the sarin gas attack on the Tokyo subway system in March 1995 by a group of religious fanatics called the Aum Shinrikyo. Twelve people died in that attack, but had the dispersal devices worked as planned, the death toll would have been higher. Al-Qa'ida leaders were impressed and saw the attack as a model for achieving their own ambitions. (In retrospect, the Tokyo attack also foreshadowed al-Qa'ida's interest in subway and railway systems, which later manifested itself in attacks in Madrid on March 11, 2004; in London on July 7, 2005; and a planned attack against the New York City subway in fall 2003 that was called off by Ayman al-Zawahiri in the last stages of preparation—"for something better.")

In February 2001, in the U.S. District Court, Southern District of New York, Usama bin Ladin was tried in absentia and others were tried in person for their involvement in the 1998 bombing of the U.S. embassies in Kenya and Tanzania. It was here that al-Qa'ida's pursuit of WMD became clear: one of the key witnesses in that trial, Jamal Ahmad al-Fadl, described how, as far back as 1993, he helped Bin Ladin try to obtain uranium in Sudan, to be used in some type of a nuclear device. Al-Qa'ida, al-Fadl testified, was willing to pay $1.5 million to acquire an unknown quantity of uranium. His testimony ended without resolution. Perhaps this was the first of many experiences for al-Qa'ida in which the group was scammed by opportunists, or perhaps the offer was real. We may never know. The important point is that the group was actively attempting to acquire nuclear material in the early 1990s. They were willing to do what needed to be done, and pay whatever it would cost, to get their hands on fissile material. In the face of such steely resolve, the only responsible course of action would be to do whatever was necessary to rule out any possibility that terrorists could get their hands on fissile material.

Bin Ladin's statements in 1998 regarding his religious obligation to obtain WMD were not made in a vacuum, either. That was the same year that Pakistan first tested a nuclear weapon. The expertise and material for fulfilling UBL's dream lay across the border from his Afghan sanctuary. We received fragmentary information from an intelligence service that, also in 1998, UBL had sent emissaries to establish contact with the nuclear scientist A. Q. Khan's network. Over decades, A.Q. had built an international network of suppliers of nuclear capability for sale to rogue states. According to the intel, A. Q. Khan had rebuffed several of UBL's entreaties, although it was not clear why. However, this new reality of the potential collaboration between a well-organized proliferation network and a terrorist group would ultimately reshape our understanding of the WMD threat, and the nature of our response to it.

Shortly before 9/11, a friendly intelligence service chanced across information that a Pakistani nongovernmental organization (NGO) called Umma Tameer-e-Nau (UTN) had been formed to establish social-welfare projects in Afghanistan. However, the information suggested that UTN had another purpose: they hoped to lend their expertise and access to the scientific establishment in order to help build chemical, biological, and nuclear programs for al-Qa'ida. (NGOs can be a convenient vehicle for providing cover for terrorist organizations, as they have legitimate reasons to traffic in expertise, material, and money.) The leadership of UTN was made up of retired Pakistani nuclear scientists, military officers, engineers, and technicians. Its founder and chairman, Sultan Bashirrudan Mahmood, was the former director for nuclear power at Pakistan's Atomic Energy Commission. Mahmood was thought of as something of a madman by many of his former colleagues in the Pakistan nuclear establishment. In 1987 he published a book called *Doomsday and Life After Death: The Ultimate Faith of the Universe as Seen by the Holy Quran*. It was a disturbing tribute to his skewed view of the role of science in jihad. The book's basic message—from the leader of a group that had offered WMD capabilities to al-Qa'ida—was that the world would end one day soon in the fire of nuclear holocaust that would usher in judgment day and thus fulfill the prophecies of the Quran.

Mahmood's associates in UTN may not have embraced his apocalyptic vision, but they shared his extremist tendencies. Chaudiri Andul Majeed, a prominent nuclear engineer who retired from the Pakistani Institute of Nuclear Science and Technology in 2000, agreed to play a key role in assisting Mahmood in his plans to share WMD with the Taliban and UBL. We also knew that UTN enjoyed some measure of support from Pakistani military officers opposed to President Musharraf, notably the former director of the Pakistani intelligence service, Gen. Hamid Gul.

It appeared that UTN's contacts with the Taliban and al-Qa'ida may have been supported, if not facilitated, by elements within the Pakistani military and intelligence establishment.

I instructed the Directorate of Operations to press all of our contacts worldwide to find out anything we could about the people and organizations with WMD that might be willing to share expertise with al-Qa'ida and other terrorist groups. We did not limit our inquiries to friends. We also spoke to the Libyans, who confirmed that they had rejected overtures from UTN peddling nuclear expertise. Ben Bonk, the deputy chief of CTC, held a clandestine meeting with Musa Kusa, the head of the Libyan intelligence service, to try to elicit what he could about Tripoli's familiarity with al-Qa'ida. During their conversation, Bonk asked if Kusa had ever heard of UTN. "Yes," the Libyan replied, "they tried to sell us a nuclear weapon. Of course, we turned them down." This information confirmed separate reporting from another intelligence service that UTN had approached the Libyans with an offer to provide chemical, biological, and nuclear expertise. Kusa's words rang true because, unbeknownst to him, we knew Libya did not need UTN since they had already secured the services of an upscale supplier of WMD services—the A. Q. Khan proliferation network.

CIA passed our information on UTN to our Pakistani colleagues, who quickly hauled in seven board members for questioning. The investigation was ill-fated from the get-go. The UTN officials all denied wrongdoing and were not properly isolated and questioned. In fact, they were allowed to return home after questioning each day. Pakistani intelligence interrogators treated the UTN officials deferentially, with respect befitting their status in Pakistani society. They were seen as men of science, men who had made significant contributions to Pakistan. Our officers read the question etched in the faces of their Pakistani liaison contacts: Surely, such men cannot be terrorists? It was a problem we would encounter time and time again as we began tracing WMD networks and leads that emerged in the Middle

East, Asia, Africa, Australia, and in North and South America. There was no question al-Qa'ida sought scientific expertise on a global scale. The question I needed an answer to urgently was whether they had already succeeded.

A Western intelligence service came to us in the fall of 2001 with a remarkable piece of information that helped break the case open. A source had told them that in August 2001, just weeks before the 9/11 attacks, UTN officials Mahmood and Majeed met with Usama bin Ladin and Ayman al-Zawahiri in Afghanistan. There, around a campfire, they discussed how al-Qa'ida should go about building a nuclear device. CIA pressed the Pakistanis to confront Mahmood and Majeed with this new information. We put the Libyan information on the table. We also passed new information that had been collected by other intelligence services. To no avail.

Then 9/11 struck, and there was no slowing down in this pursuit. The stakes were too high to accept the lack of progress that the Pakistanis were making. In late November 2001, I briefed the president, vice president, and national security advisor on the latest intelligence, our concerns, and the likelihood we would be unable to resolve this issue satisfactorily without intervention by the president. I brought along with me my WMD chief, Rolf Mowatt-Larssen, and Kevin K., our most senior WMD terrorism analyst. During the ensuing conversation, the vice president asked if we thought al-Qa'ida had a nuclear weapon. Kevin replied, "Sir, if I were to give you a traditional analytical assessment of the al-Qa'ida nuclear program, I would say they probably do not. But I can't assure you they don't." The vice president then made a comment that in my view has since been misinterpreted: "If there's a one percent chance that they do, you have to pursue it as if it were true."

I am convinced the vice president did not mean to suggest, as some have asserted, that we should ignore contrary evidence and that such a policy should be applied to all threats to our national security. On the contrary, the vice president understood instinctively that WMD must be managed differently because the implications were unique—such an attack would change history. We all felt that the vice president understood this issue. There was no question in my mind that he was absolutely right to insist that when it came to discussing weapons of mass destruction in the hands of terrorists, conventional risk assessments no longer applied; we must rule out any possibility of terrorists succeeding in their quest to obtain such weapons. We could not afford to be surprised.

The president directed that I go to Pakistan the next day and share our concerns with President Musharraf. We did not know how far UTN had gone in providing assistance to al-Qa'ida, but any fireside chat between Pakistani nuclear officials and the al-Qa'ida leadership about a nuclear weapon posed grave concerns. A U.S. Air Force 707 that at one time had served as Air Force One flew Rolf, Kevin, and me to Pakistan. During the long, restless flight, I wrote out my intended talking points on a yellow legal pad, drawing from updated information that I was receiving from Langley on the plane. Some leads were beginning to emerge concerning UTN connections to the United States, and in other countries. I intended to lay it all out for Musharraf; there was no option other than full transparency to help him make the required decisions to resolve our concerns.

We arrived in the middle of the night. After a short rest, I reviewed my plan with our senior officer in Pakistan and discussed with him the next steps he would have to take with Pakistani intelligence after I left the country—assuming we could win Musharraf's cooperation. Our senior officer stressed that our hosts were tense; they were unsure of the nature of this unusual visit for which they had received barely one day's notice. He pointed out that although things were calm in the capital city of Islamabad, the threat level was high and no one was quite sure what might happen next in those uncertain weeks that followed 9/11. The

U.S. ambassador, Wendy Chamberlin, later joined us, and we were whisked away in a heavily armed motorcade for the short but tense ride to the presidential palace.

After a few pleasantries, I explained to President Musharraf that I had been dispatched by the U.S. president to deliver some very serious information to him. I launched into a description of the campfire meeting between Usama bin Ladin, al-Zawahiri, and the UTN leaders. "Mr. President," I said, "you cannot imagine the outrage there would be in my country if it were learned that Pakistan is coddling scientists who are helping Bin Ladin acquire a nuclear weapon. Should such a device ever be used, the full fury of the American people would be focused on whoever helped al-Qa'ida in its cause."

Musharraf considered my words carefully but opened with the response we had expected: "But Mr. Tenet, we are talking about men hiding in caves. Perhaps they have dreams of owning such weapons, but my experts assure me that obtaining one is well beyond their reach. We know in Pakistan what is involved in such an achievement."

I knew that among his expert advisors was A. Q. Khan, someone who had long been under investigation for his illicit nuclear proliferation efforts. However, I didn't want the discussion to veer off toward Khan at this point. There would be another day for that topic. The issue at hand was UTN, and they were quite a different matter.

"Mr. President, your experts are wrong," I said. I told him that the current state of play between weapon design and construction and the availability of the needed materials made it possible for a few men hidden in a remote location—if they had enough persistence and money, and black enough hearts—to obtain and use a nuclear device. I turned the briefing over to Rolf, who proceeded to explain in detail how plausible the threat had become, and how our thinking had changed in terms of dealing with it. When he finished there was a brief uncomfortable silence in the room. President Musharraf was clearly reflecting on this new information. Responding with quiet confidence, he asked why we had assumed al-Qa'ida would look to Pakistan for such assistance. He recalled information he had been briefed on about "loose nukes" in Russia and the availability of nuclear material in the former Soviet Union as a more likely source of material and assistance. Still, I sensed that we had made our case.

"Let me tell you, sir," I said, "what steps we need to take." I laid out a series of steps that required immediate action. I counseled him to look at certain elements in the Pakistani military and intelligence establishment. In addition to asking for a more vigorous investigation of UTN, I suggested it might be a good time for Pakistan to perform a thorough inventory of its nuclear material. If any had gone missing, both he and I needed to know. "Can I report to President Bush that we can count on you?" I asked. "Yes, of course," he replied.

Even though we were on the ground for fewer than twenty-four hours, a picture of our big 707 with the words "United States of America" emblazoned across the fuselage had quickly appeared in the Pakistan media. With the war across the border in Afghanistan only a few weeks old and fighting still raging, U.S. and Pakistani officials were worried that terrorists might be waiting somewhere just beyond the end of the runway with a surface-to-air missile ready to bring down this symbol of American power. On takeoff, the crew executed a climb steeper than anything I imagined an old 707 could pull off. We had been advised to pull down the window shades in the darkened cabin for security reasons, but I could not refrain from lifting mine. If our plane was going to be attacked, I wanted to see it coming. Fortunately, the departure was uneventful, and I relaxed as we crossed the snowcapped Himalayas in brilliant sunlight.

By the time I got back to Washington, it was clear that President Musharraf was true to his word. Pakistani authorities had redoubled their efforts in questioning the UTN leadership. They were methodically running down all the leads we had passed. With the arrival of a team of U.S. experts, they conducted polygraph investigations of the key UTN members and eventually obtained confessions that added important new details to the story. Mahmood confirmed all we had heard about the August 2001 meeting with Usama bin Ladin, and even provided a hand-drawn rough bomb design that he had shared with al-Qa'ida leaders. He told his interrogators that he had discussed the practicalities of building a weapon. "The most difficult part of the process," he told Bin Ladin, "is obtaining the necessary fissile material." "What if we already have the material?" Bin Ladin replied. This surprised Mahmood. He said he did not know if this was a hypothetical question or if Bin Ladin was seeking a design to use with fissile material or components he had already obtained elsewhere.

According to the account, an unidentified senior al-Qa'ida leader displayed a canister for the visitors that may or may not—the account was frustratingly vague—have contained some kind of nuclear material or radioactive source. This al-Qa'ida operative shared his ideas of building a simple firing system for a weapon using commercially available supplies. Over the next several months, we ran down every lead and turned over every rock in an effort to make a judgment as to whether UTN had provided WMD to al-Qa'ida. We followed a number of serious U.S. leads. It appears we had disrupted the organization in the early stages of its efforts to ply trade with al-Qa'ida. CIA, FBI, and dozens of foreign partners had worked together in unprecedented ways in an effort to prove a negative, as best as one can do so. This effort was a success in terms of working out a new modus operandi to deal with the new threats that had emerged in the wake of 9/11. What we did not know then, and do not know now, is how many other groups like UTN are out there.

The cause for my lightning trip to Pakistan was not an aberration but part of an emerging series of nuclear-related threats. At the same time, our threat matrix was carrying unsubstantiated rumors from several reliable foreign intelligence services that some sort of small nuclear device had been smuggled into the United States and was destined for New York City. The

Department of Energy quietly dispatched detection equipment to New York to possibly detect an unexpected source of radiation before such a device could be detonated. It was a pattern that would repeat itself over time. Adding fuel to the fire, detained al-Qa'ida senior paramilitary trainer Ibn al-Shaykh al-Libi had provided the Egyptians with information that he later recanted, that al-Qa'ida had collaborated with Russian organized crime to import into New York "canisters containing nuclear material." We could not rule out that these vague, unsubstantiated streams of information were only partially right, and that Washington might be the intended target. It did not matter whether al-Qa'ida was indeed planning a WMD attack or a large-scale conventional attack, as many feared in those days and weeks after 9/11. In this period of high threat, the decision was made that the vice president and the president should not be in the same location, if at all possible. For the sake of continuity of government, the vice president was spending a lot of time at an "undisclosed location." Anyone who mocks the practice of securing the national leadership in times of crisis has not shared the reality of the threats we handled on a daily basis. None of us had any doubt that we were engaged in a war.

Our fears of imminent attack did not go away as 2001 slid into 2002.

Suleiman Abu Ghaith, a cleric of Kuwaiti origin and spokesman for al-Qa'ida, posted a statement on the Internet in June 2002 saying that "Al-Qa'ida has the right to kill four million Americans, including one million children, displace double that figure, and injure and cripple hundreds and thousands." Ghaith's rationale for such grisly figures was based on some sort of sick math extrapolating his estimates for the number of Muslims killed and wounded at the hands of the United States over the years. It would have been easy to dismiss his ranting as the hyperbole of a deranged man. But we had to consider the possibility that Abu Ghaith was attempting to justify the future use of weapons of mass destruction that might greatly exceed the death toll of 9/11.

Such weapons could be nuclear. They could be biological. They could be an unconventional massive attack on our infrastructure. But any attack would have to be big to deliver on al-Qa'ida's persistent promises to "destroy our economy."

To do so, they would need to develop a plan as intricate as the 9/11 plot, most likely planned over a long period of time by sleeper cells operating in the United States. We began what became an endless search for any leads to individuals who might fit this description. There turned out to be no shortage of radicalized Muslims who had been educated in American universities, who spoke flawless English, and who had the capability and perhaps the motive to hurt this country. Two individuals in particular represented this breed. There would be others to follow, who came to our attention in an endless stream of investigations by CIA and FBI.

Muhammed Bayazid, also known as Abu Rida al-Suri, and Mubarak al-Duri had attended the University of Arizona in the 1980s. As students, they became radicalized along with others who identified with the "jihadists" who fought against the Soviet occupation of Afghanistan. Bayazid and al-Duri attended prayer group meetings with students who would become al-Qa'ida associates—men like Wadi al-Hage, who was later linked to the 9/11 plot. With such friends, it was no surprise when we learned that al-Duri and Bayazid had joined Usama bin Ladin after he had relocated from Afghanistan to Sudan in the early 1990s.

A review of both men's dossiers revealed that they shared indicators of WMD concern. Bayazid, a Syrian, was trained as a physicist, and al-Duri, an Iraqi, was an agronomist. Both men enjoyed direct ties to Bin Ladin and helped manage his business interests in Sudan. Both men had developed business connections to Sudanese WMD-related entities, and both had established businesses that could have served as dual-use front companies for developing nuclear and biological weapons. After Bayazid's name surfaced in connection with al-Qa'ida's attempt to purchase uranium in Sudan, FBI sent agents to Sudan to interview the two men. The agents reported back that, although their suspicions were great, they were unable to develop sufficient grounds for a case against either man that would justify an extradition request.

At one of our five o'clock meetings in mid-2002, a frustrated Rolf Mowatt-Larssen suggested that if we couldn't arrest the two men then perhaps we could get them to "flip"— to change sides, in intelligence and law enforcement jargon. I sent Rolf off to Africa with orders to approach the two American-trained scientists with the mandate to try to save lives rather than take them.

It was an unusual assignment and one that we thought best undertaken with the cooperation of the local intelligence service. Rolf found the locals willing to listen to our proposal—he requested their assistance to talk to both men separately, in a neutral location. There would be no compulsion, no threats, only persuasion. Rolf explained the stakes for all of us, if the road to any future nuclear or biological attack against the United States were paved through this country. The local intelligence officer stroked his beard, smiled, and said, "I understand American threats very well. And so I know this is not a threat. It is a standard to which you would hold any country . . . cooperation on such a question is sensible to preserve civilization as we both know it . . . for this reason, I will agree to your request."

The encounters were revealing. There would be no reconciliation, no common ground or shared sense of decency and humanity with the two al-Qa'ida associates. On the contrary, they articulated the hatred, the need for revenge, that they shared. Rolf appealed to both men to agree to disagree on our differences, and to focus on a narrow area of common interest, a shared sense of moral purpose to do whatever was in our means to prevent the escalation of a war that, if left unchecked, would result in the indiscriminate deaths of thousands of innocent women and children. After a long, brooding silence, one man replied in soft, sure tones, "No . . . I think it is legitimate to kill millions of you because of how many of us you have killed." Rolf looked deeply into his cold, dark eyes—Rolf now understood Abu Ghaith's math.

The concern about al-Qa'ida's interest in WMD was more than academic. We had long worried about the security of nuclear material from the former Soviet Union. Whenever we asked the Russians for assurances that nothing of theirs had gone missing, we would receive a perfunctory response that everything was "under control." President Putin had been more candid not long after 9/11, when President Bush showed his own briefing on UTN and asked Putin point blank if Russia could account for all of its material. Choosing his words carefully, the Russian president said he was confident he could account for everything—under his watch. He was unwilling to vouch for the period before that, during Yeltsin's regime. It was a deliberately ambiguous response but, nonetheless, one that suggested we needed to pay especially close attention to smuggling incidents in the early years following the breakup of the USSR.

From the end of 2002 to the spring of 2003, we received a stream of reliable reporting that the senior al-Qa'ida leadership in Saudi Arabia was negotiating for the purchase of three Russian nuclear devices. Saudi al-Qa'ida chief Abu Bakr relayed the offer directly to the al-Qa'ida leadership in Iran, where Sayf al-Adl and Abdel al-Aziz al-Masri (described as al-Qa'ida's "nuclear chief" by Khalid Sheikh Mohammed) were reportedly being held under a loose form of house arrest by the Iranian regime. The al-Qa'ida leadership had obviously learned much from their ventures into the nuclear market in the early 1990s. Sayf al-Adl told Abu Bakr that no price was too high to pay if they could get their hands on such weapons. However, he cautioned Abu Bakr that al-Qa'ida had been stung by scams in the past and that Pakistani specialists should be brought to Saudi Arabia to inspect the merchandise prior to purchase.

As soon as I got wind of al-Qa'ida negotiations to purchase nuclear components in Saudi Arabia, I contacted the Saudi ambassador to the United States, Prince Bandar, and gave him all the details we had.

Like most people when first exposed to this threat, Bandar was incredulous. He questioned both the capability of al-Qa'ida to obtain a device and their willingness to use one within the kingdom. "Look," I said, "we don't know if they intend to detonate a device inside your country or just use Saudi Arabia as a transit point. But in either case, you have big trouble." I explained that Saudi and U.S. intelligence had recent information from clerics favorable to al-Qa'ida debating the wisdom of attacking the Saudi royal family. They were discussing in vague terms the morality under the Quran of using new weapons that did not discriminate among their victims. "Even if they don't go after the Saudi leadership," I impressed upon Bandar, "a nuke going off in the middle of your major oil distribution facility would devastate your economy and ours. Al-Qa'ida would like nothing better." Visibly shaken by the implications of the gathering threat, Bandar agreed and persuaded his government to track down and arrest al-Qa'ida within the kingdom. It was another turning point in Saudi resolve to deal with the extremist threat as a problem affecting their own survival.

From the spring through the summer of 2003, with unprecedented CIA assistance, the Saudis staged a remarkable series of preemptive actions that thwarted a number of terrorist attacks in the kingdom, and which gutted the al-Qa'ida leadership in Saudi Arabia in the process. Although al-Qa'ida had maintained its predilection for mounting conventional attacks, for the first time we uncovered clear indications of their interest in using cyanide weapons in future attacks. Cyanide had been found in a terrorist safe house.

Across the straits in Bahrain, we learned that terrorists with strong Saudi extremist connections had been planning to conduct a cyanide gas attack on the New York City subway system. The extremists had created a clever homemade dispersal device called the "mobtaker"—Arabic roughly translated as "invention"—a lethal device that could be constructed entirely from readily available material. Although the Bahrain cell operated independently of al-Qa'ida, they followed the unwritten protocol between extremists by requesting permission from al-Qa'ida central leadership to conduct the attack. Chillingly, word came back from Ayman al-Zawahiri himself in early 2003 to cancel the operation and recall the operatives, who were already staged in New York—because "we have something better in mind."

There was endless speculation at the highest levels as to the proper interpretation of al-Zawahiri's cryptic comment. We still do not know what he meant. However, we do know that the "mobtaker" cyanide device was not sufficiently inspiring to serve al-Qa'ida's ambitions. For that, the group consulted with several radical Saudi clerics in an effort to obtain Quranic justification—a "fatwa"—that would legitimize the use of weapons of mass destruction. Even Safar al-Hawali, a radical cleric who had written an open letter to President Bush after 9/11, reportedly balked at lending his name to such a fatwa. The terrorists found their cleric, however, in Shaykh Nasir bin Hamid al- Fahd, who helpfully gave al-Qa'ida just what they needed. In a document published in May 2003 called "A Treatise on the Legal Status of Using Weapons of Mass Destruction Against Infidels," al-Fahd argued that a large number of civilian deaths, numbering in the millions, would be justifiable if they came as part of an attack aimed at defeating an enemy.

Following the al-Qa'ida attacks in Riyadh in May 2003, the Saudis captured several top al-Qa'ida leaders responsible for planning the assaults. Arrested along with them was Shaykh Nasir bin Hamid al-Fahd. In custody, he confirmed that al-Qa'ida had been negotiating for the purchase of Russian devices, but he claimed ignorance regarding the nature of these devices and whether al-Qa'ida had in fact obtained them. After about six months in custody, al-Fahd appeared on Saudi television rescinding his fatwa and expressing regret for the error of his religious interpretation.

Having done all that was possible to neutralize any threats in Saudi Arabia, we turned our attention to the al-Qa'ida leadership in Iran. We pursued learning more about al-Qa'ida's interest in WMD through every means available to us. Many al-Qa'ida operatives had something to say about the organization's interest in WMD. Many would also quickly recant much of what they told us. Despite the considerable uncertainties, we were concerned about what we were able to corroborate from other information available to us. One senior al-Qa'ida operative told us that Mohammed Abdel al-Aziz al-Masri, who had been detained in Iran, managed al-Qa'ida's nuclear program and had conducted experiments with explosives to test the effects of producing a nuclear yield. We passed this information to the Iranians in the hope that they would recognize our common interest in preventing any attack against U.S. interests.

Our inability to determine the fate of the Russian devices presented great concern, not only for me but for the White House. I took Rolf to a meeting with the president and Condoleezza Rice in the early summer of 2003, at the height of the Saudi takedowns and the threat stream related to possible attack planning in the United States. The president was unusually pensive. He asked me how the Russians were doing in the war on terrorism. I told him their contribution was a disappointment—they were preoccupied with Chechnya and were not players in the global war against terrorism, certainly not as we had defined it. Clearly frustrated, the president asked Condi Rice what needed to be done to engage the Russians and get to the bottom of the current threat. She recommended that I call Defense Minister Ivanov, explain the president's concerns, and obtain Ivanov's assurances that our respective intelligence agencies would intensify their work to resolve the WMD threats.

Defense Minister Ivanov was receptive to our concerns and agreed immediately to receive CIA's representatives in Moscow. I instructed Rolf to travel to Moscow and coordinate meetings with Russian intelligence. At the old KGB headquarters in Moscow, under a watchful portrait of former KGB chairman Andropov,

Rolf pressed our Russian counterparts to work with us in ways that would have been unfathomable during the cold war. Heads nodded as all sides agreed that our two countries' national security interests were closer than one might think. Having moved past the promising opening remarks, however, it soon became evident that even high-level pressure had not prepared them for the intimate forms of concrete cooperation required to deal with the WMD threat. In the final analysis, it was still a game of spy versus spy. Both sides had spilled too much blood for too many years to expect a breakthrough on such an issue. As expected, the Russians took copious notes and asked penetrating questions regarding the information we had come to share. But the conversation became awkward as we began asking questions. The Russians could not shed light on reports we had received of missing material from the former Soviet Union. They did not recognize the names of former Soviet scientists who had reportedly collaborated with al-Qa'ida. They refused to delve into any matters related to the security of their nuclear facilities and nuclear weapons, including reports sourced to Russian officials concerning possible thefts of Russian "suitcase nukes."

As disappointed as Rolf was upon returning to Washington, he advised me that it would have been unreasonable to expect much more from the Russians on such sensitive internal security matters. If we were to improve the quality of our intelligence interaction we would need a fundamental shift in policy. At the time of my retirement, we were still trying to cross that bridge.

As luck would have it, not long after this meeting we obtained the proof we had hoped did not exist concerning the availability of fissile material for sale. In the summer of 2003, we learned that officials had arrested an individual crossing the border from Georgia to Armenia carrying a small amount of highly enriched uranium (HEU). Although the amount of material seized was far short of that required for a nuclear weapon, we could no longer ignore the fact that organized crime, smuggling networks, and corrupt officials inside nuclear facilities were working in concert to find a customer—any customer—willing to pay the going rate for such merchandise. Although this particular shipment was interdicted, I am not convinced we can rule out the possibility that a terrorist group might one day purchase enough fissile material to construct a viable nuclear device.

As much as we were worried about nuclear plots, we were also feverishly trying to get everything we could on Bin Ladin and his lieutenants' attempts to obtain biological and chemical weapons. Their interest in crude poisons and toxins—cyanide, botulinum, ricin, and the like has been well established.

Abu Musab al-Zarqawi, a senior al-Qa'ida associate, made a name for himself by running a chemical and poisons laboratory and training facility in the northern Iraqi town of Khurmal from May 2002 through early 2003. Al-Zarqawi established his ruthless reputation early on by testing the lethality of cyanide he had developed in Khurmal on a hapless associate—the poison worked, and the unsuspecting extremist died an agonizing death. Al-Zarqawi had brought his lieutenants with him from his days when he ran a training camp for jihadists, in Herat, Afghanistan. He was able to forge ties between Algerians, Moroccans, Pakistanis, Libyans, and other Arab extremists located throughout Europe. Over several months of tireless link analysis we identified al-Zarqawi–connected terrorist cells in more than thirty countries.

This loose association of groups planned a string of poison plots across Europe that began to mature in December 2002. The coordinated disruption of this European-based network represented one of the great successes of the post-9/11 war on terrorism. A global coalition of more than two dozen countries shared intelligence information on a near real-time basis. Numerous operatives and couriers were captured. Plots were disrupted in the United Kingdom, France, Spain, and Italy, among others, and lives were saved. We were able to keep the president, vice president, and other senior administration officials constantly updated as to the threats and our unfolding responses.

Shortly after the invasion of Iraq, al-Zarqawi's camp in Khurmal was bombed by the U.S. military. We obtained reliable human intelligence reporting and forensic samples confirming that poisons and toxins had been produced at the camp. As for al-Zarqawi's fate, information from a source indicated that he may have escaped to Baghdad, where he planned to lead an insurgency against U.S. forces. (Zarqawi went on to play a leading role in the insurgency until his death in mid-2006.)

Another key al-Qa'ida connection to biological weapons was Yazid Sufaat, the Jemaah Islamiya associate who hosted the first operational meeting of the 9/11 hijackers at his apartment in Kuala Lumpur, Malaysia, in January 2000. In fact, Sufaat had provided commercial cover for Zacarias Moussaoui's trip to the United States. Sufaat was also the self-described "CEO" of al-Qa'ida's anthrax program. U.S. educated and with a Malaysian military background, Sufaat had impeccable extremist credentials. In 2000 he had been introduced to Ayman al-Zawahiri personally, by Hambali, as the man who was capable of leading al-Qa'ida's biological weapons program.

Al-Qa'ida spared no effort in its attempt to obtain biological weapons. In 1999 al-Zawahiri had recruited another scientist, Pakistani national Rauf Ahmad, to set up a small lab in Khandahar, Afghanistan, to house the biological weapons effort. In December 2001 a sharp WMD analyst at CIA found the initial lead on which we would pull and, ultimately, unravel the al-Qa'ida anthrax networks. We were able to identify Rauf Ahmad from letters he had written to Ayman al-Zawahiri. Later, we uncovered Sufaat's central role in the program. We located Rauf Ahmad's lab in Afghanistan. We identified the building in Khandahar where Sufaat claimed he isolated anthrax. We mounted operations that resulted in the arrests and detentions of anthrax operatives in several countries.

The most startling revelation from this intelligence success story was that the anthrax program had been developed in parallel to 9/11 planning. As best as we could determine, al-Zawahiri's project had been wrapped up in the summer of 2001, when the al-

Qa'ida deputy, along with Hambali, were briefed over a week by Sufaat on the progress he had made to isolate anthrax. The entire operation had been managed at the top of al-Qa'ida with strict compartmentalization. Having completed this phase of his work, Sufaat fled Afghanistan in December 2001 and was captured by authorities trying to sneak back into Malaysia. Rauf Ahmad was detained by Pakistani authorities in December 2001. Our hope was that these and our many other actions had neutralized the anthrax threat, at least temporarily.

But of all al-Qa'ida's efforts to obtain other forms of WMD, the main threat is the nuclear one. I am convinced that this is where UBL and his operatives desperately want to go. They understand that bombings by cars, trucks, trains, and planes will get them some headlines, to be sure. But if they manage to set off a mushroom cloud, they will make history. Such an event would place al-Qa'ida on a par with the superpowers and make good Bin Ladin's threat to destroy our economy and bring death into every American household. Even in the darkest days of the cold war, we could count on the fact that the Soviets, just like us, wanted to live. Not so with terrorists. Al-Qa'ida boasts that while we fear death, they embrace it.

We have learned that it is not beyond the realm of possibility for a terrorist group to obtain a nuclear weapon. I have often wondered why this is such a hard reality for so many people to accept. In a scene in a book called *American Prometheus*, by Kai Bird and Martin Sherwin, in 1946 the father of the U.S. atomic bomb, J. Robert Oppenheimer, describes the specter of nuclear terrorism. Asked in a closed Senate hearing room "whether three or four men couldn't smuggle units of an atomic bomb into New York and blow up the whole city," Oppenheimer responded, "Of course it could be done, and people could destroy New York." The surprised senators then asked, "What instrument would you use to detect an atomic bomb hidden somewhere in the city?" Oppenheimer replied, "A screwdriver [to open each and every crate or suitcase]." Oppenheimer instinctively understood what we learned the hard way: that nuclear terrorism was then, and remains now, a terrifying possibility, and extraordinarily hard to stop.

The terrorists are endlessly patient. The first plans to attack the World Trade Center were made a decade before the Twin Towers fell. The plot to bring down aircraft traveling between the United Kingdom and United States that was thwarted in the summer of 2006 parallels Project Bojinka. How hard is al-Qa'ida willing to work and how long are they willing to wait to pull off the ultimate attack? What was the attack Ayman al-Zawahiri described as "something better" when he called off the 2003 attack on the New York City subway?

One mushroom cloud would change history. My deepest fear is that this is exactly what they intend.

Being sworn in as Director of Central Intelligence by FBI Director Louis Freeh, July 11, 1997. Wife Stephanie Glakas-Tenet and son John Michael holding the bible. *(Official FBI Photo)*

At a White House meeting with President Bill Clinton. Seated are Chief of Staff John Podesta and Secretary of Defense Bill Cohen. *(Official White House Photo)*

With one of the great historical figures in modern Middle East history, His Majesty King Hussein of Jordan. *(Author's personal collection)*

With former president George H. W. Bush and Mrs. Bush at a ceremony renaming CIA headquarters in honor of President Bush, April 26, 1999. *(Official CIA Photo)*

A new generation of leader in the Middle East, His Majesty King Abdullah II of Jordan. *(Author's personal collection)*

With President George W. Bush addressing CIA workforce in Agency headquarters lobby, March 20, 2001. *(Official CIA Photo)*

In the White House bunker on September 11, 2001, watching as the president addresses the nation on television. Behind me is Richard Clarke, and to the right is the vice president's wife, Mrs. Lynne Cheney. *(Official White House Photo)*

At Camp David, Maryland, briefing the president, Condoleezza Rice, and Andrew Card on CIA operations against al-Qa'ida in Afghanistan, September 30, 2001. *(Official White House Photo)*

In a meeting with Attorney General John Ashcroft *(right)* and FBI Director Robert Mueller *(left)*, October 29, 2001. *(Official White House Photo)*

An overhead view of a presidential daily briefing in the Oval Office. *Clockwise from bottom left:* President George Bush, National Security Advisor Condoleezza Rice, White House Chief of Staff Andrew Card, me, a CIA briefer, and Vice President Dick Cheney. *(Official White House Photo)*

With John McLaughlin, who served as deputy director of Central Intelligence from 2000 to 2004 and was acting CIA director from July to September 2004. *(Official White House Photo)*

Making a point to Secretary of Defense Donald Rumsfeld in the White House Situation Room, January 16, 2002. *(Official White House Photo)*

Having a discussion with White House Chief of Staff Andrew Card, May 11, 2002. John McLaughlin is in the background. *(Official White House Photo)*

In a meeting with the president in the White House Situation Room, May 20, 2002. On the president's left are Secretary of State Colin Powell, Secretary of Defense Donald Rumsfeld, and Chairman of the Joint Chiefs of Staff General Richard Myers. To the president's right are Vice President Dick Cheney, White House Chief of Staff Andrew Card, and me. *(Official White House Photo)*

Briefing the president in the Oval Office, June 11, 2002. *(Official White House Photo)*

On the phone at Camp David, September 7, 2002. *(Official White House Photo)*

An example of the leaflets released over Iraq just prior to and in the first few days of the war; millions of such documents were dropped on Iraqi military units. U.S. officials later argued that the Coalition Provisional Authority proclamation to disband the Iraqi Army had no effect because the units had "self-demobilized." In fact, they were following U.S. instructions, and many would have come back after the war to provide security if they had been ordered to do so. *(U.S. Department of Defense)*

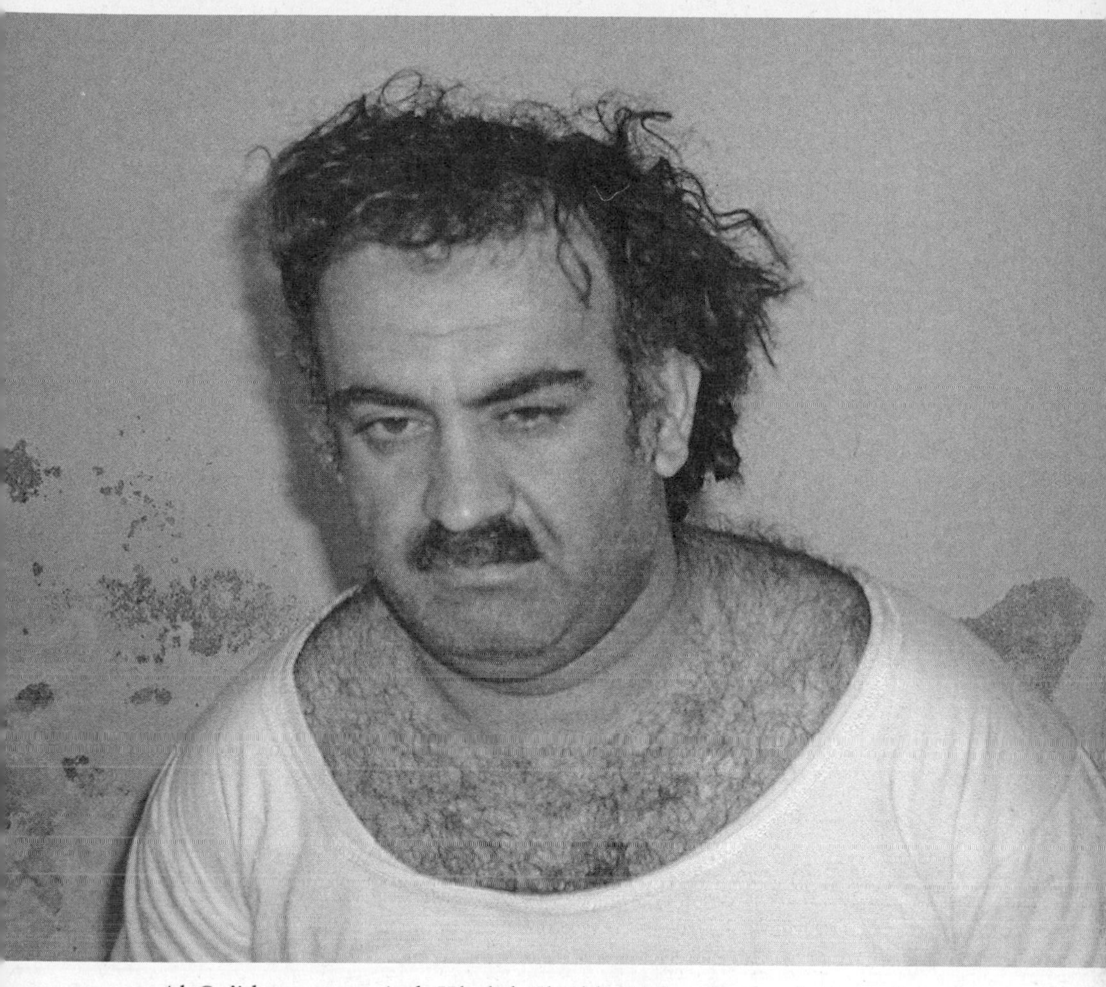

Al-Qa'ida mastermind Khalid Sheikh Mohammed minutes after he was captured in Pakistan on March 1, 2003. *(AP Images)*

Secretary of State Colin Powell addressing the U.N. Security Council on the case for going to war in Iraq, February 5, 2003. *(AP Images)*

With FBI Director Bob Mueller at my CIA farewell ceremony, July 8, 2004. *(Official CIA Photo)*

With eight other former Directors of Central Intelligence, August 16, 2005 *(from left to right)*: Porter Goss, John Deutch, Robert Gates, William Webster, Stansfield Turner, James Woolsey, James Schlesinger, George H. W. Bush. *(Official CIA Photo)*

CIA's senior management team in July 2004 just prior to my departure. *(Official CIA Photo)*

In front of a container of P2 centrifuge casings turned over to the United States by the government of Libya. The P2 photo was taken December 9, 2005, at an event commemorating the third anniversary of Libya renouncing its WMD programs. The Libyans ordered more than 10,000 such casings. Only 1,200 are needed to produce enough enriched uranium for a Hiroshima-size nuclear weapon each year. *(Manuel G. Gillispie, Oak Ridge National Laboratory)*

CHAPTER 15

The Merchant of Death and the Colonel

It is not always easy. Your successes are unheralded—your failures are trumpeted. . . . But I am sure you realize how important is your work, how essential—it is and how, in the long sweep of history, how significant your efforts will be judged.

—President John F. Kennedy at CIA headquarters, November 28, 1961

Almost a half century later, President Kennedy's words still ring true. The problem is often of the intelligence community's own creation. We are reluctant to talk publicly about our successes. Sometimes it is even useful to have positive accomplishments misperceived as failures, to throw foreign governments and rogue organizations off the scent.

A couple of successful operations that took place during my tenure, however, *did* receive some limited positive public attention. The dismantling of the A. Q. Khan proliferation network and the disarming of Libya's WMD programs are classic examples of the kinds of work that can and must be done by American intelligence if we are to avoid a catastrophic future. A. Q. Khan's nuclear proliferation network was a project we focused on during my entire seven-year tenure as DCI. Our efforts against this organization were among the closest-held secrets within the Agency. Often I would brief only the president on the progress we were making.

Dr. Abdul Qadeer Khan, a metallurgist, was the father of the Pakistani nuclear weapons program. A. Q. Khan, as he is known, studied in Europe and earned a Ph.D. in Belgium in 1972. He worked in the nuclear energy industry in the Netherlands and returned to Pakistan in 1976 to help his country compete with India, which had just detonated its first nuclear device. Khan stole from his European bosses blueprints and information that would give Pakistan a jump-start in entering the nuclear age. (Indeed, Khan was convicted in absentia of nuclear espionage, in a Dutch court in 1983, but the verdict was overturned on a technicality two years later.)

During the 1970s and '80s, Khan led an aggressive effort to build a uranium-enrichment effort. So revered was he for his efforts that Pakistan eventually renamed its research facility the Khan Research Laboratory (KRL) in his honor.

In 1979 the United States suspended military and economic assistance to Pakistan over concerns about the country's attempts to make weapons-grade uranium. By the late 1980s and early 1990s, reports began to surface in the media and elsewhere that Pakistan had succeeded in producing enough fissile material to make its own bomb.

For many years, there were rumors and bits of intelligence that Khan was sharing his deadly expertise beyond Pakistan's borders. His range of international contacts was broad—in China, North Korea, and throughout the Muslim world. In some cases, there were indications that he was trading nuclear expertise and material for other military equipment—for example, aiding North Korea with its uranium-enrichment efforts in exchange for ballistic missile technology. It was extremely difficult to know exactly what he was up to, or to what extent his efforts were conducted at the behest and with the support of the Pakistani government. Khan was supposedly a simple government employee with only a modest salary. Yet he lived a lavish lifestyle and had an empire that kept expanding dramatically.

Although CIA struggled to penetrate proliferation operations and learn about the depth of their dealings, there is a tension when investigating these kinds of networks. The natural instinct when you find some shred of intelligence about nuclear proliferation is to act immediately. But you must control that urge and be patient, to follow the links where they take you, so that when action is launched, you can hope to remove the network both root and branch, and not just pull off the top, allowing it to regenerate and grow again.

In the late 1990s, the section within CIA's Counterproliferation Division (CPD) in charge of this effort was run by a career intelligence officer who once told me that as a child the officer read a book on the bombing of Hiroshima and was awestruck by the devastation that a nuclear bomb could deliver. The book described how the blast from the thirteen-kiloton "Little Boy" bomb, which killed an estimated seventy thousand people, burned the image of three people's shadows onto a wall. The individuals themselves were vaporized. That mental picture was seared into the officer's consciousness and became part of the officer's motivation, years later, to work to keep nuclear weapons from falling into the wrong hands.

The small unit working this effort recognized that it would be impossible to penetrate proliferation networks using conventional intelligence-gathering tactics. Security considerations do not permit me to describe the techniques we used.

Patiently, we put ourselves in a position to come in contact with individuals and organizations that we believed were part of the overall proliferation problem. As is so often the case, our colleagues in British intelligence joined us in our efforts and were critically important in working against this target.

We discovered the extent of Khan's hidden network, which stretched from Pakistan, to Europe, to the Middle East, to Asia. We pieced together a picture of the organization, revealing its subsidiaries, scientists, front companies, agents, finances, and manufacturing plants. Our spies gained access through a series of daring operations over several years.

What we learned from our operations was extraordinary. We confirmed that Khan was delivering to his customers such things as illicit uranium centrifuges. A. Q. Khan was the mastermind behind proliferation efforts as far afield as North

Korea, Iran, and South Africa. We briefed the president on what we had found.

"Mr. President," one of our officers said, "with the information we've just gotten our hands on—soup to nuts—about uranium enrichment and nuclear weapons design, we could make CIA its own nuclear state."

By mid-2003 we had learned quite a bit about locations where Khan's network was producing equipment for uranium enrichment for some of his clients, and we were considering taking action against those sites. Doing so, however, might have dealt a temporary setback to Khan's scheme but would not have prevented it from springing up again somewhere else. We therefore came up with a bold solution that involved a series of carefully orchestrated approaches to the network.

What we uncovered proved that Khan and his associates were selling the blueprints for centrifuges to enrich uranium, as well as nuclear designs stolen from the Pakistani government. The network sold uranium hexafluoride, the gas that in the centrifuge process can be transformed into enriched uranium for nuclear bombs. Khan and his associates provided Iran, Libya, and North Korea with designs both for Pakistan's older centrifuges and for newer, more efficient models. The network also made available to these countries components and, in some instances, complete centrifuges. Khan and his associates used a factory in Malaysia to manufacture key equipment. Other parts were obtained by network operatives based in Europe, the Middle East, and Africa. Khan's deputy—a man named B. S. A. Tahir—ran a computer business in Dubai and used that as a front company for the Khan network, acting as the network's chief financial officer and money launderer.

We had the goods on A. Q. Khan and his cohorts and we had reached a point where we had to act, but there were still some important matters to resolve. It remained unclear to what extent Khan's dealings were known and supported by his own government. It was our job to find out.

Pakistan's president Musharraf had heroically stepped up in the aftermath of 9/11 and helped us fight al-Qa'ida and the Taliban. Now I was about to ask him to help take on a man who had, almost single-handedly, turned Pakistan into a nuclear power and was viewed as a national hero in his country.

You don't make those kinds of requests over the phone, and you certainly don't make them in front of large groups of people. As it turned out, Musharraf was coming to New York City to attend the UN General Assembly, and I requested a one-on-one session with him for September 24, 2003. We met in his hotel suite. It was what we in the intelligence business called a "four eyes" meeting—just the two of us. No handlers, no note takers.

I started by thanking him for his courageous support in the war on terrorism and told him I was now going to give him some bad news. "A. Q. Khan," I said, "is betraying your country. He has stolen some of your nation's most sensitive secrets and sold them to the highest bidders." I went on. "Khan has stolen your nuclear weapons secrets. We know this, because we stole them from him."

I pulled out of my briefcase some blueprints and diagrams of nuclear designs stolen from the Pakistani government. I'm not a nuclear physicist, and neither is President Musharraf, but I had been briefed well enough by my team that I could point out markings on the drawings that would prove that these designs were supposed to be in a vault in Islamabad and not a hotel room in New York.

I pulled out a blueprint of a Pakistani P1 centrifuge design. "He sold this to Iran." Then I produced a design for the next-generation P2 centrifuge. "He has sold this to several countries." Without pause, I laid before President Musharraf another document. "These are the drawings of a uranium processing plant that he sold to Libya."

There could be no doubt about the size and scope of the problem.

Although he later described this as one of the most embarrassing moments of his presidency, Musharraf betrayed no emotion to me. I always found him to be a cool customer, someone who seems to be taking in every word you are saying.

I told him that I knew that since March 2001 he had tried to restrict A. Q. Khan's international travel. Then I gave him a lengthy list of dozens of foreign trips Khan had undertaken despite the restriction. Even as we spoke, Khan was on an international sales trip.

"Mr. President," I said, "if a country like Libya or Iran or, God forbid, an organization like al-Qa'ida, gets a working nuclear device and the world learns that it came from your country, I'm afraid the consequences would be devastating."

I suggested a few steps we could take jointly to find out the full extent of Khan's corruption and to put an end to it once and for all.

President Musharraf asked a few questions and then simply said, "Thank you, George, I will take care of this."

Not long after returning to Pakistan, President Musharraf twice narrowly averted being killed in al-Qa'ida–inspired assassination attempts.

In December word leaked of a major investigation going on regarding the activities of the Khan Research Laboratory. On January 25, 2004, Pakistani investigators announced that Khan had provided unauthorized technical assistance to Iran's nuclear program in exchange for tens of millions of dollars. Six days later, Khan was dismissed from his position as "science advisor" to Musharraf, to allow the investigation to continue. Then, in early February, the Pakistani government announced that Khan had signed a confession admitting to having aided Iran, Libya, and North Korea with designs and equipment for their nuclear weapons programs.

Khan appeared on national television in Pakistan on February 4 and, speaking in English, made a three-minute speech. "I take full responsibility for my actions and seek your pardon," he said. He expressed the deepest "sense of sorrow, anguish and regret," saying that his actions were taken "in good faith" but were "errors in judgment." He portrayed his actions as entirely his own. "There was never, ever any kind of authorization for these activities from the government."

The next day Musharraf pardoned him but placed him under permanent house arrest. While we would have preferred to see Khan face trial, and wanted to have U.S. and International Atomic Energy Agency (IAEA) investigators extensively question him about his dealings, the outcome was still a major victory.

In the new world of proliferation, nation states have been replaced by shadowy networks like Khan's, capable of selling turnkey nuclear weapons programs to the highest bidders. Networks of bankers, lawyers, scientists, and industrialists offer one-stop shopping for those wishing to acquire the designs, feed materials, and manufacturing capabilities necessary for nuclear weapons production. With Khan's assistance, small, backward countries could shave years off the time it takes to make nuclear weapons.

A small group of our intelligence officers, working closely with our British allies, patiently pursued the Khan network for close to a decade. They succeeded brilliantly. On my next-to-last day as DCI, I went down to the small office and presented medals to the officer leading the effort and the entire team.

What we don't know is how many networks similar to Khan's may still be out there—operating undetected—and offering deadly advice and supplies to anyone with the cash to pay for them. In the current marketplace, if you have a hundred million dollars, you can be your own nuclear power.

The Khan network was closely intertwined with another major intelligence success. Through the work of U.S. and British intelligence, Libya, long a pariah state, had its weapons of mass destruction programs neutralized without the firing of a shot.

CIA had been having clandestine meetings with senior Libyan officials since 1999. Our efforts were designed to try to resolve issues regarding terrorism and to learn what we could from the

Libyans about various Islamic terrorist groups. These meetings, conducted with our British colleagues, were held in several European cities. The Libyan delegation was led by Col. Muammar al-Gadhafi's chief of intelligence, Musa Kusa, who got a master's degree from Michigan State University in 1978. Illustrative of the surreal world in which we had to operate, CIA officers found themselves exchanging pleasantries with the man who, by some accounts, was the mastermind behind the Pan Am 103 bombing in December 1988 that killed 270 people.

These contacts continued for several years, up through the time that a Scottish court convicted one Libyan intelligence officer of complicity in the airline bombing and acquitted another. Libya's cooperating with the Scottish tribunal, and other acts, were signs—admittedly faint—that Libya might be looking for a way off the terrorism limb they had climbed out on more than twenty years previously.

Following the 9/11 attacks, Colonel Gadhafi publicly condemned the terrorists actions, calling them "terrible," and announced that the Libyan people were ready to send humanitarian aid to America. That was an interesting sign.

We exchanged some terrorist tracking data with Libya in the aftermath of September 11, but our focus was on pursuing al-Qa'ida in Afghanistan, and so the contacts subsided for a while. Then, in March of 2003, an envoy from Colonel Gadhafi made an informal approach to British officials. He said that Gadhafi was thinking about giving up his WMD programs and asked whether should Libya do so, would the West be willing to ease sanctions on his country.

A senior British intelligence official flew to the United States just as the war in Iraq was starting. I met with him the next day. Five days later, I joined President Bush and British prime minister Blair at Camp David. Blair was accompanied by my counterpart, Sir Richard Dearlove. A "spy's spy," Sir Richard is one of the most skilled and talented intelligence officers I have ever worked with. Extraordinarily thoughtful and articulate, he had instant credibility with political leaders on both sides of the Atlantic.

Although the bulk of our time was spent talking about Iraq, we also discussed Gadhafi's surprising initiative. Here we were, just days after launching an invasion of Iraq that was inspired, at least in part, by our concerns about Saddam's nuclear, biological, and chemical programs, and out of the blue, another rogue state wanted to talk about the possibility of coming clean on its own programs.

We debated what Gadhafi's motivation might be. It seemed to us that the Libyans had come to the realization that they had gotten nothing out of their very expensive flirtation with WMD. They were struggling to find their place in the world—Libya was the odd man out in both the Arab and African worlds. You also couldn't discount the effect that 150,000 U.S. troops positioned around Iraq might have on focusing the mind.

Whatever the impetus, this was certainly an opportunity that we could not dismiss lightly. I returned from Camp David and called into my office Jim Pavitt and Steve Kappes, the top two officers in our clandestine service. I briefed them on the opening with Libya and told them that it needed to be handled at a high level and with the utmost discretion. Pavitt and I were up to our ears with Operation Iraqi Freedom, but we had the perfect candidate in Kappes. Steve is one of the most capable case officers I have been privileged to know. Fluent in Russian and Farsi, he had handled some of the toughest assignments that the Agency had to offer. I put the project in his hands and got back to worrying about Iraq. Together Kappes and a senior British counterpart were given the mission for their respective services. They set up a meeting with the Libyans to determine if they were really serious about renouncing their WMD programs.

Kappes and his British colleague flew to a European city in mid-April. Initially, the plan was for them to meet the Libyan intelligence chief Musa Kusa and Libyan diplomat Fouad Siltni in their hotel over breakfast. Steve and his colleague selected a table that allowed them to keep an eye on the entire hotel restaurant. Just before the appointed meeting time, two young men of Middle Eastern extraction walked in. Kappes noted that they had the air of security professionals about them. Moments later, Ehud Barak, the former Israeli prime minister, walked into the restaurant. Clearly, this was not a discreet enough environment for sensitive discussions. While Kappes kept an eye on the Israelis, the Brit intercepted the Libyans and took them up to a meeting room on the hotel's top floor. Kappes soon joined them.

Once settled in, Musa Kusa, a tall, well-dressed man, launched into a lengthy canned speech about Libya's position. We had decided not to give the Libyans any written material from the United States in that first meeting, but Kappes conveyed the president's desire that Libya take the necessary steps to return to "the family of nations."

That first meeting lasted more than two hours. After some discussion, Musa Kusa essentially admitted that his country had violated just about every international arms control treaty it had ever signed. Then he said that they wanted to relinquish their weapons programs, that we should trust them to do so, and he asked for a sign of good faith from us.

Steve and his British colleague explained the "trust but verify" concept made famous by President Reagan and said that there would be no signs of good faith from either of our two countries until we could get experts on the ground in Libya and verify the extent of the Libyan holdings, and assure ourselves that the programs were being dismantled.

When Steve returned from the trip, I took him to the Oval Office one morning to brief the president on what had transpired. Although nothing definitive had been accomplished, we had the prospect of making a real breakthrough. We and the Brits started assembling teams of WMD experts who might go to Libya to inspect their programs.

But the Libyans started dragging their feet. They weren't ready for foreigners to go poking around their weapons programs, it seemed. I flew to London in mid-May to meet with my counterparts. One of the topics of discussion was how to jump-start the process. Later that month, Kappes and a senior British officer invited the Libyans to a meeting in a European capital. Gadhafi's son Saif al-Islam attended along with Musa Kusa.

Saif started to play the role of tough-guy negotiator, telling Steve and his British colleague what the Libyans expected from us before anything would happen on their end. Steve and the Brit allowed the leader's son to go on for a while and then cut him off. "Look," Steve said, "you need to understand that none of that is going to happen. We aren't going to make any concessions until we get our people on the ground and confirm that everything you are telling us about your stockpiles and your intentions is true. Please go back and tell your father that."

Several more months passed without progress on the Libyan side. Another meeting was held in August, this time without Gadhafi's son. Musa Kusa invited Steve and his British colleague to come to Libya and meet with Gadhafi himself. President Bush instructed us to make no promises until we saw solid proof of Libyan intentions and evidence that their decision was irreversible.

Steve and his British colleague flew into Tripoli in early September. As is typical in the Middle East, the promised meeting was delayed several times while they waited at a hotel on the edge of the Mediterranean. Musa Kusa warned them that the first few minutes of the meeting with Gadhafi might be "a little rough."

Finally, in the early evening, they were summoned. Musa Kusa himself drove them to Gadhafi's office. Along the way, he found time to work into the conversation that this was the same location that the U.S. had bombed in 1986, allegedly killing one of Gadhafi's adopted daughters.

They were ushered into Gadhafi's large office. Two huge globes sat astride either end of a large desk that featured a modern personal computer. (Steve would learn that Gadhafi spent hours surfing the Web, to keep up with developments in the outside world.) The leader was wearing expensive Italian loafers and a gaudy shirt with a map of Africa emblazoned on it. After brief introductions, the visitors took seats and Musa Kusa put his head down, as if he knew what was coming, and the interpreter pulled out his pad. Gadhafi immediately launched into a loud and colorful diatribe, slamming the West, and the United States in particular, for every misdeed imaginable. The interpreter had great difficulty keeping up with the Arabic words as they flew off Gadhafi's tongue.

Then, at about the seventeen-minute mark in the tirade, Musa Kusa's head came up as if he could tell that the rant was about to end. Sure enough, Gadhafi ran out of steam, took a breath for the first time, and smiled. "Nice to see you. Thanks for coming," he said. And then he got down to business.

We want to "clean the file," he kept saying. Everything is on the table. At one point there was mention of Libya's WMD programs, and that set off Gadhafi, who claimed that he did not *have* WMD programs. A discussion followed of exactly what a "weapon of mass destruction" was, and then they moved on. At another point, someone mentioned that the United States and Britain would want to conduct "inspections" of Libyan weapons facilities. Again Gadhafi was outraged, but eventually it became clear that if our side called them "visits" instead of "inspections," there might not be such a big problem.

The meeting lasted for about two and a half hours. It ended without any conclusion other than Gadhafi saying, "Work things out with Musa Kusa." On the way out of the office, however, the visitors were informed that Gadhafi's son Saif, who had not been present in his father's office, wanted to meet with them right away. They were driven to Saif's beach house, where the staff had apparently taken the living room furniture and placed it outside on the Mediterranean sand. By now it was around midnight, and they enjoyed a very late dinner and informed the Libyan leader's son about the state of play.

On returning to the United States, I took Kappes down to brief the president once again. I knew Steve would neither oversell nor undersell the situation. He gave the president his assessment that the Libyans had multiple reasons for wanting to do a deal now. Their fear of Islamic extremists is as large as ours, he explained. If they can find a way to get back into the good graces of the West, the Libyans could send their brightest kids to American colleges, and they could attract major oil companies to help foster the economic prosperity that was eluding them. Still, he said, the Libyan's track record was such that they would likely get cold feet before the deal was done.

The matter was still extraordinarily closely held. I briefed Colin Powell about what we were up to, and we told Rich Armitage and Bill Burns, the State Department's chief official working on Middle Eastern affairs. If this effort ultimately succeeded, it would be their job to work on normalizing relations with Libya.

Then, in the fall of 2003, elements of our two success stories—A. Q. Khan and Libya—merged. Through our operations against the Khan network, we learned that a ship of German registry, called the BBC *China*, was carrying centrifuge parts bound for Libya. After it passed through the Suez Canal, we worked to have the ship diverted to the Italian port of Taranto, where it arrived on October 4. There inspectors found precisely manufactured centrifuge parts in forty-foot containers listed on the ship's manifest as simply "used machine parts."

While we were delighted that we had intercepted the shipment, we were reluctant to make too big a deal of it at the time, hoping that we could use the incident to drive home to the Libyans that we knew all about their plans and to give them greater incentive to renounce all their WMD.

The Brits dispatched their senior officer to inform Gadhafi before the seizure hit the press. The Libyans claimed that the shipment had been arranged long before the current secret negotiations began and that the people responsible for monitoring it didn't know about an impending decision to renounce WMD.

Few U.S. government officials were aware of the "back-channel" negotiations taking place with the Libyans. Some prominent people who were not aware of the secret talks wanted to trumpet the seizure. We learned that then undersecretary of state for arms control, John Bolton, planned to hold a press conference to cite the incident as a great success for the president's "Proliferation Security Initiative," a two-year-old program to foster international cooperation on limiting illicit arms shipments. In truth, catching the BBC *China* had almost nothing to do with that program. We were concerned that if U.S. officials launched into the typical and well-deserved Libya-bashing language, Gadhafi might cancel the whole deal out of embarrassment.

We called Rich Armitage, one of the few State Department officials aware of our ongoing efforts, and got him to direct Bolton to stand down. The order was understandably mystifying to Bolton and resulted in his calling Kappes and chewing him out for not coming directly to him.

After the Libyans finally gave us the blessing for inspection teams to visit their country, a handful of CIA weapons experts flew from the United States to the United Kingdom to pick up their British counterparts. On October 19 they traveled to Tripoli in an unmarked airplane. The sight of a jet labeled "United States of America" landing there was something neither the Libyans nor we were ready to explain. Just before touching down, the aircrew told Steve Kappes that Tripoli was refusing to grant landing rights. No one knew whether this was a bureaucratic screwup or if the Libyans had once again gotten cold feet, so Steve told the crew to tell the tower to call Musa Kusa if they had any questions about their arrival. Within minutes, landing approval was granted. Steve thought it was a good thing that the Libyans were keeping the team's arrival under wraps. But, as the plane taxied toward the terminal, Steve looked out the aircraft window and saw a marching band taking up positions. It turned out that there was no reason to worry—the band was present to greet some other arriving dignitary—and the CIA plane parked at a remote location.

Just as we had kept the Libyan initiative a closely held secret in the United States, it was an especially big secret in Gadhafi's country, too. Kappes, his British counterpart, and their teams were taken to a compound where a large gathering of local officials had been assembled. Kappes could tell that the Libyans had no guidance on what to say to the visitors. They appeared frightened and may have thought that the whole exercise was a loyalty test by the Great Leader to see who could keep the tightest jaw. Slowly, over a period of days, the Libyans finally figured out that they were supposed to reveal what they knew, and that this was not some kind of trick.

On October 21, after two days of limited progress, Gadhafi asked that Kappes meet him alone. Back in his big office, the colonel proceeded to launch into another signature rant. After a while he stopped and asked if the United States would really fulfill its commitments if he renounced his WMD programs. "Yes sir, the president is a man of his word," Steve told him. "But if he feels his word has been dishonored . . . well, he is a very serious-minded man." Gadhafi just kept repeating that he wanted to "clean the file, clean the file."

After a few days, things bogged down again. So Steve and his British colleague used the tried-and-true "pack our bags" routine. They ordered the weapons inspectors to pack up and called for their aircraft to come collect them. Musa Kusa sighed. "You guys are such a pain," he said, but then ordered increased openness, and the bags were unpacked.

Progress was slowly being made, with the Libyans showing the U.S. and British inspectors how far along they had been on various weapons programs. In many cases, the Libyans tried to conceal parts of their programs, not knowing how much we already knew. They'd show us their Scud B missiles, and we would say, "Fine, now where are your Scud Cs?"

When our inspectors were shown a storage facility for highly toxic chemicals, they were stunned. The surprise was not that the Libyans possessed the deadly chemicals, but that they were stowed in large plastic jugs and the Libyans' sole safety precaution was to hold their noses when they entered the facility. The Americans quickly backed out and donned complete body-covering chemical defense suits before reentering the storehouse.

The process of inventorying the various programs took several months. The Libyans were most uncooperative on the nuclear account, however. They had no idea how much we already knew about their program.

In late November 2003, Steve and his British colleague invited Musa Kusa to a meeting. "Look," they said, "we know you guys purchased a centrifuge facility." About this time the Libyans realized that there was no turning back. Having started to tell us about their programs, they had to complete the effort, given what we already knew.

In fact, we knew virtually all there was to know about their program due to our operation against the Khan network. It was like playing high-stakes poker and knowing your opponent's cards. In this case, the stakes were the complete and peaceful disarmament of a nuclear weapons project that would eventually have given the colonel a nuclear weapons capability.

Sometimes we knew more than the Libyans themselves did. At one point we told them, "Hey, we know you guys paid a hundred million dollars for all that stuff from A. Q. Khan." There was a puzzled silence on the other side. "A hundred million? We thought the price was two hundred million!" Apparently, someone had made a heck of a profit on the side.

By mid-December enough progress had been made that the deal would soon become public. Even that was a carefully orchestrated dance; Gadhafi would first announce to his own people that he had decided to renounce his WMD programs. Then Prime Minister Blair was to make public comments welcoming the news, to be followed by remarks from President Bush. The timing was tightly negotiated for December 19. And then, at the last minute, word came from Libya that the colonel wanted to delay. Uh oh, we thought. He is about to pull the rug out from under this deal. But the explanation turned out to be a simple one. The Libyan national soccer team was playing on television that night, and Gadhafi didn't want to annoy the fans by breaking into the coverage of an important game with an announcement about something most Libyans didn't care about, weapons of mass destruction.

PART III

CHAPTER 16

Casus Belli

One of the great mysteries to me is exactly when the war in Iraq became inevitable. In the period after 9/11, just as in the months before it, I was singularly obsessed with the war on terrorism. My many sleepless nights back then didn't center on Saddam Hussein. Al-Qa'ida occupied my nightmares—not *if* but *how* they would strike again. I was wracking my brain for things we could do to delay, disrupt, or—God willing—prevent an attack. Looking back, I wish I could have devoted equal energy and attention to Iraq. Given all the mistakes that would eventually be made, Iraq deserved more of my time. But the simple fact is that I didn't see that freight train coming as early as I should have.

Not that there weren't rumblings from the very beginning of the Bush administration. Many of the incoming senior officials had been heavily involved with Iraq when they were last in government. Not long before the inauguration, Dick Cheney had asked departing defense secretary William Cohen to give the incoming president a full and complete briefing on Iraq and the options involved. To me, it was both natural and appropriate to want to bring the new president up to speed on what continued to be a thorny issue for the United States. Our air crews were patrolling Iraq's no-fly zones at considerable risk. Meanwhile, the UN sanctions against Saddam were steadily eroding.

From the beginning, too, it was evident that the vice president intended to take an active interest in the workings of CIA and in the intelligence we turned out. Many media accounts, and indeed some of the court filings in the Libby case (in which the vice president's former chief of staff was found guilty of perjuring himself regarding the Valerie Plame Wilson leak matter), have contended that there was some kind of war between CIA and the Office of the Vice President. If there was a war, it was one-sided and we were noncombatants. At the time, I viewed the vice president as enormously supportive of intelligence, helping us get the resources we needed. Because of his past service in government, he knew a lot about our business and was never shy about asking tough questions. I welcomed them. Tough questions should never be a problem—so long as you don't change the answer from what you believe to what you think the inquisitor wants to hear. And we never did.

Sure, some of our analysts, junior and senior, chafed at the constant drumbeat of repetitive queries on Iraq and al-Qa'ida. Jami Miscik, our senior analyst, came to me one day in mid-2002 complaining that several policy makers, notably Scooter Libby and Paul Wolfowitz, never seemed satisfied with our answers regarding allegations of Iraqi complicity with al-Qa'ida. I told her to tell her analysts to "quit killing trees." If the answer was the same as the last time we got the question, just say "we stand by what we previously wrote." But if there was any evidence of collaboration between Saddam and terrorist organizations, it was important to know, just as it was important to know if there was a nexus between terrorism and WMD, another of the vice president's deep concerns.

The focus on Iraq by senior Bush officials predated the administration. Paul Wolfowitz, Doug Feith, and Richard Perle were among eighteen people who had signed a public letter from a group they named "The Project for the New American Century" calling for Saddam's ouster. It is often forgotten, but regime change in Iraq was also the explicitly stated policy of the Clinton administration, and was the goal of the Iraq Liberation Act, passed by Congress in 1998. One hundred million dollars was appropriated to the State Department for the express purpose of seeking an end to Saddam's regime. This policy emerged in the aftermath of a failed 1996 covert-action program and was announced to the world. Most important, the U.S. government's intention to bring about regime change in Baghdad was proclaimed to the long-suffering people of Iraq. America's promise to topple Saddam remained the law of this land from halfway through Bill Clinton's second term right up until U.S. troops invaded in March 2003.

At the start of the Bush administration, Secretary Powell in particular pushed the notion of introducing "smart sanctions." In meetings early in 2001, he noted that the United States was getting killed in the court of public opinion by the incorrect impression that UN sanctions were causing the starvation of Iraqi babies. To restore our public image, Powell urged new sanctions that would more clearly be focused on military-related procurement. Other senior administration officials argued that this would only increase Saddam's opportunities to evade the sanctions, refill his coffers, and restore his weapons programs. Powell did eventually gain approval for the "smart sanctions," but this was rapidly overtaken by other efforts within the administration.

On February 7, 2001, little more than two weeks into the new administration, Condi Rice chaired a Principals Committee meeting in the White House that focused on Iraq. My deputy, John McLaughlin, sat in for me that day. Like many meetings in the early days of the Bush administration, this one appeared to be intended to gather information and to assign bureaucratic missions so that a government-wide policy could later be developed.

The topic of Iraq faded into the background during that spring and summer—at least for me—as plenty of other issues demanded my attention. The forcing down of a Navy EP-3 by China in April, an event now almost completely forgotten, caused eleven days of intense concern. And I spent a good part of early June in the Middle East trying to come up with a work plan that would stabilize the security situation between the Israelis and Palestinians. But Saddam wasn't being ignored.

Within our Directorate of Operations, the Iraq Operations Group (IOG) was planning for any covert actions that might be ordered inside Iraq or on the periphery of the country. In August 2001, we appointed a new head of IOG (whom I can't name because he is still under cover). An articulate, passionate, smart, and savvy Cuban American, this officer used to tell people that he was in this country as the result of one failed U.S. covert action, the Bay of Pigs, and that he didn't plan to preside over another. To make sure that didn't happen, he conducted a review of the lessons learned from our long, not-too-happy history of running operations against Iraq since the end of the Gulf war in 1991. The principal message taken away from the review was that Saddam was not going to be removed via covert action alone. As much as some would wish for an "immaculate deception"—some quick, easy, and cheap solution to regime change in Iraq—it was not going to happen.

A number of otherwise savvy senior government officials and media pundits concluded in early 2002 that the CIA was simply unwilling to take on so difficult a job. That wasn't the case at all. Rather, our analysis concluded that Saddam was too deeply entrenched and had too many layers of security around him for there to be an easy way to remove him. Whenever we talked to Iraqis, either expatriates or those still living under Saddam's rule, the reaction was always: "CIA, you say you want to get rid of Saddam. You and whose army? If you are serious about this, we want to see American boots on the ground." My own aversion to a CIA go-it-alone strategy was based both on our estimate of the chance of success (slim to none) and my belief that our plate was already overflowing with missions in the war on terrorism.

There was another, unstated, reason why the "silver bullet" option was never going to fly. Even if we had managed to take Saddam out, the beneficiary was likely to have been another Sunni general no better than the man he replaced. Such an outcome would not have been consistent with the administration's intent that a new Iraq might serve as a beacon of democracy in the Middle East.

After 9/11, everything changed. Many foreign policy issues were now viewed through the prism of smoke rising from the World Trade Center and the Pentagon. For many in the Bush administration, Iraq was unfinished business. They seized on the emotional impact of 9/11 and created a psychological connection between the failure to act decisively against al-Qa'ida and the danger posed by Iraq's WMD programs. The message was: We can never afford to be surprised again. In the case of Iraq, if sanctions eroded and nothing were done (and the international community had little patience for maintaining sanctions indefinitely), we might wake up one day to find that Saddam possessed a nuclear weapon, and then our ability to deal with him would take on an entirely different cast. Unfortunately, this train of thought also led to some overheated and misleading rhetoric, such as the argument that we don't want our "smoking gun to be a mushroom cloud."

There was never a serious debate that I know of within the administration about the imminence of the Iraqi threat. (In truth, it was not about imminence but about acting before Saddam did.) Nor was there ever a significant discussion regarding enhanced containment or the costs and benefits of such an approach versus full-out planning for overt and covert regime change. Instead, it seemed a given that the United States had not done enough to stop al-Qa'ida before 9/11 and had paid an enormous price. Therefore, so the reasoning went, we could not allow ourselves to be in a similar situation in Iraq. Even without a 9/11, however, the skepticism that had greeted Powell's "smart sanctions" proposal revealed a pretty clear split between its proponents and those who thought we needed a more robust approach to pressuring Saddam. Still, had 9/11 not happened, the argument to go to war in Iraq undoubtedly would have been much harder to make.

Whether the case could have been made at all is uncertain. But 9/11 did happen, and the terrain shifted with it.

My odd encounter with Richard Perle in front of the West Wing in mid-September 2001 was just the first hint of things to come. It was not an isolated incident. I recently talked with a senior military officer who happened to be in Europe when the attacks of 9/11 occurred. Struggling to get a flight back to the United States, he made his way to the U.S. airbase at Mildenhall, England, where he bumped into another temporarily stranded senior official, Doug Feith. They caught a ride aboard an Air Force tanker, one of the few planes permitted to transit the closed airspace of the United States. Onboard the flight, the military officer told Feith that al-Qa'ida was responsible for the previous day's attacks and a theater-wide campaign would need to be launched against them starting in Afghanistan. To his amazement, Feith said words to the effect that the campaign should immediately lead to Baghdad. The senior military officer strongly disagreed. During meetings at Camp David the weekend following the terrorist attacks, Paul Wolfowitz in particular was fixated on the question of including Saddam in any U.S. response. He spoke of Iraq in the context of terrorism alone. I recall no mention of WMD. The president listened to Paul's views but, fairly quickly, it seemed to me, dismissed them. So did I. Rumsfeld did not seem nearly as consumed with the Iraqi connection as was his deputy, and he did not join in this portion of the debate in any meaningful way. When an informal vote was taken on whether to include Iraq in our immediate response plans, the principals voted four to zero against it, with Don Rumsfeld abstaining.

I am sure that Wolfowitz genuinely believed that there was a connection between Iraq and 9/11. I am also certain that he felt deeply that the first step toward altering the face of the Middle East for the better began with leadership change in Iraq. But again, for me, Iraq was not uppermost in my mind. In the weeks following the attacks of 9/11, we quadrupled the size of CIA's Counterterrorism Center, made massive shifts in personnel and money, and closed down and scaled back operations in many parts of the world to support the offensive that was being launched against al-Qa'ida. It wasn't just that we wanted revenge against Bin Ladin. More important was the fact that there were clear, unmistakable signs that the United States might be hit again, even signs that the next attack would dwarf 9/11 in violence and casualties. If someone had told me to quit paying so much attention to terrorism in the months following September 11 and to start boning up on Iraq instead, I would have stared at them in disbelief.

To be sure, a number of people were fixated on Iraq, and a number of decisions and actions during the late fall of 2001 and into early 2002 created a momentum all their own. One of CIA's senior Middle East experts recently told me of a meeting he had in the White House a few days after 9/11. A senior NSC official told him that the administration wanted to get rid of Saddam. Our analyst said, "If you want to go after that son of a bitch to settle old scores, be my guest. But don't tell us he is connected to 9/11 or to terrorism because there is no evidence to support that. You will have to have a better reason." The National Security Council staff held meetings in the White House Situation Room with increasing regularity to discuss Iraq. Many of the meetings were so-called Deputies Committee meetings, or DCs, usually attended by the second in command from the various agencies. Others involved the Principals Committee, or PC. Although I went to some of the PC meetings, I frequently delegated the task to my long-suffering deputy, John McLaughlin. The DCs were already his burden.

Before long, the NSC staff started hosting another series of meetings that included representatives from State, Defense, the Joint Chiefs, the Vice President's Office, Treasury, and CIA, in addition to the NSC. These meetings had no formal title but were informally called "small group" meetings. Usually held twice a week over lunch, these get-togethers were frustratingly unproductive from the point of view of those who attended. After a while, McLaughlin started bringing along senior CIA analysts and operations officers to backbench him. Then he quit going altogether and had his seconds moved up to the front-row seats.

In talking now to those who did attend, I'm told that the sessions, in retrospect, seemed odd. A presidential decision on going to war was always alluded to by the NSC in hypothetical terms, as though it were still up in the air and the conferees were merely discussing contingencies. Sometimes there would be lengthy debates over such arcane details as how quickly after the war began could we replace Iraq's currency and whose picture should be on the dinar; the old currency had Saddam's mug on it. In none of the meetings can anyone remember a discussion of the central questions. Was it wise to go to war? Was it the right thing to do? The agenda focused solely on what actions would need to be taken if a decision to attack were later made. What never happened, as far as I can tell, was a serious consideration of the implications of a U.S. invasion. What impact would a large American occupying force have in an Arab country in the heart of the Middle East? What kind of political strategy would be necessary to cause the Iraqi society to coalesce in a post-Saddam world and maximize the chances of our success? How would the presence of hundreds of thousands of U.S. troops, and the possibility of a pro-West Iraqi government, be viewed in Iran? And what might Iran do in reaction? In looking back, there seemed to be a lack of curiosity in asking these kinds of questions, and the lack of a disciplined process to get the answers before committing the country to war. And in hindsight, we in the intelligence community should have done more to answer those questions even though not asked. One of our senior analysts subsequently told me that the impression given was that the issue of "should we go to war" had already been decided in meetings at which we were not present. We were just called in to discuss the "how" and occasionally the "how will we explain it to the public."

There was never any doubt of the military outcome, but there was precious little consideration, that I'm aware of, about the big picture of what would come next. While some policy makers were eager to say that we would be greeted as liberators, what they failed to mention is that the intelligence community told them that such a greeting would last for only a limited period. Unless we quickly provided a secure and stable environment on the ground, the situation could rapidly deteriorate.

In addition to the "small group" meetings at the White House, the Pentagon hosted similar meetings referred to as the "Executive Steering Group" meetings, or ESGs, generally attended by officials one echelon below those going to the "small group" sessions downtown. But once again, reports coming back to CIA headquarters said that the meetings started out talking about what actions would need to be taken "if we went to war," and quickly segued into discussions of what should happen "when we went to war," without stopping for any debate on "should we."

Over the past couple of years, I have asked various people who were in senior positions at CIA at the time, "When did you know for sure that we were going to war in Iraq?" The answers are instructive. Those involved in assembling support for the U.S. military had the sense from early in the Bush administration that war was inevitable. By and large, the analysts whom I have talked to—the ones who were following Saddam's weapons programs or who were examining possible links between Iraq and al-Qa'ida—came much later to the conclusion that we were going to war.

Richard Haass, the former director of policy planning at the State Department, has said that Condi Rice told him in July of 2002 that "the decisions were made," and unless Iraq gave in to all our demands, war was a forgone conclusion.

In May of 2002, my counterpart in Great Britain, the head of MI-6, Sir Richard Dearlove, traveled to Washington along with Prime Minister Blair's then national security advisor, David Manning, to take Washington's temperature on Iraq. Sir Richard met with Rice, Hadley, Scooter Libby, and Congressman Porter Goss, who was then the chairman of the House Intelligence Committee.

In the spring of 2005 some documents dating back to July 2002 were leaked to the British press. The documents, which came to be known as "the Downing Street Memos," reported on a "perceptible shift" in the attitude in Washington, saying that military action was now seen as "inevitable." One memo records "C," the designation the Brits use for the head of the British Secret Intelligence Service, as saying that "intelligence and facts were being fixed around the policy."

Sir Richard later told me that he had been misquoted. He reviewed the draft memo, objecting to the word "fixed" in particular, and corrected it to reflect the truth of the matter. He said that upon returning to London in July of 2002, he expressed the view, based on his conversations, that the war in Iraq was going to happen. He believed that the momentum driving it was not really about WMD but rather about bigger issues, such as changing the politics of the Middle East.

Dearlove recalled that he had a polite but significant, disagreement with Scooter Libby, who was trying to convince him that there was a relationship between Iraq and al-Qa'ida. Dearlove's strongly held view, based on his own service's reporting, which had been shared with CIA, was that any contacts that had taken place between the two had come to nothing and that there was no formal relationship. He believed that the crowd around the vice president was playing fast and loose with the evidence. In his view, it was never about "fixing" the intelligencc itself but rather about the undisciplined manner in which the intelligence was being used.

In a memo that Doug Feith, the undersecretary of defense for policy, sent to John McLaughlin on September 6, 2002, he forwarded a cable summarizing his comments at a recent conference in Berlin attended by U.S., British, French, and German officials. The cable quotes Feith as having told the gathering that "war is not optional." "At stake," he reportedly said, "is the survival of the United States as an open and free society." The summary went on to say that Feith told his colleagues that U.S. action was based on self-defense. "So with regard to Iraq, the question of whether one can prove a connection with Iraq and the September 11th attack is not (repeat not) of the essence." One of the foreign attendees apparently agreed, saying that we should not get caught up in the "legalisms about clear evidence of imminent threat," given Saddam's history of deception.

While we at CIA were intensely focused on al-Qa'ida, and others in the administration were obsessed with Iraq, there was a third subset of people who seemed to have *Iran* on their minds. A strange series of events brought this to our attention. In late December 2001, the U.S. ambassador to Italy, Melvin Sembler, told CIA's senior man responsible for Italy that Michael Ledeen, an American conservative activist, was in Rome, along with some DOD officials, talking to the Italians about secret contacts with Iranians. Ledeen had figured prominently in the Iran-Contra scandal in the 1980s and had introduced Manucher Ghorbanifar, an Iranian middleman, con man, and fabricator, to Oliver North. Ledeen's latest mission was news to us.

A few weeks later, on January 14, 2002, a senior representative of Italian intelligence was in Washington and visited me. He asked me what I knew about U.S. government officials exploring contacts with Iranians. I shot a look at other members of my staff in the meeting. It was clear that none of us knew what he was talking about. The Italian quickly changed the subject.

On February 1, 2002, Ambassador Sembler told our senior officer in Italy that he was getting questions from the State Department about the DOD visitors, who apparently were Larry Franklin and Harold Rhode of Doug Feith's staff. The ambassador said there were reports that the two men were talking about a twenty-five-million-dollar program to support Iranians who opposed the Tehran regime. We still had no idea what this was about, but what we were hearing sounded like an off-the-books covert-action program trying to destabilize the Iranian government. Without the appropriate presidential authorities, normally run through the CIA, and without congressional notification, such a program might well be illegal. This started to give the appearance of being "Son of Iran-Contra."

I picked up the phone and called Steve Hadley and asked him what the hell was going on. Hadley appeared to know something about this initiative. He reminded me that he had mentioned to me in early December 2001 that DOD might meet with some Iranians in Europe who had terrorist threat information. True, but there'd been no mention of anything like this; no discussion of Ledeen, Ghorbanifar, or Iranian opposition. I remember being uncomfortable about the previous discussion and didn't understand why CIA wasn't being asked to get directly involved. But if there was information available about a threat to U.S. interests, I wasn't going to let bureaucratic reasons stand in the way of our getting the details. But what I was hearing now was something entirely different. Hadley asked me if Paul Wolfowitz hadn't called me before to explain all this. My answer was no.

Steve sent me a memo he had received from Michael Ledeen dated January 18, 2002. In the memo, Ledeen talked about how he had arranged the meeting with Iranian officials who were "in violent opposition to the regime." It also said that Pentagon officials suggested that the initiative working with these people be "managed entirely by DOD personnel" and that "the Iranians have stipulated that they are totally unwilling to deal with anyone from CIA, but they are quite comfortable with Pentagon officials."

I was furious. Don't these guys remember the past? I thought. I called Hadley after reviewing the Ledeen memo. "Steve," I said, "this whole operation smells." I followed up with a memo of my own on February 5, 2002, strongly recommending he immediately get on top of the matter.

When Colin Powell found out, he hit the roof. Powell had become national security advisor in 1987 to help clean up the first Iran-Contra mess; he didn't want to be around for another one. Powell contacted Condi Rice and told her that the issue needed to be taken care of immediately and that if it were not, he would raise the matter directly with the president.

Hadley told John McLaughlin in mid-February that the situation had been resolved and that Ledeen was out of the picture. John asked for a written response to my earlier note, but none was ever received.

On July 11, 2002, a senior CIA officer was told by the ambassador to Italy that Ledeen had called him to say he would be returning to Rome the next month to "continue what he had started." Our Rome rep met with his Italian counterparts and asked them not to provide any assistance to Ledeen unless the ambassador or CIA requested that they do so. A senior CIA lawyer contacted his NSC opposite number and asked whether anyone at the NSC had authorized Ledeen's visit. If not, he suggested, CIA might have to file a "crimes report" with the Justice Department, a requirement when we learn of a possible violation of the law.

About two weeks later, the NSC lawyer contacted CIA to say that Steve Hadley had called Ledeen in and "read him the riot act," telling him to "knock it off." In light of that, he said, they didn't see any need for a crimes report.

There was a series of Ledeen-inspired inquiries that would come in over the transom, via Congress, the White House, DOD, and elsewhere. The common thread was that he had urgent and highly sensitive information and would like to talk about a reward. These tips led nowhere.

On August 6, 2003, after the United States had ousted Saddam, Ledeen contacted DOD with word that he had a source who knew that a significant amount of enriched uranium was buried in Iraq some thirty to forty meters deep, underneath a riverbed, but that some of it had been moved to Iran. Ledeen told a DOD official that he had already briefed Scooter Libby and John Hannah of the vice president's staff, and he intended to share the info with the Senate Intelligence Committee staff but would not tell CIA. Like most Ledeen tips, this one proved worthless.

Two days later, on August 8, word leaked to the media about Ledeen and Ghorbanifar's earlier meetings with Pentagon officials, possibly to discuss regime change in Iran. Various White House and DOD officials admitted that, yes, there were some meetings, but nothing came of them. I called Condi Rice and urged the NSC staff, once again, to get to the bottom of the matter. "If you don't," I said "this will all end up on the president's desk, and he will take the blame." Condi mentioned that after the first meeting in Rome, the DOD officials had "accidentally bumped into" the Iranians again in Paris, while crossing the street or some such thing. "Condi," I said, "in this line of work there is no such thing as an accidental meeting."

Later that month, in one of my weekly NSC meetings, once again I raised my concerns about what was going on and that the NSC needed to get to the bottom of the matter. I reiterated to Steve Hadley that we had no intention of meeting with Ghorbanifar. CIA had issued a "burn notice" (a formal declaration that a source is deemed to be untrustworthy) on him nearly two decades previously, and we had no reason to revise our opinion of his credibility. DOD opened an investigation into the contacts between their staff and Ghorbanifar. I do not know the outcome.

Ultimately, the Ledeen follies on Iran were a distraction from the administration's main focus: Iraq. Back in May 2002, the NSC expressed interest in putting out an unclassified publication that would lay out some of what we knew—or thought we knew—about Iraq's WMD programs. The National Intelligence Council, or NIC, had produced a similar document that the Clinton administration used to help justify the December 1998 Desert Fox bombing campaign. The NIC stepped up to the plate again, and the assignment went to Paul Pillar, one of the national intelligence officers. As is common with projects such as this, the drafting proceeded only intermittently. There was discussion of releasing the draft as a U.S. government "white paper"—one that would not carry the seal of any one agency—but ultimately the document was put on the shelf after the NSC seemed to lose interest in it.

Separately, in the summer and fall of 2002, the NSC asked John McLaughlin to have the Agency assemble its intelligence on Saddam's WMD programs and his human rights record, and outline what we believed about Iraq's connection to terrorism. While these efforts were going on in the background, the public debate was roiling. On August 15, 2002, Brent Scowcroft, who had served as national security advisor under President Ford and the first President Bush, and was then chairman of George W. Bush's Foreign Intelligence Advisory Board, published a hard-hitting Op-Ed piece in the *Wall Street Journal* titled "Don't Attack Saddam." In the article, Scowcroft argued that an attack would divert U.S. attention from the war on terrorism. It is no surprise that the advice was not well received at 1600 Pennsylvania Avenue. As moderate voices joined in the debate for caution in Iraq, the Bush administration pledged to listen carefully to the various sides, but its rhetoric seemed considerably ahead of the intelligence we had been gathering across the river in Langley.

I was surprised, for example, when I read about a speech Vice President Cheney gave to the Veterans of Foreign Wars on August 26, 2002, in which he said, "Simply stated, there is no doubt that Saddam Hussein now has weapons of mass destruction. There is no doubt he is amassing them to use against our friends, against our allies, and against us." Later in the speech, the vice president would tell the VFW, "Many of us are convinced that [Saddam] will acquire nuclear weapons fairly soon."

The speech caught me and my top people off guard for several reasons. For starters, the vice president's staff had not sent the speech to CIA for clearance, as was usually done with remarks that should be based on intelligence. The speech also went well beyond what our analysis could support. The intelligence community's belief was that, left unchecked, Iraq would probably not acquire nuclear weapons until near the end of the decade.

In his VFW speech, the vice president reminded the audience that during the first Gulf war, the intelligence community underestimated Iraq's progress toward building a nuclear weapon. No doubt that experience had colored the vice president's view of U.S. intelligence gathering ever since, but it also had a profound impact on my views and those of many of our analysts. Given Saddam's proclivity for deception and denial, we, too, were haunted by the possibility that there was more going on than we could detect.

The VFW speech, I suspect, was an attempt by the vice president to regain the momentum toward action against Iraq that had been stalled eleven days earlier by Scowcroft's Op-Ed piece. I have the impression that the president really wasn't any more aware than we were of what his number-two was going to say to the VFW until he said it. But if the speech was meant mostly as a wake-up call, it was a very loud one.

In the aftermath of Iraq, I was asked by Senator Carl Levin at a hearing before the Senate Select Committee on Intelligence, on March 9, 2004, if I should have intervened when I heard officials make public comments that went beyond our intelligence. It was a fair question. Clearly, decision makers are entitled to come to their own conclusions with regard to policy. Intelligence is an important part of the decision-making process but hardly the sole component. Policy makers are allowed to come to independent judgments about what the intelligence may mean and what risks they will tolerate. What they cannot do is overstate the intelligence itself. If they do, they must clearly delineate between what the intelligence says and the conclusions they have reached. In fairness to the vice president, prior to the production of the National Intelligence Estimate in October 2002, we at CIA had written pieces in key publications, such as the President's Daily Brief, that were very assertive about Iraq's WMD programs. However, none that I can recall put Iraq's acquisition of a nuclear weapon on the time line suggested in the VFW speech. Perhaps when policy makers who remember previous history, such as the vice president, read "overly assertive" analysis, their views are quickly hardened.

Policy makers have a right to their own opinions, but not their own set of facts. I had an obligation to do a better job of making sure they knew where we differed and why. The proper place to make that distinction is in one-on-one discussions with the principals, and I did so on a number of occasions. No one had elected me to go out and make speeches about how and where I disagreed on thorny issues. I should have told the vice president privately that, in my view, his VFW speech had gone too far. Would that have changed his future approach? I doubt it, but I should not have let silence imply agreement. We did a much better job of pushing back when it came to desires on the part of some in the administration to overstate the case on possible Iraq connections to al-Qa'ida.

On Friday afternoon, September 6, 2002, a week after the vice president delivered his VFW speech, the president's National Security Team gathered at Camp David and remained overnight for meetings about Iraq the next day. In advance, the NSC staff sent around thick briefing books packed with background information for the participants to read. One paper toward the front of the book listed things that would be achieved by removing Saddam—freeing the Iraqi people, eliminating WMD, ending threats to Iraq's neighbors, and the like.

Toward the middle of the book was a paper that discussed in general terms how Iraq would be dealt with following Saddam's removal. The paper said that we would preserve much of Iraq's bureaucracy but also reform it. An appendix listed for the attendees certain lessons learned from the occupations of Germany and Japan after World War II. Near the back of the book, at Tab P, was a paper CIA analysts had produced three weeks earlier. Dated August 13, 2002, it was titled "The Perfect Storm: Planning for Negative Consequences of Invading Iraq." The paper provided worst-case scenarios that might emerge from a U.S.-led regime-change effort. The summary said that following an invasion:

> The US will face negative consequences with Iraq, the region and beyond which could include:
>
> - Anarchy and the territorial breakup of Iraq;
> - Regime-threatening instability in key Arab states;
> - A surge of global terrorism against US interests fueled by deepening Islamic antipathy toward the United States;
> - Major oil supply disruptions and severe strains in the Atlantic alliance.

It's tempting to cite this information and say, "See, we predicted many of the difficulties that later ensued"—but doing so would be disingenuous. The truth is often more complex than convenient. Had we felt strongly that these were likely outcomes, we should have shouted our conclusions. There was, in fact, no screaming, no table-pounding. Instead, we said these were *worst case*. We also, quite accurately, labeled them scenarios. We had no way of knowing then how the situation on the ground in Iraq would evolve. Nor were we privy to some of the future actions of the United States that would help make many of these worst-case scenarios almost inevitable.

The Perfect Storm paper ended with a series of steps the United States could take to help reduce the chance of some of these negative consequences taking hold, including diplomatic initiatives to enhance the chances of Arab-Israeli peace. Promoting the notion that, although we were acting militarily in Iraq, we remained committed to an equitable resolution of this critical issue, which would have great resonance in the Islamic world, we advised. It was important that we be able to show the Arab world that we could make war and peace at the same time.

The meeting on Saturday morning, September 7, sparked considerable debate about the wisdom of trying to revive a UN inspection regime. Colin Powell was firmly on the side of going the extra mile with the UN, while the vice president argued just as forcefully that doing so would only get us mired in a bureaucratic tangle with nothing to show for it other than time lost off a ticking clock. The president let Powell and Cheney pretty much duke it out. To me, the president still appeared less inclined to go to war than many of his senior aides.

A week later, on Saturday, September 14, Steve Hadley convened another meeting in the White House Situation Room, attended by second-echelon officials from the NSC, State Department, DOD, and CIA. The agenda was titled, "Why Iraq Now?" Bob Walpole, the national intelligence officer for strategic programs, was among those present. He recalls telling Hadley that he would not use WMD to justify a war with Iraq. Someone, whom he did not know at the time but now recognizes as Scooter Libby, leaned over to another participant in the meeting and asked, "Who is this guy?"

Walpole explained to Hadley that the North Koreans were ahead of Iraq in virtually every category of WMD. Bob knew that we had recently discovered Pyongyang's covert program to produce highly enriched uranium, and he correctly assumed this would become public knowledge soon. "When that gets out, you guys will have a devil of a time explaining why you are more worried about a country that might be working on nuclear weapons rather than one that probably already has them and the wherewithal to deliver them to the U.S.," he told the group.

Someone suggested that the confluence with terrorism made Iraq a bigger threat. Two other CIA analysts present spoke up, saying that a much stronger case could be made for Iran's backing of international terrorism than could be made for Iraq's. They recall Doug Feith saying that their objections were just "persnickety."

CHAPTER 17

"The One Issue That Everyone Could Agree On"

The United States did not go to war in Iraq solely because of WMD. In my view, I doubt it was even the principal cause. Yet it was the public face that was put on it.

The leaders of a country decide to go to war because of core beliefs, larger geostrategic calculations, ideology, and, in the case of Iraq, because of the administration's largely unarticulated view that the democratic transformation of the Middle East through regime change in Iraq would be worth the price. WMD was, as Paul Wolfowitz was quoted as saying in *Vanity Fair* in May 2003, something that "we settled on" because it was "the one issue that everyone could agree on."

In early September 2002, with a vote looming on authorizing the use of force in Iraq, CIA came under pressure from members of the Senate Select Committee on Intelligence to produce a written assessment of Iraq's WMD programs. Specifically, they wanted a National Intelligence Estimate to aid their deliberations regarding whether or not to authorize the president to take the nation to war.

NIEs are intended to provide senior policy makers with the consensus of the American intelligence community on a given subject and to portray honestly dissenting and alternative views. Typically, NIEs require several months of preparation and jawboning by CIA, DIA, NSA, INR, DOE, NGA, and other agencies.

An NIE on Iraq should have been initiated earlier, but at the time I didn't think one was necessary. I was wrong. While there was no decision to go to war yet, the clock had begun to tick. We had not done an NIE specifically on Iraqi WMD in a number of years, but we had produced an array of analysis and other estimates that discussed Iraqi weapons programs, in the context of broader assessments on ballistic missiles and chemical and biological weapons. We all believed we understood the problem. In hindsight, even though policy makers were not showing much curiosity, that was the time we should have initiated a new series of analytical reports on Iraqi WMD and other issues regarding the implications of conflict in Iraq. This was my responsibility. But back then, I was consumed with al-Qa'ida—the people really trying to kill us—and I didn't pay enough attention to another gathering storm.

On September 9, 2002, Senator Richard Durbin of Illinois wrote me urging that I direct the production of an NIE and also an unclassified summary to explain the issue to the American public. The next day, Senator Bob Graham, chairman of the Senate Select Committee on Intelligence, followed that up with a letter requesting the production of an NIE "on the status of Iraq's programs to develop weapons of mass destruction and delivery systems, the status of the Iraqi military forces, including their readiness and willingness to fight, the effects a U.S.-led attack on Iraq would have on its neighbors, and Saddam Hussein's likely response to a U.S. military campaign designed to effect regime change in Iraq."

I reluctantly agreed and, on September 12, 2002, directed the National Intelligence Council staff to initiate a crash project to produce an NIE on the "status of and outlook for Iraq's weapons of mass destruction programs." The NIE was to answer two key questions on nuclear weapons: Did Saddam have them, and if not, when could he get them? I expected no surprises.

Like those of us in the intelligence community, the NSC staff questioned whether an NIE was needed. Steve Hadley thought that the data were already available in other documents.

Because of the impending vote on the use of force, scheduled for early October, a production process that normally stretched for six to ten months had to be truncated to less than three weeks. Even that was not fast enough for some unsympathetic members of Congress who wanted the NIE delivered almost instantly. Senator Graham went so far as to make statements to the press chastising us for foot-dragging. Not satisfied with the demands for *this* NIE, some senators were also pressing us to do another one evaluating the effectiveness of planned U.S. covert and military actions in Iraq. Assessing U.S. plans has never been a function of a National Intelligence Estimate. We were startled to have to explain this to a committee charged with overseeing intelligence—but that didn't stop the drumbeat.

The press of business and the shortened time available to produce the document meant we were headed uphill from the beginning. Had we started the process sooner, I am confident we would have done a better job highlighting what we did and didn't know about Saddam's WMD programs, and we would have sorted out some of the inconsistencies in the document. The lack of time, however, did not relieve us of the responsibility to get the information right. The flawed analysis that was compiled in the NIE provided some of the material for Colin Powell's February 5, 2003, UN speech, which helped galvanize public support for the war.

Some observers have gone so far as to suggest that our Iraq NIE is evidence that senior members of the intelligence community, like some senior policy makers, were hell bent on war. The truth is just the opposite. The person in charge of managing the NIE was Bob Walpole, the national intelligence officer for strategic programs. Not your typical bureaucrat—he's a Mormon bishop who often comes to work on a motorcycle—Bob is both a brilliant analyst and one of the most unlikely people to be accused of being a war hawk that you could imagine. When he was given the mission of coordinating the NIE, he came to me quite concerned. "I just don't believe in this war," he said with considerable angst. "Some wars are justifiable, but not this one."

"Look," I told him, "we don't make policy. Our job is to tell the people who *do* what we know and what we think. It's up to them to decide what to do about it."

"All right," Bob sighed, but I could tell he wasn't happy with the prospect. Nonetheless, in the weeks ahead he would spend many nights sleeping in his office to get the job done.

Because of the time pressures, analysts lifted large chunks of other recently published papers and replicated them in the Estimate. Twelve previous intelligence community publications formed the spine of the NIE. To meet the deadline, on September 23 a quickly assembled draft was sent around to intelligence community agencies for review. A day-long coordination meeting with intelligence community analysts was held two days later. The next day, a draft incorporating the analysts' changes and comments was sent back to the various Agency leaders.

On Tuesday, October 1, senior representatives of all the contributing agencies met with me to discuss, debate, and approve a final document. This is a standard part of the NIE process—the meeting was called the National Foreign Intelligence Board, or NFIB—but the narrow time frame, combined with often highly technical material, pushed standard procedures to the breaking point. Consider, for example, the controversial issue of aluminum tubes.

In early 2001, Iraq had been caught trying to clandestinely procure sixty thousand high-strength aluminum tubes manufactured to extraordinarily tight tolerances. The tubes were seized in the Middle East. The Iraqi agent tried in vain to get the tubes released, claiming they were to be used in Lebanon to make race car components. Whatever their intended use, under UN sanctions, Saddam was prohibited from acquiring the tubes for any purpose. All agencies agreed that these tubes could be modified to make centrifuge rotors used in a nuclear program. CIA analysts believed that these tubes were intended for the enrichment of uranium. Others thought they were intended to make rockets.

To test the theory, CIA brought together a "red team" of highly experienced experts from the Oak Ridge National Laboratory—people who had actually built centrifuges. Their assessment was that the tubes were more suited for nuclear use than for anything else. The Department of Energy's representative at the NFIB delivered his agency's assessment that the tubes were probably not part of a nuclear program. He was not a technical expert, however, and, despite being given several opportunities, he was unable to explain the basis of his department's view in anything approaching a convincing manner. About all we could take away from his statement was that DOE did not disagree with the assessment that Saddam was trying to revive or "reconstitute" his nuclear weapons program—a program that was within months of producing a weapon when it was interrupted at the time of the first Gulf war. Although the U.S. Army's National Ground Intelligence Center was not represented at the meeting, their view that it was highly unlikely that the tubes were intended for rockets gave added impetus to those who believed the tubes had a nuclear purpose.

With more time, I'm certain we would have delayed a decision on the aluminum tubes until greater clarification emerged—we were staring at a jumbled mess, basically—but in the end, the majority of agencies believed that the tubes were part of the evidence of nuclear reconstitution. But there was certainly no unanimity of thought.

The dissenting views were clearly and extensively laid out in the report. Not only did the Estimate make that point, but Colin Powell would go on to underline it in his UN speech the following February.

Perhaps the most widely misunderstood section of the NIE dealt with yellowcake, an element that can be enriched to make nuclear weapons–grade uranium. The Estimate included an account of Saddam's reported attempts to procure yellowcake from the African nation of Niger, taken from a September 2002 paper by the Defense Intelligence Agency. That account, told in a few paragraphs on page twenty-four of the document, was not a major pillar of the NIE. The Estimate noted that Saddam already had access to large amounts of yellowcake in Iraq—550 tons of it, enough to produce as many as 100 nuclear weapons. This yellowcake was supposed to be under seal by international inspectors, but that was at best a flimsy wall of protection.

Although it would loom large in subsequent criticisms of the NIE, the Niger yellowcake was not among the half dozen reasons cited why all agencies, with the exception of the State Department's Bureau of Intelligence and Research (INR), believed that Iraq was resuming its nuclear weapons program. Even INR wrote in the NIE that it believed Iraq was pursuing "at least a limited effort" to "acquire nuclear weapons related capabilities" and that the evidence indicated "at most a limited reconstitution effort."

We assessed that Saddam did not have a nuclear weapon and that if he had to make his own fissile material, he probably would not be able to do so until 2007 to 2009. However, we indicated in the NIE that we had only moderate confidence in that judgment. We also indicated that INR thought that, although Saddam clearly *wanted* nuclear weapons, there was inadequate evidence to prove that he had an ongoing integrated and comprehensive program to develop them.

If Saddam could obtain fissile material elsewhere, it would not be hard for the regime to make a weapon within a year. After all, we believed that some terrorist groups could do so if they came into possession of the all-important highly enriched uranium or plutonium.

None of the intelligence agencies challenged the judgments regarding Saddam's chemical and biological weapons programs. The NIE said that Saddam was continuing and expanding his ballistic missile efforts in contravention of UN sanctions. The missile assessment turned out to be on target.

Contrary to popular misconception, the NIE also gives full

voice to those agencies that wanted to express alternative views. Dissenting opinions are not relegated to footnotes and, indeed, often appear in boxes with special colored backgrounds to make them stand out. These make up an unprecedented sixteen pages of the ninety-page NIE. Agency heads had approval of not only the language that is used to express their reservations, but also where those reservations are displayed in the document.

What isn't emphasized, however, is the poor human access to Saddam's WMD programs and the limitations of our knowledge. It would have been helpful to have clarified that the use of the words "we judge" and "we assess" meant we were making analytical judgments, not stating facts. As the founding father of CIA analysis, Sherman Kent, wrote in the *Foreign Service Journal* in 1969, "Estimating is what you do when you do not know."

A careful reading of the NIE gives a more nuanced impression of its comments than the public has been led to believe. The phrase "we do not know" appears some thirty times across ninety pages. The words "we know" appear in only three instances. Unfortunately, we were not as cautious in the "Key Judgments," a five-page summary at the front of the document. The Key Judgments is written with language that, especially on chemical and biological weapons, is too assertive and conveys an air of certainty that does not exist in the rest of the paper. The nuance was lost.

The first key judgment states, "We judge that Iraq has continued its weapons of mass destruction/WMD programs in defiance of UN resolution and restrictions." Characterized as a "judgment," that's not bad, but the second sentence drops uncertainty regarding chemical and biological weapons: "Baghdad has chemical and biological weapons as well as missiles with ranges in excess of UN restrictions; if left unchecked, it probably will have a nuclear weapon during this decade. (See INR alternative view at the end of these Key Judgments.)" Although the missile statement is accurate and the nuclear judgment has its caveats ("if left unchecked" and "probably") and the reference to INR's alternative views, the chemical and biological judgments are stated as facts. They were *not* facts and should not have been so characterized.

The second key judgment states clearly that *"We lack specific information on many aspects of Iraq's WMD programs."* The problem was that statement followed a boldface assessment stating that *"we judge that we are seeing only a part of Iraq's WMD efforts, owing to Baghdad's vigorous denial and deception efforts. Revelations after the Gulf War starkly demonstrate the extensive efforts undertaken by Iraq to deny information."*

The absence of evidence and linear thinking, and Iraq's extensive efforts to conceal illicit procurement of proscribed components, told us that a deceptive regime could and would easily surprise us. It was never a question of a known imminent threat; it was about an unwillingness to risk surprise.

More troubling to me than the technical issues, over which experts can disagree, were instances where information from possible fabricators was included in the NIE. The most notorious example of bad information came from a German-run source dubbed "Curve Ball," whose information about mobile biological production trailers would figure largely both in the NIE and in Colin Powell's February 2003 speech to the UN. Curve Ball's information made its first appearance in our December 2000 NIE on biological weapons, where we stated, "new intelligence acquired in 2000 provides compelling information about Iraq's ongoing BW activities . . . and causes us to adjust our assessment upward of the BW threat posed by Iraq . . . the new information suggests Baghdad has continued and expanded its offensive BW program by establishing a large scale, redundant and concealed BW capability." At the time, Curve Ball's reporting was given added credence by the UN discovery in 1995 of Iraqi military documents pertaining to a secret mobile fermentation project.

We also had trouble with information from sources we used to validate what we heard from Curve Ball. For example, the Estimate contains information obtained in March 2002 from an

Iraqi defector, a former Iraqi major by the name of al-Asaaf, who had been referred to the Defense Intelligence Agency by the Iraqi National Congress. DIA had concerns about al-Asaaf's story regarding Iraq's mobile BW program, its interest in "dirty bombs," and its work on proscribed long-range missiles. The Iraqi major passed a DIA polygraph, but those who administered it felt that he had been "coached" in his story. Soon, much of what he had to say to DIA also appeared in a May 2002 article in *Vanity Fair*. The Iraqi National Congress arranged al-Asaaf's interview with the publication. The fact that his information was being peddled as if it were a PR campaign should have set off alarm bells.

DIA officials eventually concluded that the man was unreliable and was quite possibly feeding the United States fabricated information. But senior DIA officials sat through the hour-and-a-half NFIB meeting without ever mentioning that possibly bogus information was being cited in the Estimate we were all evaluating. Perhaps they didn't recognize their own information when they saw it, but that strains credulity.

DIA is not alone in bearing responsibility for the error. In July 2002, the National Intelligence Council staff did a study of the value, or lack thereof, of intelligence provided by the INC and cited this same source, al-Asaaf, as a possible fabricator. Three months later, they, too, failed to mention the matter as the NFIB reviewed the draft Estimate. I subsequently learned that some CIA analysts were also aware of al-Assaf's fabrication and failed to notice its inclusion in the NIE.

Although not mentioned in the Estimate, my views about Iraq's pursuit of WMD were greatly influenced by a very sensitive, highly placed source in Iraq. Little has been publicly said about this source. Indeed, at the time the NIE was being produced, because of the sensitivity of the source, most of the analysts involved were not even aware of the source's existence. The reporting, as it continued to stream in after the production of the Estimate, however, gave those of us at the most senior level further confidence that our information about Saddam's WMD programs was correct.

This source reported that production of chemical and biological weapons was taking place, biological agents were easy to produce and to hide, and prohibited chemicals were also being produced at dual-use facilities. This source stated that a senior Iraqi official in Saddam's inner circle believed, as a result of the UN inspections, that Iraq knew the inspectors' weak points and how to take advantage of them. The source said there was an elaborate plan to deceive inspectors and ensure that prohibited items would never be found.

Every once in a while, doubts would creep in about why so much of our evidence was indirect or why it had been so long since inspectors had found something. Right about then, this source would pop up with something incredibly specific that would not only affirm our intelligence but eliminate the doubts we might be having.

Sometimes a single source can make all the difference. Oleg Penkovsky was a single source whose reporting proved indispensable in helping the United States get through the Cuban Missile Crisis forty years earlier.

In many ways, we were prisoners of our own history. The judgments we delivered in the NIE on Iraq's chemical, biological, and nuclear weapons programs were consistent with the ones we had given to the Clinton administration. Yet by 2002, we made some leaps based on technical analysis that led us to assume that Saddam had more capability, particularly with regard to chemical weapons, than we later learned was warranted.

Inevitably, the judgments were influenced by our underestimation of Iraq's progress on nuclear weapons in the late 1980s and early 1990s—a mistake no one wanted to repeat.

Martin Indyk, with whom I served on the NSC staff early in the Clinton administration and who went on to be assistant secretary of state for Near Eastern Affairs, illustrated for me the mind-set that we were all operating with in the mid-1990s. Martin and I were convinced that Iraq had weapons of mass destruction because Saddam had an entire organization dedicated to concealing them. "We observed how they operated," Martin said. "Saddam refused to account for the material that was missing from the previous war, and logically it did not make sense, since if he would just come clean he could get out of sanctions and we would be screwed."

"I remember going to bed at night," Martin recalled, "expecting to wake up the next morning and find that UNSCOM would go to the secret site and catch Saddam red-handed. We'd wake up in the morning and there was nothing there. There was never anything there. With the benefit of hindsight, we should have thought: wait a minute, if we never find it, maybe it is not there. I didn't think about the possibility that Saddam was bluffing us."

I did not believe he was bluffing, either. With the quality of UN inspections growing weaker over time, the political will to maintain sanctions fading, and Saddam's coffers ballooning through the Oil-for-Food program, I had little doubt in my own mind what Saddam was up to. I believed he had WMD, and I said so.

From then on, after UNSCOM's departure, we had to rely more on analysis and extrapolation of more nuanced technical data. We divorced technical analysis from our understanding of Iraqi culture, however, and this hurt us in central ways. We failed, for example, to factor in how the regime's harsh treatment of its citizens would make truthful reporting to superiors on the status of weapons programs less likely. We did not fully consider the impact of nearly a decade of international sanctions, UNSCOM inspections, continuous overflights, and U.S. military actions. Yet Saddam gave us little reason to believe that he had changed his stripes or his trajectory.

Nevertheless, in 2002, to conclude that Saddam was not pursuing WMD, our analysts would have had to ignore years and years of intelligence that pointed in the direction of active programs and continuing evidence of aggressive attempts on Iraq's part to conceal its activities. Even with more time, could analysts have concluded that Saddam had no weapons programs, or even the ability to quickly surge to produce the weapons themselves? I doubt it.

In retrospect, we got it wrong partly because the truth was so implausible. We knew plenty of countries that were working on WMD programs and desperately trying to conceal that fact. But we had no previous experience with a country that did not possess such weapons but pretended that it did. Saddam made a speech in June 2000 in which he said you cannot expect Iraq to give up the rifle and live only with a sword when his neighbor [Iran] had a rifle. After his capture in December 2003, Saddam was asked by George Piro, an FBI Special Agent, what he had meant by that statement. Saddam said that he had two audiences in mind. One was the UN Security Council, as Saddam wanted the disarming of Iraq to be part of a broader disarming of the Middle East. The other audience was Iran. Saddam then said, "You guys just don't understand. This is a rough neighborhood."

There is another factor that few people outside the intelligence community would recognize or credit, and that is how the remedy for one so-called intelligence failure can help set the stage for another. Following the controversy over some of our missile analysis in the mid-1990s, a commission headed by Donald Rumsfeld had taken us to task for not leaning forward more boldly and imaginatively in projecting missile development in countries such as Iran and North Korea. In response, we began to give more weight in these assessments to what *could* occur, rather than stopping with what we confidently knew. This is perhaps another way of saying, connect all the dots in order to warn adequately. I have often wondered if this was the prevailing sentiment among analysts as we did our Iraq work. Did it push us to be more assertive than we should have been?

Saddam was a genius at what the intelligence community calls "denial and deception"—leading us to believe things that weren't true. But he was a fool for not understanding, especially after 9/11, that the United States was not going to risk underestimating his WMD capabilities as we had done once before. The irony is that he could have allowed UN inspectors free run of the country—and if they found nothing, UN sanctions would have melted. In that case, he might be alive and living in a palace today. Without sanctions, he would be well on his way to possessing WMD. Before the war, we didn't understand that *he* was bluffing, and he didn't understand that *we were not*.

When we finally did complete the nineteen-day Estimate late on the evening of October 1, the document was rushed to Capitol Hill with the ink still wet on its covers.

The morning of October 2, 2002, twelve hours after we had delivered the NIE to Congress, the Senate Select Committee on Intelligence held a closed-door hearing to discuss its contents. My staff had informed the committee several days earlier that I would be unable to attend because I was required to be at the White House at the same time, ironically to meet with other congressional leaders. In my place, I sent John McLaughlin and Bob Walpole, the lead national intelligence officer for the NIE, to brief members. The members, though, seemed to have forgotten that I had advised them I could not be present. Several were upset about my absence and about the NIE having been delivered so late the night before, around 10:30 P.M. Their anger was misplaced—McLaughlin and Walpole were well qualified to respond to the senators' interests.

With great difficulty, McLaughlin persuaded the committee to go ahead with the hearing, and then only by promising that he and I would return to brief the senators again two days later. We did that on Friday, October 4. That closed-door session was very contentious.

One senator asked us how our views differed from those of our

British allies, who had just published their own white paper days before. Bob Walpole cited two points of divergence. First, he said, we differed by a few months with the British on how quickly Saddam could make a nuclear weapon. Second, we differed with the Brits on intelligence suggesting that Saddam had been trying to obtain uranium from Africa. Senator Kyl pointed out that there was reference to yellowcake in our Estimate. Walpole said, yes, we mention it as a possibility, but only after we say that we are much more worried about the 550 tons of yellowcake that Saddam already has access to inside Iraq. Even then, Walpole pointed out, yellowcake is not mentioned in the Key Judgments or in our unclassified paper.

As soon as we delivered the classified Estimate to the Hill, calls began for us to instantly produce an unclassified version. This, too, was virtually impossible in the time allotted, but our efforts to be accommodating led to another major error. Someone came up with the bright idea of taking an unclassified white paper that the NIC had drafted months before on the same subject, and had sat unpublished on a shelf, and modifying it for this purpose. Doing so would be far faster than trying to come up with an unclassified version of the NIE. But there's a saying that "if you want it bad, you get it bad," and that was precisely what we got.

In an effort to meld the white paper and the NIE, analysts took the Key Judgments from the NIE, declassified them, and stuck them on the front of the white paper. Because they are written from the point of view of the entire intelligence community, NIEs are replete with statements such as "we assess that" and "we judge that." The white paper had been crafted in a different style, and in merging the two documents, those responsible opted for the latter style. Out went the "we"s, and what remained were bolder assertions, such as "Saddam has." The classified NIE already had too few cautionary "we judge"s in the Key Judgment section. Now, with a few strokes of a keyboard, the unclassified paper—the only one most Americans would ever see—came out sounding far too assertive, even though it did note that there were differences among specialists over issues such as the aluminum tubes and UAVs. The moral to the story is that white papers should never be written before a classified estimate has been completed.

Following McLaughlin and Walpole's October 2 appearance before the Senate Select Committee on Intelligence, several Democratic senators demanded that a few sentences from the testimony be declassified and cleared for public release. The senators also wanted released some language that was contained in the classified NIE but not in the unclassified white paper.

On October 7, McLaughlin signed a letter to them on my behalf containing the words they were seeking from the NIE:

> *Baghdad for now appears to be drawing a line short of conducting terrorist attacks with conventional or C.B.W. [chemical and biological weapons] against the United States.*
>
> *Should Saddam conclude that a U.S.-led attack could no longer be deterred, he probably would become much less constrained in adopting terrorist actions. Such terrorism might involve conventional means, as with Iraq's unsuccessful attempt at a terrorist offensive in 1991, or C.B.W.*
>
> *Saddam might decide that the extreme step of assisting Islamic terrorists in conducting a W.M.D. attack against the United States would be his last chance to exact vengeance by taking a large number of victims with him.*

The letter also authorized the release of some dialogue between Senator Carl Levin and John McLaughlin, who testified in a closed hearing. The witness said that the likelihood of Saddam's initiating a terrorist attack in the foreseeable future was, in our judgment, "low," but that if Saddam felt cornered, the chances of his using WMD were "pretty high."

Democratic members of the committee released the letter to

the media almost immediately, provoking a flurry of stories. The articles suggested that the letter contradicted President Bush's assertion on the imminent threat posed by Iraq and implied that the use of force by the United States would only increase the likelihood that Saddam would either use WMD himself or share it with terrorists. The articles prompted a frantic call from Condi Rice. She wanted me to "clarify" the issue right away. So, at her request, I spoke with a *New York Times* reporter who was working on the story. "There was no inconsistency in the views in the letter and those of the president," I told the reporter. The sentence seized upon in the letter was about a judgment call as to whether and when Saddam might use WMD and whether he might share them with terrorist organizations. We labeled our views as "low confidence" judgments—in other words, we were not very sure we had a good idea what Saddam would do if cornered.

In retrospect, I shouldn't have talked to the *New York Times* reporter at Condi's request. By making public comments in the middle of a contentious political debate, I gave the impression that I was becoming a partisan player. That certainly wasn't my intention.

The intelligence reports and analysis used over the years on the WMD issue, and repeated in the NIE, were flawed, but the intelligence process was not disingenuous nor was it influenced by politics. Intelligence professionals did not try to tell policy makers what they wanted to hear, nor did the policy makers lean on us to influence outcomes. The consistency of our views on these weapons programs was carried forward to two presidents of different political parties who pursued vastly different courses of action. Even though the daily reports the president saw in the run-up to the production of the NIE were uneven and assertive in tone, and at times more assertive on some issues than the NIE, they were a reflection of honest analysis.

Policy makers also have the responsibility to challenge the analysis they receive. Their uncritical attitude in this case was highlighted by a question posed by Brent Scowcroft in a recent speech: "What happens when the intelligence community provides intelligence that policy makers want to hear?" He could have added: *particularly when war and peace hang in the balance.*

An NIE had never been relied upon as a basis for going to war, and, in my view, the decision to invade Iraq was not solely predicated on this one. But if we had done a better job in all our analysis and in this NIE, war critics would have had a harder time today implying that "the intelligence community made us do it."

The notion that we somehow cooked the books on the Iraq NIE is only part of current mythology. Maybe the greater exaggeration is the profound effect the NIE supposedly had on decision makers. In a little-remembered article in April 2004, the *Washington Post* reported, "No more than six senators and a handful of House members who did not serve on the house and senate Intelligence Committees read beyond the five-page National Intelligence Estimate executive summary, according to several congressional aides responsible for safeguarding the classified material." The full NIE ran some ninety pages.

Some who later rightly criticized the NIE had previously made their own public statements that went beyond what was in the Estimate. Senator Jay Rockefeller, the respected ranking Democrat on the Senate Intelligence Committee, said on the floor of the Senate on October 10, 2002, that "There is unmistakable evidence that Saddam Hussein is working aggressively to develop nuclear weapons and *will likely* have nuclear weapons within the *next five years*" (emphasis added). The first sentence of the Key Judgments of the October 2002 estimate itself says only that *"if left unchecked,"* Saddam *"probably"* will have a nuclear weapon *"during this decade"* (emphasis added).

Rather than being "unmistakable," the evidence was a matter of some dispute among analysts, a point made clear in pages of dissenting opinions in the NIE. Rockefeller went on to remind his colleagues of the same history that caused our analysts much concern. He said, "We also should remember we have always underestimated the progress Saddam has made in development of weapons of mass destruction."

Congress was not alone in its lack of genuine interest in the NIE before the war. Senior administration officials in the NSC, Department of Defense, and elsewhere had also put the document at the bottom of their reading lists. Everyone seemed to think they knew either what was in the document or what ought to be in it.

Few people may have read the NIE, but in no way does this excuse the many shortcomings of our Iraqi analysis over the years, in the Estimate or in the testimony we presented to Congress. Misinformation and misimpressions go to the heart of our credibility, our mission, even our reason for being.

Given what we knew then, the NIE should have said:

> *We judge that Saddam continues his efforts to rebuild weapons programs, that, once sanctions are lifted, he probably will confront the United States with chemical, biological and nuclear weapons within a matter of months and years. Today, while we have little direct evidence of weapons stockpiles, Saddam has the ability to quickly surge to produce chemical and biological weapons and he has the means to deliver them.*

We should have said, in effect, that the intelligence was not sufficient to prove beyond a reasonable doubt that Saddam had WMD. The evidence was good enough to win a conviction in a civil suit but not in a criminal case. Would we have gone to war with such conclusions? I don't believe the war was solely *about* WMD, so probably yes. But more accurate and nuanced findings would have made for a more vigorous debate—and would have served the country better.

In the spring of 2004, during one of my final appearances before the House Permanent Select Committee on Intelligence, Congressman Norm Dicks commented on the NIE. Norm is a longtime friend of the intelligence community and of mine personally, yet he had harsh words that day. Regarding the Estimate, and the faith he had in me, he said, "We depended on you, and you let us down." For me, it was one of the lowest moments of my seven-year tenure, because I knew he was right.

CHAPTER 18

No Authority, Direction, or Control

Mr. President," I said one morning in March 2003, "the vice president wants to make a speech about Iraq and al-Qa'ida that goes way beyond what the intelligence shows. We cannot support the speech, and it should not be given."

The Iraq WMD issue had been around for years. People believed they knew it backward and forward. There was no raging debate within the administration about our conclusions. But there *was* debate, intense focus, and, in the eyes of some analysts, pressure regarding the question of Iraq's relationship with al-Qa'ida and complicity in 9/11. We could go as far as outlining contacts between Iraq and al-Qa'ida going back a decade, to Bin Ladin's time in the Sudan, to Abu Musab al-Zarqawi finding safe haven in Iraq, and to at least a dozen Egyptian Islamic Jihad operatives who showed up in Baghdad in the spring and summer of 2002. We could cite training that may have been provided, particularly regarding chemical and biological weapons. But one thing is certain, we consistently told the Congress and the administration that the intelligence did not show any Iraqi authority, direction, or control over any of the many specific terrorist acts carried out by al-Qa'ida.

Let me say it again: CIA found absolutely no linkage between Saddam and 9/11. At best, all the data in our possession suggested a plausible scenario where the "enemy of my enemy might be my friend," that is, two enemies trying to determine how best to take advantage of each other. In the world of terrorism, nothing is ever very clear, and the murkiness of the intelligence required an exhaustive effort to run down every lead to satisfy ourselves that there was no state complicity with al-Qa'ida's actions on 9/11.

We told the president what we did on Iraq WMD because we believed it. However, we did not bend to pressure when it came to a possible past Iraq–al-Qa'ida connection. The absence of such a connection would have been impossible for others to disprove following an invasion, unlike WMD, which were either there or not. Those who say that we cooked the books or knowingly let the administration say things that we knew to be untrue are just wrong.

People often forget what it was like after 9/11. A senior analyst put it this way, "Intelligence is central to the Bush administration. Every single day it was the discipline around which they started their day. And then after 9/11, the first attack on American soil of any magnitude in sixty years, they were in fear. In fairness to them, people do not understand how goddamn dangerous we thought it was. The absence of solid information on additional threats was terrifying."

It took us a while to understand how important the Iraq connection was for some in the administration, but we learned quickly. The vice president and others pushed us hard on this issue, and our answers never satisfied him or some of our other regular "customers." Paul Wolfowitz and Scooter Libby, for example, were relentless in asking us to check, recheck, and re-recheck. Wolfowitz's strong views on the matter were no secret. He even wrote a blurb for Laurie Mylroie's 2000 book, *Study of Revenge: Saddam Hussein's Unfinished War Against America*, in which he said the book "argues powerfully" that the perpetrator of the 1993 World Trade Center attack was actually "an agent of Iraqi intelligence," and it asks, if that is true, what that would tell us about Saddam's ultimate ambitions.

The truth was that CIA was not initially prepared for the intense focus that the administration put on the Iraq–al-Qa'ida relationship. We had devoted little analytic attention to it prior to September 11. We were instead consumed with the very hot war with Sunni extremists all over the world. People were coming to kill us. We had no preconceived conclusions on the Iraq–al-Qa'ida connection—unlike our certainty on Iraqi WMD—and it would require us to start from the bottom up, do a zero-based review and look at the whole issue dispassionately. On one level, this was a blessing.

It was also a curse, because initially, and for some time, our answers to the elaborate, nuanced, and voluminous questions the administration asked were inconsistent and incomplete, and often had to be revisited. Early on we probably did not inspire much confidence in policy makers who knew their brief and knew where they wanted to end up. Senator Fritz Hollings once said that going to a press conference with Vice President Hubert Humphrey was like jumping into a swimming pool with Olympic champion Mark Spitz. Well, that was what it was like briefing Dick Cheney, Scooter Libby, and Paul Wolfowitz on this subject. They were smart, tough, and had command of the data. Initially, we did not. But over time, that changed in a dramatic way.

The first time I recall a briefing at our headquarters on Iraq and al-Qa'ida was in September of 2002. The briefing was a disaster. Libby and the vice president arrived with such detailed knowledge on people, sources, and timelines that the senior CIA analytic manager doing the briefing that day simply could not compete. We weren't ready for this discussion. We determined that from that moment on we would have multiple lower-level subject-matter expert analysts—people who knew a lot about a narrow range of topics—meet with them.

By November of 2002, we were ready for another visit from the vice president and his team. There was extensive preparation, practice sessions called "murder boards," and total collaboration between regional and terrorist analysts. The November meeting was described by a participant this way:

Scooter Libby approached it like an artful attorney. An analyst would make a point and Libby would say, okay this is what you say. But there are these other things happening. So if this were true, would it change your judgment? And the analyst would say, well if that was true, it might. And Libby would say, well if that's true, what about this? And six "if that were trues" later, I finally had to stop him and say, "Yes, there are other bits and pieces out there. We've looked at these bits and pieces in terms of the whole. And the whole just does not take us as far as you believe. And everything else is just speculation. That was a push by policy makers to see how far we would go."

Some analysts viewed this kind of grilling as being pressured, but most did not. Their view was, if a country is about to go to war, policy makers are going to ask tough questions to understand all the elements of the issue. One senior analyst said to me, "Were they trying to push us and drive us? Absolutely. By the questions they asked and by the way they asked the questions again and again with changed nuances. They were trying to pull out every last iota of what we might say that supported where they wanted to go. But they are policy makers. It is our job to consider what they say, think about it, and write what we think. We stuck to our guns."

The truth is we were not ready to take a position on the Iraq–al-Qa'ida question because there were differing views within the Agency about how to think about the issue. The division was between analysts who focused on specific regions and those who specialized in terrorism. This uncertainty was played out earlier in the year on June 21, 2002, when we produced the paper "Iraq & al-Qa'ida: Interpreting a Murky Relationship." In contrast with every other type of analysis, because of the nature of the threat, terrorism analysis by design takes weaker information and makes more aggressive conclusions, sometimes from information that regional analysts might discard.

The "Murky Relationship" paper was an academic exercise. Its "scope note" at the beginning explained that the paper was an effort to see what our conclusions might be if the most forward leaning explanations of our intelligence turned out to be true. The note read in part: "This intelligence assessment responds to senior policy maker interest in a comprehensive assessment of Iraqi regime links to al-Qa'ida. Our approach is *purposefully aggressive in seeking to draw connections*, on the assumption that any indication of a relationship between these two hostile elements could carry great dangers to the United States" (emphasis added).

Regional analysts who focus on geographic areas believed that fundamental distrust stemming from stark ideological differences between Saddam and Usama bin Ladin, and the potential fear that Islamic extremism posed for Iraq, significantly limited the cooperation that was suggested by the reporting. The terrorism analysts who specialized in the broad range of terrorism and who wrote the paper took note of the ideological differences but believed to be credible the reporting that suggested a deeper relationship. The paper made clear that there were no conclusive signs between Iraq and al-Qa'ida with regard to terrorist operations. Yet it posited that there were enough data with regard to safe haven, training, and contacts to at least require us to be very concerned. Jami Miscik, our chief analyst, believed that the analysis should be published because of the risks to the United States, and it was.

In our shop, many saw this as almost too aggressive. Some analysts involved complained informally to an ombudsman, whom we had earlier appointed to deal with claims of politicization, that we had gone too far in coming up with our "murky" conclusion. As described to me by a senior analyst, "Barry [the ombudsman] sat us down and said: 'Grow up. This is not politicization. This is misunderstanding and hurt feelings.' The two groups need to sit down and hash it out."

Despite the fact that some of our analysts felt we had gone too far, many in the administration, such as Paul Wolfowitz and Scooter Libby, believed that the "Murky Relationship" paper had not gone far enough. Within a couple months the classified document was being mocked in a Jim Hoagland column in the

Washington Post. Hoagland's piece led with the rhetorical slap: "Imagine that Saddam Hussein has been offering terrorist training and other lethal support to Osama bin Ladin's al-Qa'ida for years. You can't imagine that? Sign up over there. You can be a Middle East analyst for the Central Intelligence Agency."

Other senior administration officials questioned our preliminary analysis. The Senate Intelligence Committee later uncovered an internal Pentagon memo sent to both Paul Wolfowitz and Doug Feith saying that although the facts represented in the "Murky Relationship" paper were good, CIA's analysis attempted to "discredit, dismiss, or downgrade" much of the reporting, and our interpretations "should be ignored."

During the late summer of 2002, we started working on a more comprehensive paper that would explain what we knew and suspected about Iraq's involvement with terrorism. While we could not make the al-Qa'ida connection, there was no doubt that Saddam was making large donations to the families of Palestinian suicide bombers and was known to be harboring several prominent terrorists, including Abu Nidal, a ruthless killer responsible for attacks on El Al ticket counters in Rome and Vienna in 1985, resulting in 18 deaths and injury to 120 people. Saddam also gave refuge to one of the individuals still being sought for the first World Trade Center bombing.

We were still gathering material for the comprehensive paper when we received an offer from a Pentagon group working under Doug Feith to share with us their observations on the case for a connection between Iraq and terrorism. Although the suggestion was a bit odd—since it was coming from people in the policy shop, not people in intelligence positions—we agreed to hear them out. A small group of Pentagon officials showed up at CIA headquarters on August 15, 2002.

Present from the Pentagon were Feith; Richard Haver, a longtime civilian intelligence professional who had worked for Dick Cheney in the first Bush administration; Vice Admiral Jake Jacoby, the head of the Defense Intelligence Agency; and several others from Feith's shop. Haver had been the one who dropped by my office in late December 2000 and hinted broadly that I was soon to be replaced by Don Rumsfeld. That had not been his only off-the-mark assessment. Shortly before September 11, 2001, he gave a speech at the National Security Agency during which he told the audience that the intelligence community was spending far too much time on terrorism.

Attending the meeting on our side with me were Ben Bonk, the deputy chief of our Counterterrorism Center; several analysts from Ben's staff; and a number of analysts from CIA's Directorate of Intelligence who were coordinating the forthcoming Iraq terrorism paper.

Feith's team, it turned out, had been sifting through raw intelligence and wanted to brief us on things they thought we had missed. Trouble was, while they seemed to like playing the role of analysts, they showed none of the professional skills or discipline required. Feith and company would find little nuggets that supported their beliefs and seize upon them, never understanding that there might be a larger picture they were missing. Isolated data points became so important to them that they would never look at the thousands of other data points that might convey an opposite story.

The woman on Feith's team gave the presentation, which was titled "Iraq and al-Qa'ida—Making the Case." She started out by saying that there should be "no more debate" on the Iraq–al-Qa'ida relationship. "It is an open-and-shut case," she said. "No further analysis is required." This statement instantly got my attention. I knew we had trouble on our hands.

The briefing slides she used were equally self-certain. One slide said that Iraq and al-Qa'ida had a "mature, symbiotic relationship." Wrong. There was nothing in the intelligence to suggest a mature, symbiotic relationship. Another slide said there were "some indications of possible Iraq coordination with Al Qaeda specifically related to 9/11." By this point, the "Atta in Prague" story, which CIA had brought forward after 9/11, was eroding.

I listened for a few more minutes, trying to be polite, before saying, "That's very interesting." This was one of my rare moments of trying to be subtle. What I was really thinking was, This is complete crap, and I want this to end right now.

Shortly thereafter I excused myself and pulled Jacoby aside. As an active-duty vice admiral and head of DIA, he worked for both Don Rumsfeld and me. Reverting to my normal blunt self, I told him, "This is entirely inappropriate. You get this back in intelligence channels. I want analysts talking to analysts, not people with agendas." Pentagon officials were later quoted anonymously in the media describing the same meeting but claiming that "the scales fell away" from the CIA's eyes when we saw their fine analysis. In fact, their analysis had little, if any, impact on us whatsoever.

Only much later did we learn that "Team Feith" had been going around briefing officials at the White House, the NSC, and the Office of the Vice President with a story similar to the one we found so weak in August. In these briefings they added an extra slide titled "Fundamental Problems with How Intelligence Community Is Assessing Information." The slide went on to complain that we were being too picky and applying a standard of proof that we would not normally require. But we weren't too impressed with their work, either, especially their willingness to blindly accept information that confirmed preconceived notions. We came to call their approach "Feith-based analysis."

When the Pentagon's inspector general issued a report in February 2007 calling some of Feith's efforts "inappropriate" (which to my mind is the kindest thing you could say about it), Feith shot back. He said peddling his alternative intelligence was simply an exercise in "good government." Nonsense. This was an example of bad government. Policy makers are entitled to their own opinions—but not to their own set of facts. Feith's charts mischaracterized the intelligence. If policy makers want to reach their own judgments they can do so, so long as they say, "The views I am about to express are not supported by the DCI and his analysts." And Feith should have had the courage to tell us that his opening slide, shown to the White House, said in essence that CIA analysis stinks.

Our second paper on Iraq and al-Qa'ida, published in September 2002, was shared with only a small number of senior officials. As often happens, shortly after that report went out, new intelligence came in suggesting that there might have been greater contact regarding training between Iraq and al-Qa'ida. In light of that, we set out to vet and add these new details for an upgraded report that would be given wider dissemination among administration and congressional officials than the first, closely held document. Agency analysts went so far as to show a draft to Feith's team and to ask if they had any comments or objections to it. Feith's staffers said they did "but would make their views known through other channels." In retrospect this was a pretty clear warning that we were being second-guessed and undermined.

By December the revision was done, and we titled the report, "Iraqi Support of Terrorism." I asked, out of courtesy, that a copy of the draft be forwarded to the White House before it was shared with other senior officials. We were explicit in saying that we were not soliciting proposed edits; we just didn't want the administration surprised when we issued the paper. Despite those caveats, a series of calls from the White House continued to pour in asking us to revise or withdraw the paper. John McLaughlin was on the receiving end of one such call, from a testy Scooter Libby asking for more revisions. The answer was no—we would make no more revisions. Jami Miscik received the brunt of those calls. She, too, stood firm. Jami believed that she had pushed her analysts to ensure they employed every analytic best practice and that no solid reporting had been ignored. But she would not go beyond where the intelligence took us.

After Steve Hadley called Jami from the NSC, wanting to engage her in yet another discussion on the paper, she stormed into my office and said she would resign before she would delay or amend the paper again. Completely supportive of her, I picked up my white secure telephone and punched Hadley's number. "Steve," I said, "knock this off. The paper is done. It is finished. We are not changing it. And Jami is not coming down there to discuss it anymore."

Message received. A day or two later Jami was at the White House—for an entirely different reason—and got word that the president wanted to see her. He clearly had heard about the flap and asked if "his guys" had "stepped over the line." Not wanting to prolong the controversy, Jami told me that she assured the president that it was nothing that we couldn't handle.

On January 28, 2003, the paper was published. So what did it say? Our analysts believed that there was a solid basis for identifying three areas of concern with regard to Iraq and al-Qa'ida: safe haven, contacts, and training. But they could not translate this data into a relationship where these two entities had ever moved beyond seeking ways to take advantage of each other.

The intelligence told us that senior al-Qa'ida leaders and the Iraqis had discussed safe haven in Iraq. Most of the public discussion thus far has focused on Zarqawi's arrival in Baghdad under an assumed name in May of 2002, allegedly to receive medical treatment. Zarqawi, whom we termed a "senior associate and collaborator" of al-Qa'ida at the time, supervised camps in northeastern Iraq run by Ansar al-Islam (AI).

AI, a radical Kurdish Islamic group, was closely allied to al-Qa'ida. Kurdish Islamists and al-Qa'ida had come together in the summer of 2000 to create a safe haven for al-Qa'ida in an area of northeastern Iraq not under Iraqi government control, in the event Afghanistan was lost as a sanctuary. The area subsequently became a hub for al-Qa'ida operations. We believed that up to two hundred al-Qa'ida fighters began to relocate there in camps after the Afghan campaign began in the fall of 2001. The camps enhanced Zarqawi's reach beyond the Middle East. One of the camps run by AI, known as Kurmal, engaged in production and training in the use of low-level poisons such as cyanide. We had intelligence telling us that Zarqawi's men had tested these poisons on animals and, in at least one case, on one of their own associates. They laughed about how well it worked. Our efforts to track activities emanating from Kurmal resulted in the arrest of nearly one hundred Zarqawi operatives in Western Europe planning to use poisons in operations. What was even more worrisome was that by the spring and summer of 2002, more than a dozen al-Qa'ida–affiliated extremists converged on Baghdad, with apparently no harassment on the part of the Iraqi government. They had found a comfortable and secure environment in which they moved people and supplies to support Zarqawi's operations in northeastern Iraq.

More al-Qa'ida operatives would follow, including Thirwat Shihata and Yussef Dardiri, two Egyptians assessed by a senior al-Qa'ida detainee to be among the Egyptian Islamic Jihad's best operational planners, who arrived by mid-May of 2002. At times we lost track of them, though their associates continued to operate in Baghdad as of October 2002. Their activity in sending recruits to train in Zarqawi's camps was compelling enough.

There was also concern that these two might be planning operations outside Iraq. Credible information told us that Shihata was willing to strike U.S., Israeli, and Egyptian targets sometime in the future. Shihata had been linked to terrorist operations in North Africa, and while in Afghanistan he had trained North Africans in the use of truck bombs. Smoke indeed. But how much fire, if any?

Could we prove that this was Iraqi complicity with Zarqawi and the two Egyptian Islamic Jihad operatives? No. Do we know just how aware Iraqi authorities were of these terrorists' presence either in Baghdad or northeastern Iraq? No, but from an intelligence point of view it would have been difficult to conclude that the Iraqi intelligence service was not aware of their activities. Certainly, we believe that at least one senior AI operative maintained some sort of liaison relationship with the Iraqis. But operational direction and control? No.

In the laborious exercise undertaken by analysts to understand the history of a potential Iraq–al-Qa'ida relationship, they went back and documented the basis of a variety of sources—some good, some secondhand, some hearsay, many from other intelligence services. There were, over a decade, a number of possible high-level contacts between Iraq and al-Qa'ida, through high-level and third-party intermediaries. Our data told us that at various points there were discussions of cooperation, safe haven, training, and reciprocal nonaggression.

During the mid-1990s, Sudanese national Islamic Front Leader Hasan al-Turabi reportedly served as a conduit for Bin Ladin between Iraq and Iran. Turabi in this period was trying to become the centerpiece of the Sunni extremist world. He was hosting conferences and facilitating the travel of North Africans to Hezbollah training camps in the Bekaa Valley, in Lebanon. There was concern that common interests may have existed in this period between Iraq, Bin Ladin, and the Sudanese, particularly with regard to the production of chemical weapons. The reports we evaluated told us of high-level Iraqi intelligence service contacts with Bin Ladin himself, though we never knew the outcome of these contacts.

A senior al-Qa'ida detainee told us in 2002 that he believed it unlikely that Bin Ladin would ally himself with Baghdad and thereby compromise al-Qa'ida's mission and independence. He also said that several of Bin Ladin's lieutenants had urged cooperation with Iraq, believing that the benefit of possible training, safe haven, and help with al-Qa'ida's WMD efforts outweighed any risks to al-Qa'ida's independence. According to the detainee, Saddam became more interested in al-Qa'ida after the East Africa and *Cole* bombings. But certainly by that time, al-Qa'ida had demonstrated its prowess to conduct conventional attacks, and was well established in its sanctuary in Afghanistan.

The one possible connection that analysts viewed as most disturbing was training. There were solid reports from senior al-Qa'ida members that raised concerns about al-Qa'ida's enduring interest in acquiring chemical and biological expertise from

Iraq. In the public debate that has since occurred, this has now all come down to the recantation of an individual named Ibn Sheikh al-Libi. A senior military trainer for al-Qa'ida in Afghanistan, al-Libi was detained in late 2001 and transferred into military custody in Afghanistan in early January of 2002. At the time, he was the highest ranking al-Qa'ida member in U.S. custody.

We believed that al-Libi was withholding critical threat information at the time, so we transferred him to a third country for further debriefing. Allegations were made that we did so knowing that he would be tortured, but this is false. The country in question understood and agreed that they would hold al-Libi for a limited period, and then return him to U.S. military custody, where he would be registered with the International Committee of the Red Cross.

In the course of questioning while he was in U.S. custody in Afghanistan, al-Libi made initial references to possible al-Qa'ida training in Iraq. He offered up information that a militant known as Abu Abdullah had told him that at least three times between 1997 and 2000, the now-deceased al-Qa'ida leader Mohammad Atef had sent Abu Abdullah to Iraq to seek training in poisons and mustard gas. Another senior al-Qa'ida detainee told us that Mohammad Atef was interested in expanding al-Qa'ida's ties to Iraq, which, in our eyes, added credibility to the reporting.

Then, shortly after the Iraq war got under way, al-Libi recanted his story. Now, suddenly, he was saying that there was no such cooperative training. Inside CIA, there was sharp division on his recantation. It led us to recall his reporting, and here is where the mystery begins.

Al-Libi's story will no doubt be that he decided to fabricate in order to get better treatment and avoid harsh punishment. He clearly lied. We just don't know when. Did he lie when he first said that al-Qa'ida members received training in Iraq or did he lie when he said they did not? In my mind, either case might still be true. Perhaps, early on, he was under pressure, assumed his interrogators already knew the story, and sang away. After time passed and it became clear that he would not be harmed, he might have changed his story to cloud the minds of his captors. Al-Qa'ida operatives are trained to do just that. A recantation would restore his stature as someone who had successfully confounded the enemy. The fact is, we don't know which story is true, and since we don't know, we can assume nothing.

The additional context I had to consider was this: the kind of training al-Qa'ida may have been pursuing with Iraq in the chemical and biological arena was part of a larger, more robust and compartmented WMD program that al-Qa'ida was pursuing and continues to pursue. It is a program sanctioned and directed by the senior leadership. Would they have sought to attain building blocks from more sophisticated programs? My view at the time was that it was completely possible.

Did we look at Zarqawi's operations at the lower-level poisons facility in northeastern Iraq as part of al-Qa'ida's intention to both use these lesser capabilities and also obscure their more important and lethal programs? Of course, you can pull out the al-Libi recantation and say, "You see, this was all hyped." Yet if you ignore the Iraqi context we were operating in with regard to al-Qa'ida's pursuit of WMD capability, you end up missing the larger and more important picture. This was my mind-set. Run it all down, put all the concerns on the table, and give everybody your best judgment.

There was more than enough evidence to give us real concern about Iraq and al-Qa'ida; there was plenty of smoke, maybe even some fire: Ansar al-Islam; Zarqawi; Kurmal; the arrests in Europe; the murder of American USAID officer Lawrence Foley, in Amman, at the hands of Zarqawi's associates; and the Egyptian Islamic Jihad operatives in Baghdad. But for some in the administration, it was never enough. They had pushed the data farther than it deserved. They made command linkages where we could not see them. They sought to create a connection between Iraq and the 9/11 attacks that would have made WMD, the United Nations, and the international community absolutely irrelevant. The first problem is that case was never, ever true. The second problem is that in trying to make more out of the case, advocates ended up undermining the case we had. People just stopped listening.

It was during this period that we dealt with another high-profile issue. Reports dating back to late 2001 alleged that one of the 9/11 hijackers, Mohammed Atta, might have met with Ahmad Khalil al-Ani, a member of the Iraqi intelligence service, in Prague just months before the 2001 attack. The White House, Department of Defense, and CIA were all intensely interested in the allegation. If it could be shown that Iraq was an active participant in the planning for the 9/11 attacks, there would be no question regarding an immediate effort to oust Saddam.

We devoted extraordinary effort to the issue but could never find any convincing evidence that the visit had happened. In fact, over time the intelligence suggesting such a meeting eroded. Proving something *didn't* happen is problematic, but in this case, we and the FBI concluded that such a meeting was highly unlikely. Nonetheless, we kept being asked to reinvestigate the matter, and while doing so, we kept hearing highly placed officials, including the vice president, say on television that it was "pretty well confirmed" that the visit had occurred. By May of 2002, FBI and CIA analysts voiced increased skepticism that these meetings had taken place. The case for the meetings continued to weaken from that time forward.

It is my understanding that, in 2006, new intelligence was obtained that proved beyond any doubt that the man seen meeting with the member of the Iraqi intelligence service in Prague in 2001 was *not* Mohammed Atta.

A second possible linkage to 9/11 and Iraq involved an Iraqi national named Shakir who worked at the airport in Kuala Lumpur as a part-time facilitator for Arab visitors, a job he had obtained through an Iraqi embassy employee. In January 2000, Shakir facilitated the travel of 9/11 hijacker Khalid al-Mihdhar from the airport. Shakir's immediate departure from Malaysia one week after helping al-Mihdhar through the airport, and previous travel and contacts with extremists, raised red flags. After months of exhaustive analytic work, we could not establish that Shakir was an Iraqi agent.

The Iraq–al-Qa'ida controversy continued, even after Saddam was long gone from power. Once U.S. forces reached Baghdad, they discovered—stacked where they could easily find them—purported Iraqi intelligence service documents that showed much tighter links between Saddam and Zarqawi and Saddam and al-Qa'ida. CIA analysts worked with the U.S. Secret Service to have the paper and ink checked and tried to verify the names and information in the documents. Time and again, documents that were supposedly produced in the early 1990s turned out to be forgeries. CIA officers interviewed Iraqi intelligence officers in Baghdad who also discounted the authenticity of the documents. It was obvious that someone was trying to mislead us. But these raw, unevaluated documents that painted a more nefarious picture of Iraq and al-Qa'ida continued to show up in the hands of senior administration officials without having gone through normal intelligence channels.

As one senior analyst put it to me, "The administration is relying too much on flawed information. These are documents found on the floor of burnt-out buildings, strewn all over the floor, and taken at face value and not being looked at by trained analysts. Trained analysts would ask questions like, 'What is the source? What do I know about the source?' 'Do they have the access that they claim?' So there is absolutely no standard of analytic tradecraft applied to any of this. Rather, it was presented to us as proof, evidence and confirmation."

On March 13, 2003, we received for our clearance review a speech that had been drafted for the vice president to give on the eve of the war. The proposed speech was sharply at odds with our paper of January 28, 2003, going far beyond the notion of Iraq as a possible training site for al-Qa'ida operatives. The speech draft came to conclusions we could not support, suggesting Iraqi complicity in al-Qa'ida operations.

This prompted a heated conversation between John McLaughlin and Scooter Libby. John subsequently provided in writing detailed reasons why we could not support the speech. "Clearly a policy maker is free to say 'given my read of the intelligence, here is what I make of it,'" John wrote, but he went on to say that the text "goes further than most of our analysts would, implying that Iraq has operational direction and control over al-Qa'ida terrorists." The next morning, just before the president's intelligence briefing, I raised the issue.

"Mr. President," I said, "the vice president wants to make a speech about Iraq and al-Qa'ida that goes way beyond what the intelligence shows. We cannot support the speech and it should not be given." Although I never learned why, the vice president chose not to give his speech.

The push to make the Iraq–al-Qa'ida connection didn't end with the start of the war. The November 24, 2003, issue of the *Weekly Standard* magazine had a lengthy article called "Case Closed," which was based on a top-secret memorandum that Doug Feith had sent Senate Intelligence Committee chairman Pat Roberts and ranking member Jay Rockefeller a few weeks before. The article claimed that much of the information in the memo contained intelligence "detailed, conclusive and corroborated by multiple sources" showing an "operational relationship" between Usama bin Ladin and Saddam Hussein going back to the early 1990s.

In fact, much of the material in the memo was the kind of cherry-picked, selective data that Feith, Libby, and others had been enamored of for so long. The Pentagon issued a press statement noting that the memo contained a lot of raw reports but claimed, inaccurately, that the intelligence community had cleared its submission to Congress.

Two months later, Vice President Cheney was in Denver and was asked about the Iraq–al-Qa'ida connection. He cited the

Weekly Standard article containing the leaked Feith memo as "your best source of information" on possible ties. I disagree. The best source of information was our January 2003 paper, which said that there was no Iraqi authority, direction, or control over al-Qa'ida.

Stretching the case continues to this day. On the eve of the fifth anniversary of September 11, the vice president appeared on NBC's *Meet the Press*. Asked about previous administration comments seeming to link Iraq to 9/11, the vice president ducked the question but referred to testimony I had given a few years before, about contacts between Iraq and al-Qa'ida. "The basis for that is probably best captured in George Tenet's testimony before the Senate Intel Committee, in open session, where he said specifically that there was a pattern of relationship that went back at least a decade between Iraq and al-Qa'ida." On *Fox News Sunday*, Condi Rice was asked a similar question and gave a similar answer. "What the president and I and other administration officials relied on—and you simply rely on the central intelligence. The Director of Central Intelligence, George Tenet, gave that very testimony, that, in fact, there were ties going on between al-Qa'ida and Saddam Hussein's regime going back for a decade. Indeed, the 9/11 Commission talked about contacts between the two."

They quoted my testimony accurately, as far as they went, but both failed to mention that, at the same time, I told them and Congress that our intelligence did not show Iraq and al-Qa'ida had ever moved beyond seeking ways to take advantage of each other. We were aware of no evidence of Baghdad's having "authority, direction and control" of al-Qa'ida operations. In other words, they told only half the story.

CHAPTER 19

Slam Dunk

Many people today believe that my use of the phrase "slam dunk" was the seminal moment for steeling the president's determination to remove Saddam Hussein and to launch the Iraq war. It certainly makes for a memorable sound bite, but it is belied by the facts. Those two words and a meeting that took place in the Oval Office in December 2002 had nothing to do with the president's decision to send American troops into Iraq. That decision had already been made. In fact, the Oval Office meeting came:

- ten months after the president saw the first workable war plan for Iraq;
- four months after the vice president's Veterans of Foreign Wars speech in which he said there was "no doubt" that Saddam had weapons of mass destruction;
- three months after the president told the United Nations that the Iraqi regime should "immediately and unconditionally forswear, disclose, and remove or destroy all weapons of mass destruction, long-range missiles, and all related material";
- several months after the U.S. military began repositioning assets to be used in war to facilities throughout the Middle East;
- two months after Congress had authorized the use of force in Iraq; and

• two weeks after the Pentagon had issued the first military deployment order sending U.S. troops to the region.

As so often happens with these matters, the context has disappeared, and all that is left are the words themselves, two words that have taken on a significance that far exceeds their import at the time. Let me set the scene.

On Saturday, December 21, 2002, I went to the White House for the usual briefing that we delivered to the president six days a week. But that day an additional meeting had also been scheduled after the morning briefing. About two and a half weeks earlier, NSC officials had asked us to start assembling a public case that might be made against Saddam regarding his possession and possible use of WMD. Although this presentation by CIA would eventually evolve into the speech that Secretary of State Colin Powell delivered to the United Nations, at the time it was not clear who the ultimate audience would be—or even who would present the case. That morning, our charge was simply to assemble materials for a briefing that might someday go public. White House staffers had made it clear that they were looking for an "Adlai Stevenson moment," a reference to Stevenson's famous UN presentation during the Cuban missile crisis, but Bob Walpole had told them that our collected intelligence was nowhere near that categorical.

In the intervening few weeks, a small team of senior analysts had pulled together the requested material. Now it was our turn to deliver it to the president, vice president, Andy Card, Condi Rice, and a few others. The presentation itself fell to John McLaughlin. A champion debater in college, John is not one to go beyond the facts or to stray into bombast. Within and beyond the Agency, John's briefings were well known for being precise, measured, and low key. He had brought some charts with him to illustrate his points and an executive assistant to help with the visual presentation.

It's important to remember both what John was doing that morning and what our charge had been. This meeting was not called for us to mull over the entire issue of Iraq and WMD. Everybody in the room—as well as the most credible intelligence services in the world—already believed that Saddam had chemical and biological weapons and was working on a nuclear program. The incomplete data declarations Saddam was giving to the UN and a stream of information from well-placed intelligence sources served only to buttress our confidence. Our job that day wasn't to prove the WMD case, or validate the claim. Our job was to lay out the information relevant to WMD that we believed (a) to be true, and (b) could be cleared for public release without doing damage to intelligence sources and methods. We weren't going to put anything in a public presentation that would jeopardize the lives or continued productivity of precious intelligence sources. Nothing John said in his briefing should have been new to anyone in the room.

Inevitably perhaps, given the high expectations, the substance of John's presentation underwhelmed the audience. This was a first cut, and as most first drafts go, it was very rough. Clearly this didn't compare to the Stevenson moment the White House was searching for. I was disappointed, too. I was sure there was more supporting data in the recently produced NIE, and I felt certain we could find a way to release some of it to the public. Worse, I felt that we had wasted the president's time by giving him an inferior briefing.

George Bush was gracious. "Nice try," he told John, but he quickly added that what he had just heard was not likely to convince "Joe Public." The president suggested that maybe we could add punch to the presentation by bringing in some lawyers who were accustomed to arguing cases before juries. At no time did he or anyone else in the room suggest that we collect more intelligence to find out if the WMD were there or not. As I said, everyone in the room already believed Saddam possessed WMD.

The focus was simply on sharpening the arguments. Some might criticize us for participating in what was essentially a marketing meeting, but intelligence was going to be used in a public presentation and it was our responsibility to ensure that the script was faithful to what we believed to be true and that it placed no sensitive intelligence sources or methods at risk.

To that end, I was asked if we didn't have better information to add to the debate, and I said I was sure we did. I wanted to convey that I thought it would be possible to declassify enough additional information—communications intercepts, satellite reconnaissance photos, sanitized human intelligence reports, and so forth—to help the public understand what we believed to be true. If I had simply said, "I'm sure we can do better," I wouldn't be writing this chapter—or maybe even this book. Instead, I told the president that strengthening the public presentation was a "slam dunk," a phrase that was later taken completely out of context and has haunted me ever since it first appeared in Bob Woodward's book *Plan of Attack*.

Whoever later described the scene to Bob Woodward painted a caricature of me leaping into the air and simulating a slam dunk, not once but twice, with my arms flailing. Credit Woodward's source with a fine sense of the ridiculous, or at least a fine sense of how to make me look ridiculous, but don't credit him or her with a deep sense of obligation to the truth. Even though I am often blunt and prone to talk with my hands, both McLaughlin and I know that this basketball pantomime never happened. In fact, neither John nor his executive assistant even remember my statement. I certainly don't deny using the term "slam dunk" or strongly believing that Saddam had WMD. But the phrase has, in my view, been intentionally misused and thus completely misunderstood by the public at large.

To double-check John's recall and mine, I asked another CIA officer who was sitting next to me in the Oval Office that morning and who had accompanied me to daily presidential briefings for nearly three years what the officer remembered about the incident. "I am sure you said 'slam dunk,' but it was no more than a passing comment. I have been with you when you are really trying to make a point, so I have a basis for comparison. The picture that has been incorrectly portrayed is: You said, 'slam dunk' and they all went, 'Well, we're done. Let's go to war.' But that's not the way it was."

In thinking about all this, I have a few tips for future CIA directors, and for anyone who aspires to participate in government at a similar level. First, you are never offstage. Anything you say can be used down the road to make someone else's point. That's the way Washington has evolved—there are no private conversations, even in the Oval Office.

Second, in a position such as mine, you owe the president exactness in language. I didn't give him that, and as a result I ended up writing the talking points for those anxious to shift the blame for Iraq away from them and onto CIA in general and me in particular.

Third, I advise future directors of the Agency to be wary of the pitfalls when engaging with policy makers on intelligence related to their policies. On the one hand, if you keep hands off, chances are the intelligence may be misused. On the other, if you engage, you run the risk of seeming to support policy even when you are striving for neutrality.

I can honestly say that we always sought to give the president our best judgments. We did not go beyond our conclusions to justify a policy. Those who feel that we were stretching the case or telling the president what he wanted to hear are simply wrong.

That said, how influential was my comment to the president's thinking? In a way, President Bush and I are much alike. We sometimes say things from our gut, whether it's his "bring 'em on" or my "slam dunk." I think he gets that about me, just as I get that about him. What's more, I think each of us regularly factored that into his understanding of what the other was saying. Other than that, I don't pretend to know what was going through his mind that Saturday morning or in the weeks afterward, but there are some hints.

That Christmas Eve, three days after our Oval Office meeting, Jami Miscik was up at Camp David providing the PDB for the president. One of my predecessors was also there, the president's father.

The first President Bush mentioned to Jami that he had heard that there had been an Iraq WMD briefing a few days earlier and that it "hadn't gone well." She later told me that she informed both presidents Bush, father and son, that while there was no "smoking gun" on Iraq WMD, she offered to review the data that had been presented a few days before. In discussing the matter further, she said she was troubled by the lack of intercepted communications one would expect to find with an active WMD program. Human intelligence in a place like Iraq is hard to get, but why there wasn't more signals intelligence was a mystery, she told them.

The second President Bush responded to Jami that there had to be better information that could be presented, but in doing so he made no mention of the "slam dunk" incident. Jami says she never heard that phrase until she read about my purported performance in Bob Woodward's book. That certainly doesn't sound to me like a seminal moment in the decision to go to war.

How is it, then, that an offhand comment made in a closed-door meeting on a Saturday morning has come to symbolize so much? I don't think it was an accident. Back in early 2001, when my old mentor Senator David Boren advised me to assist the new administration for six months before resigning, he added a cautionary note: "Be careful, you are not one of the inner circle going back to the campaign. It doesn't matter how the president may feel; if it suits that group, they will throw you overboard."

If I had cared less about carrying out the Agency's mission in a time of war, I would have heeded the caution.

From the fall of 2003 onward, the security situation in Iraq continued to deteriorate. Rather than acknowledge responsibility, the administration's message was: Don't blame us. George Tenet and the CIA got us into this mess. To this day, certain administration officials continue to use the phrase "slam dunk" as a talking point. In his September 10, 2006, appearance on *Meet the Press*, in response to a question from Tim Russert, Dick Cheney referenced me and cited "slam dunk" not once but twice. I remember watching and thinking, "As if you needed *me* to say 'slam dunk' to convince *you* to go to war with Iraq."

Like the vice president and many others, Bob Woodward has not been above using this phrase for his own ends. Shortly after the start of the Iraq war in 2003, White House communications officials had strongly urged CIA cooperation with Woodward on his latest book. We had provided some background, again at White House request, for Woodward's previous book, *Bush at War*, and the administration wanted to replicate what they saw as the PR success of that effort.

I was not at all certain that cooperating this time was a good idea. While the Afghan campaign was then a clear success, the war in Iraq was still unresolved, the hunt for WMD was ongoing, and the rising insurgency in Iraq was proving problematic. Nonetheless, we kept getting calls from the White House saying, "We're cooperating fully with Woodward, and we would like CIA to do so, too."

Accordingly, we provided some senior officials to give Woodward background information, describing our role in the preparation for and conduct of the war. We believed that there was a way, without giving away any secrets, to talk about, for example, the dangerous and vital work done by our case officers who had spent months in northern Iraq gathering intelligence prior to the war.

Woodward was in frequent contact with my spokesman, Bill Harlow, chasing down things he had heard elsewhere and trying to set up interviews. In one background session with a senior CIA official in early 2004, at which I was not present, Woodward off-handedly raised the subject of the December 21, 2002, meeting and the phrase "slam dunk." He made no special issue of it then. Nor did he request that Harlow ask me about the meeting or the context in which the words had been used.

After his manuscript had gone to print, Woodward mentioned to Harlow that there was going to be something in it that we might find a bit dicey, and he described in greater detail the supposed "slam dunk" scene. Still, he downplayed it and said it was not that big a deal. Maybe that really is how he felt, but when the book came out, following extensive excerpting in the *Washington Post*, "slam dunk" seemed to be all anyone talked about.

Reporters later told Harlow that when they called the White House for reaction to the Woodward book, administration spokesmen were quick to point out the quote. It was, after all, the perfect public-relations deflection. In a situation as complex as the war in Iraq, the public yearns for a simple explanation. Now they had one.

Woodward quotes the president in his book as saying that my "slam dunk" comment was a very important moment. I truly doubt President Bush had any better recollection of the comment than I did. Nor will I ever believe it shaped his view about either the legitimacy or timing of waging war. Far more likely, the president's staff brought up the "slam dunk" scene in the course of prepping him for the Woodward interview—quite possibly the same staff member or members who originally fed the scene to Woodward. They might even have suggested that the president work "slam dunk" into one of his answers if the question was never directly asked. Then, with all the prep work done, the memories "refreshed," Woodward was ushered into the Oval Office, the tape recorder was turned on, and the rest is now history.

I've spoken to Woodward several times since his book came out, and he, of course, doesn't think that he was used or was unfair. He believed the phrase wasn't as big a deal as some might make it. But when he was on television in 2005, defending himself over not originally reporting what he knew about the Valerie Plame and Joe Wilson incident, Woodward said he was too busy in 2003, working on his book and learning important stuff like "slam dunk."

"Slam dunk," he said, "was the basis of this incredibly critical decision the President and his war cabinet were making on, do we invade Iraq?"

I have another two-word reaction to that statement. The first word is "bull."

CHAPTER 20

Taking "the Case" Public

The last thing I ever expected was to be a member of a Greek chorus. But there I was, on international TV, a prop on the set, sitting behind Colin Powell as he spoke to the UN General Assembly on February 5, 2003. Little did I know that as Powell ran through chapter and verse of what we *thought* we knew about Saddam's WMD programs, this drama would later turn out to be a tragedy.

The speech was the end result of several months of planning, extrapolating, and negotiating. If the United States and our allies were going to win international support for an invasion of Iraq, it was going to take a compelling argument that would turn the legions of skeptics into the "coalition of the willing." The administration debated who could make such a presentation, to whom it would be given and, most important, what would be said.

On a Saturday morning shortly after Christmas 2002, John McLaughlin and Bob Walpole were attending yet another meeting at the White House. The subject turned to trying to improve upon the unsatisfactory presentation we had given a week or so before, during the "slam dunk" meeting, and how we could improve on it. The NSC staff suggested drawing from the NIE to bolster the public argument for toppling Saddam. Condi asked Walpole to summarize the Estimate's key judgments. He began doing so from memory, citing all the "we assess" and "we judge" language that appears in the document.

"Wait a minute," Condi interrupted. "Bob, if you are saying these are *assertions*, we need to know this now." That was the word she used. "We can't send troops to war based on assertions."

Walpole calmly said that the NIE was an "assessment" and that these were analytical judgments. He explained that the agencies attached certain levels of confidence to the various judgments—some matters we had high confidence in, others moderate or low—but there was a reason the document's title contained the word "estimate."

Condi asked what he meant about confidence levels. Walpole said that, for example, the analysts had "high confidence" that Saddam had chemical weapons.

"What's high confidence, ninety percent?" she asked.

"Yeah, that's about right," Bob replied.

Condi said, "That's a heck of a lot lower than we're getting from reading the PDB." After the war, as part of our lessons-learned efforts, we went back and had analysts review everything the Agency had written regarding Iraq and WMD. We had in fact been much more assertive in what we were writing for the president on some issues, such as aluminum tubes, than we had been in some of our other publications, including the NIE. Walpole told her that the strongest case for Saddam having weapons of concern was missiles. Walpole was aware that the Iraqis had recently made a declaration to the UN about their Al-Samoud missile. Our experts studied the data and had just concluded that the missile was badly designed and would not reach as far as previously feared. "But you cannot go to war over missiles that exceed authorized ranges by just a few tens of kilometers," he said.

Relying on the information that we would later learn was wrong, Walpole assured her that the next strongest case was biological weapons. While we had confidence about chemical weapons, Walpole said, that case was largely built on analytical inference. "The weakest case," he explained, "was nuclear." There were alternative views, and the agencies had only moderate confidence in the views that they expressed.

Turning to John McLaughlin, the national security advisor said, "You [the intelligence community] have gotten the president way out on a limb on this."

McLaughlin was stunned and not at all happy about being chastised. He later came back to Langley and told me about the conversation. "We've got *them* out on a limb?" he said. It wasn't, after all, the intelligence community that had been clamoring to go to war in Iraq. We had had our hands full with the war on terrorism.

On January 6, 2003, I attended another meeting in Condi's office along with McLaughlin, Walpole, and Steve Hadley. Hadley noted that the Iraq nuclear case in the proposed speech, a presentation that did not yet have an audience, was weak and needed to be "beefed up." Walpole replied that the draft was weak because the *case* was weak. That was why there were alternative views expressed on the issue in the NIE.

On January 24, 2003, at still another meeting, Hadley asked Walpole to provide him information on what Saddam needed if he were to obtain nuclear weapons. Walpole replied that that information was contained in the NIE published three months previously.

"Humor me," Hadley said. "The NIE is ninety pages. Can you just excerpt that part and send it to me?"

Walpole subsequently faxed twenty-four pages of material to Hadley for background purposes. Out of that, and out of context, White House officials much later seized on one paragraph from page twenty-four of the NIE to justify including Niger yellowcake and Saddam's nuclear weapons ambitions in the president's State of the Union speech, delivered only days later. Not only did doing so completely ignore the tenor of what we had been telling Rice, Hadley, and others in these meetings, but it also ignited the "sixteen words" flap that would come back to bite us a half year later.

By late January, Colin Powell was picked to make the case for going to war before the United Nations. His mandate was to give a speech that would tell the world why time was running out for Iraq. At one point, Condi Rice and Karen Hughes had urged Powell to speak on three consecutive days. Their vision was that he would speak one day only about Iraq and terrorism. The next day he would address Iraq and human rights. Then he would finish with a lengthy speech about Iraq and WMD. Colin wisely nixed that notion, but it was clear to everyone that this was going to be a speech of extraordinary importance.

Colin asked to come out to CIA headquarters along with several of his speechwriters and senior aides to work through the speech and make sure it was as solid as possible. Although he didn't say so explicitly, I believe one of the reasons he wanted to have the speech worked on at the Agency was the sense that, within our barbed wire–encircled headquarters compound, we were relatively free from interference from downtown.

This was an unusual role for us. We had two undesirable options from which to pick. We could let the administration write its own script, knowing that they might easily mischaracterize complex intelligence information, or we could jump in and help craft the speech ourselves. We chose the latter.

We believed Colin would use as a template for his speech a document that grew out of John McLaughlin's infamous presentation in the "slam dunk" session. Bob Walpole had sent the NSC a revised draft weeks earlier based on the NIE, as they requested. When Colin's team first arrived at CIA, they had in their hands a fifty-nine-page document on WMD with which they presumed we were familiar. Powell assumed that the White House had pulled the document together in coordination with the intelligence community. But what the White House handed him was something very different, something that we had never seen before and that had not been cleared by CIA. Powell's team kept asking us about intelligence underlying elements in the draft, and my staff found themselves repeatedly saying, "We don't know what you are talking about." Colin later told me he saw Scooter Libby at one point and asked, "What are you guys thinking, giving me a draft like that?" Libby reportedly gave him a sheepish look and said, "I wrote it as a lawyer presenting a brief." Powell said the draft looked like it was "a lawyer's brief, not an analytical product."

Eventually, those working on the speech figured out that John Hannah of the vice president's staff was quite familiar with the WMD brief. So, despite the desire to shield the information vetting from kibitzers, they had to ask Hannah to come out to Langley to explain the origins of the material in the speech draft.

Hannah arrived with a stack of raw intelligence, and each time he was asked about some item that had mysteriously appeared in the speech draft, he cited a fragment of information. Time and again, CIA analysts would explain that the information being relied on was fragmentary, unsubstantiated, or had previously been proved wrong. In the end, line after line of the speech draft was thrown out. At one point Hannah asked Mike Morell, who was coordinating the review of the speech for CIA, why the Niger uranium story wasn't in the latest draft. "Because we don't believe it," Mike told him. "I thought you did," Hannah said. After much wrangling and precious time lost in explaining our doubts, Hannah understood why we believed it was inappropriate for Colin to use the Niger material in his speech.

Some members of Secretary Powell's team who participated in assembling the speech have subsequently spoken out about the ordeal and given the impression that they were standing alone on the bulwark, keeping out the bad intelligence. That is not how CIA participants remember it. We had a number of senior intelligence professionals assigned to check the accuracy of what was being said against the intelligence reporting, and others charged with examining the reliability of the sources. Our memory is that CIA and State Department officials worked side by side to rid the draft of material that would not stand up. Our goal from beginning to end was to come up with rhetoric that was both supported by underlying intelligence and worthy of what we all hoped would be a defining moment. Despite our efforts, a lot of flawed information still made its way into the speech. No one involved regrets that more than I do. But I have often wondered whether we might have uncovered more of those flaws if our people had not had to spend two days getting the garbage out of a White House draft that we had never seen before.

The UN speech was supposed to focus mostly on WMD. Weapons of mass destruction programs were, as Colin once put it, in the UN's "in-box"—in other words, something it was concerned about and responsible for—because Saddam had so consistently ignored United Nations sanctions. The White House staff, however, seemed especially keen on including material about terrorism. In addition to their own piece on WMD, Scooter Libby had provided Powell with a forty-page paper of unknown origin entitled "Iraq's Dangerous Support for Terror," which the secretary promptly dismissed. They kept suggesting language so far over the top (for example, suggesting possible Iraqi-9/11 connections) that I finally pulled aside Phil Mudd, the then deputy chief of our Counterterrorism Center, and told him to write the terrorism piece of the speech himself.

"It is highly unusual, hell, it is practically inappropriate," I told him, for us to write a speech for policy makers. "But if we don't do it, the White House will cram some crap in here that we will never live down." Mudd wrote the terrorism portion of the speech, and he did a damn good job of it. Despite some problems, that piece of Powell's remarks stands up much better today than does the larger portion on Iraq and WMD.

The process of working on the speech was difficult right up to the end. A handful of senior CIA analysts and I went to New York on February 4 along with Powell and his staff and joined them as he continued to refine and rehearse the remarks he planned to give the next day. The one fax machine capable of sending and receiving classified material broke down, and we struggled to get last-minute information from Washington and from Powell's staff across town. I stayed up until about two o'clock the night before—actually the morning of—the presentation, working on the terrorism portion of the speech. At last, though, we were all able to agree on a text. After all the back-and-forth, we believed we had produced a solid product.

If Colin had any reservations about giving the speech, he did not tell me. Once he had agreed to undertake the mission, he was going to give it his best shot. Late in the process, Colin asked me to sit behind him at the UN. That was about the last place I wanted to be—I had been scheduled to make an overseas trip to the Middle East at the time—but Powell and his deputy, Rich Armitage, were two of my closest colleagues in the administration. If he wanted me there, I was going to be there, even if my presence was more than a little odd for a serving DCI.

Walking into the UN General Assembly on the morning of February 5 was a surreal moment for me. I sat next to John Negroponte, who at the time was the U.S. ambassador to the UN. After Colin finished what I thought was an extraordinary performance, and other council members began to speak, I left the chamber mentally and physically exhausted.

It was a great presentation, but unfortunately the substance didn't hold up. One by one, the various pillars of the speech, particularly on Iraq's biological and chemical weapons programs, began to buckle. The secretary of state was subsequently hung out to dry in front of the world, and our nation's credibility plummeted.

One particularly damning part of the speech is now so notorious that it deserves special attention. The story begins in 1998, when an Iraqi chemical engineer wandered into a German refugee camp. Within a year or so, he had earned his German immigration card by agreeing to cooperate and provide information to the German Federal Intelligence Service, or BND. The Germans gave the man his perversely prescient code name: Curve Ball.

As intelligence services generally do with their spies, the BND kept its engineer under tight wraps, but eventually shared with the U.S. Defense Intelligence Agency some of the information he was providing them. Curve Ball alleged that Iraqi scientists had a biological weapons program located in mobile laboratories that could be moved to evade UN weapons inspectors.

Because BND controlled the asset tightly and because DIA had responsibility for intelligence from Iraqi refugees in Germany, CIA was twice removed from the source. It was a situation far from ideal. The Germans would not permit either DIA or CIA to have direct access to Curve Ball. They told us that he did not speak English and that he disliked Americans. (It later turned out that his English was pretty good.) We did have one opportunity to observe him when a German-speaking U.S. doctor evaluated him during a physical. The doctor noted that the man appeared hungover and he expressed doubts about his reliability. Those doubts seem prophetic now, but I must say that if we dismissed everything we heard from sources with drinking problems, some accurate intelligence would be thrown out the window.

I've since learned that there were debates between our analysts and our intelligence collectors about the case. Some of the collectors from our Directorate of Operations didn't like the way the case "felt"—they had a gut instinct that there was something wrong with Curve Ball, but little more to go on. The analysts believed passionately that the science Curve Ball was describing was accurate—too accurate to be dismissed. There was the fine detail of Curve Ball's reporting—he clearly knew what a mobile lethal-germ lab looked like—and the ever-increasing value of his information as the search for Saddam's WMD mounted.

On balance, and in the absence of any other red flags from the Germans or DIA, Curve Ball appeared to be an invaluable asset. He wasn't. As the Silberman-Robb Commission, a presidential panel looking into Iraq intelligence shortcomings, would report in March 2005, sirens should have been going off all over the place. Whether they were or not is a matter of fierce debate.

Jim Pavitt, the then deputy director of operations and head of the clandestine service, instructed Tyler Drumheller, head of the European Division, to ask for a CIA officer to be allowed to have a face-to-face meeting with the engineer. In late September or early October 2002, Drumheller met with his German counterpart over lunch at a Washington restaurant to convey the request, but got nowhere.

Drumheller, whom I always considered to be a capable officer, now says the German told him, "You do not want to see him [Curve Ball] because he's crazy. Speaking to him would be 'a waste of time.'" The German reportedly went on to say that his service was not sure whether Curve Ball was telling the truth, that he had serious doubts about Curve Ball's mental stability and reliability. Curve Ball, he said, may have had a nervous breakdown. Further, the BND representative worried that Curve Ball was "a fabricator." According to Drumheller's account, the German cautioned, however, that the BND would publicly and officially deny these views if pressed, because they did not wish to be embarrassed.

If that is true, this is how it should have played out: What the German had to say at that lunch in late September or early October 2002 should have been immediately and formally disseminated as a matter of record in a report that would have alerted intelligence and policy officials to the potential problem with Curve Ball. A second, corresponding formal report also should have been instantly sent across the intelligence and policy communities to analysts and policy makers who had received previous Curve Ball reporting. The transmittal of these two reports would have immediately alerted experts doing the work on Iraq WMD issues across the intelligence community to a problem requiring resolution. No such report was disseminated, nor was the issue ever brought to my attention. In fact, I've been told that subsequent investigations have produced not a single piece of paper anywhere at CIA documenting Drumheller's meeting with the German. The lead analyst on this case in our Weapons Intelligence, Nonproliferation, and Arms Control Center (WINPAC) insists she was never told about the meeting.

Issuing "burn notices," as they are called, on questionable sources is how the system is supposed to work. Because this didn't happen in this instance, we're forced to rely now on the recollection of individuals as to what may or may not have been said or what did or did not occur.

In his testimony before the Silberman-Robb Commission and in interviews subsequent to publication of the commission's findings in early April 2005, Drumheller insisted that the news of the German lunch hit Langley like a small bombshell.

In an April 26, 2005, *L.A. Times* story, he was even more insistent that word of his meeting with the German had spread broadly through the Agency. He admitted not telling me personally, but he said, "Everyone in the chain of command knew exactly what was happening. . . . Literally inches and inches of documentation," including "dozens and dozens of e-mails and memos," would show that warnings had been sent to John McLaughlin's office and to WINPAC, and that Curve Ball's credibility had been seriously questioned in numerous meetings.

Drumheller has told the media in various interviews that he personally went to see John McLaughlin about the time of Colin Powell's UN speech to express concern about Curve Ball's information. He has said he doesn't remember John's exact response but that it was something to the effect of "Oh my, I hope that's not true." John is convinced that this did not happen. I have absolute confidence that had such a meeting taken place, John would have pursued the matter in the meticulous style for which he is well known. He fought steadfastly against White House attempts to stretch the evidence on Iraq–al-Qa'ida ties. He understood the importance being placed on Curve Ball's information, and he would have battled just as hard to keep Curve Ball's information out of the Powell speech had someone made the case to him that it posed problems.

If Drumheller or anyone had brought to John McLaughlin or me these doubts about Curve Ball's credibility, let alone his sanity, we would have gone to great lengths immediately to resolve the matter. Unfortunately, the first either of us learned of Tyler Drumheller's lunch with the German BND official and of the latter's supposed warnings—and his refusal to stand publicly behind them—was when we were interviewed by the Silberman-Robb Commission as it prepared its March 2005 report, two years too late to do a damn thing about it. Our senior officer in Germany at the time says Drumheller never apprised him of the luncheon conversation, nor did the Silberman-Robb Commission ever interview him. The German BND representative was asked by CIA officers in 2005 about his 2002 lunch with Drumheller. He denied ever having called Curve Ball a "fabricator" and said he only warned that he was a "single source" whose information the Germans could not independently verify.

A search of CIA records in 2005 revealed that a cable did come in to our headquarters from our rep in Germany on December 20, 2002. The cable went to Drumheller's office for action. It contained a letter addressed to me from the chief of the BND saying that Curve Ball would not agree to go public himself and that CIA would not be able to debrief him in person. It said that the Germans did not object to the public use of Curve Ball's information, as long as we protected the source. The letter went on to explain how the Germans had shared his information with at least two other foreign intelligence services and three U.S. intelligence agencies. It said they found his information "plausible" but that they could not independently verify what he was saying.

As far as I can tell, that cable never left Drumheller's desk in the European Division at Langley. Our senior officer in Berlin was expecting to get a response from me to my German counterpart, because he cabled and e-mailed our headquarters numerous times seeking one. That, too, would be standard protocol. But none was forthcoming. I had never seen the German letter but had simply been told that the German BND had cleared our use of the Curve Ball material.

On January 27, 2003, right before the Powell UN speech, our man in Germany sent another cable, this one expressing his own reservations about the source. He did so because he had received no response to his December 20 cable. Curve Ball's reporting was problematic, he said, and should be relied on only after "most serious consideration." This cable, too, went to Drumheller for action. In the three days and nights we sat at headquarters working on the secretary's speech, nobody ever told us of our senior man in Germany's reservations or of the letter from the BND chief.

Finally, frustrated at the lack of response to the December 20 cable, on the day of Colin Powell's UN speech, February 5, 2003, our Berlin rep translated the original letter from the BND chief and sent it, along with the original in German, via diplomatic pouch to headquarters. It arrived on February 26 and was delivered to Drumheller's European Division. My successor, Porter Goss, asked his staff to run down the Curve Ball story. They found in 2005 that the letter, located in the European Division, had not been formally logged in as received. Despite extensive searching, no records have been found that the letter was sent to either John McLaughlin or me.

Above and beyond the formalities, cables, and letters, though, were a number of critical break points—before, at the outset of, and during the Iraq war—when this information clearly was of vital importance. I did not believe that there could be any doubt among senior CIA officials at the time that the Agency was depending heavily on Curve Ball's information. Why so many opportunities to sound the alarm were missed is a mystery to me. Powell's UN speech was one such moment, but there were many others, such as when the National Intelligence Estimate was being written and approved. It was precisely during this time or just shortly afterward when Drumheller presumably had his revelatory lunch with the German. The issue could also have been mentioned when my staff was helping prepare my multiple testimonies before the Senate Intelligence, Foreign Relations, and Armed Services Committees. But it was not.

In May 2003, CIA and DIA issued a report following the discovery of a trailer found in Iraq that closely matched the one described by Curve Ball. We went back to the Germans, again through Drumheller's division, and had them show Curve Ball a spread of photos of trailers—much as you would display in a criminal lineup. Curve Ball picked out the picture of the trailer we found in Iraq and said, "That's it." Even then, neither Drumheller nor anyone else said to John or me, "Stop. This is a fabricator, you cannot rely on him."

In February of 2004 and in subsequent appearances before the Senate Intelligence Committee in closed session on March 4, 2004, I raised the subject of our concerns about Iraq's capability to produce biological weapons in the trailers cited by Curve Ball. Every presentation of the "evidence" for such a capability was vetted far and wide through the upper echelons of the Agency. Yet at no time did anyone in the analytic or operational chain of command come forward to tell me of the specific information supposedly imparted by the German BND to the CIA European Division chief in the fall of 2002.

In 2005 Drumheller told the Silberman-Robb Commission that he spoke with me on the telephone around midnight when I was in New York on the eve of Colin Powell's UN presentation in February 2003. In a *Frontline* special in 2006, Drumheller claimed that he said, "Boss . . . there's a lot of problems with that German reporting, you know that?" And that I replied, "Yeah, don't worry about it; we've got it." I remember no such midnight call or warning. Drumheller and I did speak very briefly earlier in the evening, but our conversation had nothing to do with Curve Ball; rather it involved getting clearance from the British to use some of their intelligence in the speech. According to a CIA memorandum for the record, in speaking to Senate Intelligence Committee staffers in 2005, Drumheller said that "way too much emphasis" was being placed on the phone call, and when asked if he could confirm that I understood what he was trying to convey in the purported phone call about Curve Ball, he responded, "No, not really."

Drumheller had dozens of opportunities before and after the Powell speech to raise the alarm with me, yet he failed to do so. A search of my calendar between February 5, 2003, the date of the Powell speech, and July 11, 2004, the date of my stepping down as DCI, shows that Drumheller was in my office twenty-two times. And yet he seems never to have thought that it might be worth telling the boss that he had reason to believe a central pillar in the case against Saddam might have been a mirage.

In fact, it seemed that just the opposite was communicated. In May 27, 2003, the head of the German BND, August Hanning, paid me a visit in Washington. My office received an e-mail from Drumheller's deputy, with a copy that went to Tyler, recommending that I be sure to thank Hanning for agreeing to allow us to use the Curve Ball material in our public discussions.

In advance of Hanning's visit, I received a memo laying out our goals for the session, a matter of course before every meeting with a foreign intelligence official. The memo was *signed by Tyler Drumheller.* The first page included a list of five suggested talking points to advance our goals. Number three, all in bold, suggests that I:

> *Thank Dr. Hanning for the Iraqi WMD information provided by the BND asset "Curve Ball." Inform Dr. Hanning that we would like to work with the BND to craft an approach to Curve Ball to secure his cooperation in locating evidence of Iraq's biological weapons (BW) programs, and about the direct involvement of Dr. Rihab Taha al-Azzawi in Iraq's mobile BW program.*

If the chief of the European Division believed that it was a mistake for us to use the Curve Ball material and knew that the Germans had warned us off it, why was he asking me to thank the Germans?

The meeting happened, and I presume I used the talking point that was suggested. In any case, Drumheller sat there through that meeting, and a lunch in Hanning's honor that followed, and never mentioned any concerns.

How can you explain these huge disconnects? Why would good men and women argue behind closed doors about Curve Ball's reliability, yet not come forward to express their concerns at an appropriate level? I've asked myself that question dozens of times. We were under enormous pressure to meet our own standards of excellence and from an administration that was moving toward war. But were we, as an institution, in some sort of meltdown? I don't believe that for a second.

The best reason I can come up with is that the people who knew that Curve Ball might be a fabricator figured that coming forward wouldn't make any difference. The rush to Baghdad wasn't going away. They would just be stepping in front of a roaring train. If that was their thinking, then their reticence is inexcusable.

But why would people be asserting things now about trying to alert me to the problems of Curve Ball—claims that have been proved untrue? Perhaps some people's recollections of "if only someone had listened to me" have become sharper than reality. I don't know. What I do know is that concerns about Curve Ball did not get disseminated far and wide through the Agency as they should have been. We allowed flawed information to be presented to Congress, the president, the United Nations, and the world. That never should have happened.

CHAPTER 21

Diplomacy by Other Means

The runaway freight train that was the war in Iraq arrived in March 2003. For CIA, this war was, in every respect, different from the one we had fought in Afghanistan. There, we had been, in military parlance, the "supported" command. In Iraq, we were "supporting." The difference is far more than semantic—it speaks to our performance in both theaters.

In Afghanistan, CIA largely came up with the plan. Indeed, we had been nurturing and refining the strategy for months before the attacks of September 11, hoping to get permission to go after al-Qa'ida in their sanctuary. With the help of a small number of Special Forces troops and overwhelming U.S. airpower, we had been able to marshal the strength of various warlords and tribal factions to oust the Taliban.

We told the administration from the very beginning that an entirely different model would have to be used for Iraq. Shortly after the Bush administration came to office, we briefed senior officials, particularly the vice president, that CIA covert action would almost certainly be unable to topple Saddam.

CIA came to this conclusion through painful experience in the mid-nineties. Our attempts to identify a Sunni military leader with the capability and following to take on Saddam's elite units proved difficult. Saddam regularly shuffled or even killed senior officers just for the sport of it, and this greatly increased the challenge of getting access to the right networks without being compromised. A combination of Saddam's ruthlessness and our own mistakes had resulted in scores of Iraqis in our employ being killed.

Covert action against Saddam in the past had not been as large or as well funded as our activities in Afghanistan against the Russians during the cold war. Some of our potential partners in the region had judged that we were not serious because of the paucity of resources devoted and because we had never committed ourselves to supporting covert action with military force. There was always the possibility that U.S. airpower might come into play once we had validated the feasibility of a potential overthrow of Saddam. In practice, the execution of such a plan was extremely difficult and unlikely.

What we learned in Afghanistan was that covert action, effectively coupled with a larger military plan, could succeed. What we were telling the vice president that day was that CIA could not go it alone in toppling Saddam; all instruments of U.S. power had to be aligned to achieve the objective. Some may have believed that by saying so, we in essence were saying that we were more than willing to hold the military's coat, thus making war inevitable. In truth, we were simply conveying the reality of our historical experience.

Thus in Iraq, unlike in Afghanistan, CIA's role was to provide information to the military about the whereabouts and capabilities of enemy forces, assess the political environment, coordinate the efforts of indigenous networks of supporters who paved the way for U.S. military advances, and conduct sabotage operations and the like. That's a more traditional role for intelligence to play, but none of it came easily.

The first action the Agency undertook in February 2002 was to resurrect the Northern Iraq Liaison Element (NILE) teams of CIA officers that had historically encamped with the Kurds in northern Iraq. Arriving in early July, they began the painstaking effort of recruiting agents, creating networks of people and tribes not only willing to gather data but also to take action. We wanted them to take aggressive actions to challenge the legitimacy of the regime wherever they could, sabotage railheads, disrupt communications nodes, attack local Ba'ath Party headquarters, and communicate their actions with the military to maximize their effectiveness.

We operated out of northern Iraq and over the borders of neighboring countries to the south and west. We gave the military full transparency to the contacts we were making, introducing U.S. Special Forces to individuals inside Iraq who held some promise in persuading military units to defect once a ground war started, either by switching sides or surrendering. In the end, while few if any units did defect, neither did many regular army units fight. And changing sides was not a very appealing option for them. Regular army forces often had Iraqi Republican Guard units behind them. They faced likely death if they advanced on U.S. military units in front of them, and almost certain death if Saddam's Special Forces to the rear felt they were not supporting the regime.

That left us with encouraging surrender, and our case officers worked with clandestine sources to deliver that message to the Iraqi army. But not too long before the war got under way, this choice was taken off the table. The reason was quite simple. The U.S. had so few forces on the ground that a successful campaign to induce capitulation would have quickly resulted in the prisoners of war outnumbering the invading army.

The fallback position was to suggest to the Iraqi military units that they simply lay down their arms and go home. The U.S. military started air-dropping leaflets bearing that message, and Iraqi soldiers took it to heart, walking away in large numbers once the shooting got under way. (Later, when he was trying to justify his controversial May 23, 2003, edict disbanding the Iraqi army, Jerry Bremer would say that the army had already disbanded itself. True enough, but the Iraqi army did so largely at the behest of the U.S. government, and certainly not in the expectation that its soldiers would be cut adrift, taking their weapons with them, often with no means to support their families.)

I visited CIA officers at several secret bases in the desert west and south of Iraq just prior to the war. The bases had been created in the middle of nowhere in large part to train and equip Iraqi tribal networks so they could reenter their country to conduct surveillance and sabotage, and send back data to the U.S. military. The officers I met with had been living in tents for months preparing for the war, and they were eager to get started. Many of them were young—quite a few were on their first tours of duty—and I was the only DCI they had ever served under. My visit was intended to give them a morale boost and to let them know that I was very proud of them and confident in their ability to meet any challenge. Privately, though, I could not help but worry that many of these young men and women could soon die.

At one of these visits, I met with a contingent led by Gen. Mohammed Abdullah Shawani, who had been chief of Iraqi Special Forces during the Iran-Iraq war. General Shawani was introduced to the Agency in 1991, quickly becoming one of the U.S. government's most critical partners in working against Saddam's regime. A physically imposing figure with the size and strength of a football offensive lineman, Shawani was a born leader with a significant following within the traditional and Special Operations elements of the regular Iraqi army. A special operator and pilot by training, he gained fame and the highest Iraqi military honors when he led a heliborne attack against an Iranian-occupied hilltop during the Iran-Iraq war.

Shawani, or "the General," as he was known to his Iraqi followers, quickly became key to developing a strong network inside Iraq for the Agency. Unfortunately, the network was compromised by Saddam's security services in the mid-1990s, resulting in the torture and execution of Shawani's three sons. Shawani continued to work tirelessly to develop agent networks within Iraq and assisted the Agency in contacting Iraqi tribal and religious leaders in the months leading up to the invasion in the spring of 2003. During the prosecution of the war, Shawani helped develop and lead the Agency-sponsored Iraqi paramilitary group known as "the Scorpions." Such was his following in the regular Iraqi military that when Shawani went to talk to a large group of Iraqi soldiers being held prisoner in Kuwait, he was immediately recognized by a number of the senior Iraqi officers, who stood at attention and saluted him.

Thanks to Shawani and many others, in the hours and days before the war got under way, CIA teams were able to slip into Iraq and meet up with established networks to try to prevent the Iraqi military from destroying the bridges crossing the Euphrates and leading into Baghdad. Others met up with agents working to prevent Saddam from torching the southern oil fields.

As war approached, our designated senior officer for Baghdad, "Charlie S.," moved to Doha, where he sat at Gen. Tommy Franks's side. Charlie became an important member of the military team. He was constantly providing information from our networks of sources about potential military targets. Sometimes he would even give advice *not* to bomb. An example: When Central Command (CENTCOM) learned where a senior Iraqi intelligence officer was hiding, the military's first reaction was to target a Tomahawk cruise missile on his coordinates. However, as our liaison to Tommy Franks's headquarters, Charlie convinced his military counterparts of the intelligence value of taking this Iraqi officer alive. Though difficult, the effort to get ground forces to this officer's location proved to be very worthwhile based on the information subsequently obtained from him.

The Northern Iraq Liaison Element (NILE) teams operated continuously in northern Iraq after July 2002, working under extremely arduous conditions, far from any military support, and in constant danger from Saddam's security forces. Nonetheless, they produced some extraordinary successes. They managed to recruit whole networks of Iraqi agents dedicated to helping us overthrow Saddam's regime.

One group of Iraqis, united by religious affiliations, was particularly important. Once we were able to convince the group's leaders that this time the United States was serious about getting rid of Saddam, and, not coincidentally, once we provided their leaders a couple million dollars to demonstrate our resolve, they began to produce highly actionable intelligence. The group secretly brought in four Iraqi military officers a week to be debriefed by the CIA NILE team. The head of the religious sect, someone our guys referred to as "the Pope," sat in on the sessions. Often those being debriefed refused to answer some questions, saying that what we were asking for was "too sensitive." Each time, "the Pope" would interrupt. "You will answer the question!" he would instruct, and they would obey. Every military officer we debriefed told us that Saddam did indeed possess WMD.

One early windfall came when a member of the group handed us a CD-ROM that was essentially a personnel roster of Saddam's Special Security Organization. We cross-checked the list against some of the names we already knew. It proved legitimate and enabled us to identify and expose several double agents that the Iraqi intelligence apparatus was trying to infiltrate into our midst.

Other members of this network gave us the locations of Iraqi missile emplacements and would tell us precisely when the batteries would be tested. Using U.S. reconnaissance aircraft, we were able to validate that the information we were given was accurate. The missiles were precisely where our sources told us we would find them. As a result, the U.S. military was able to make short work of eliminating Saddam's surface-to-air missiles when the shooting war started.

In the run-up to the war, the United States had promised to deliver a large amount of weapons to the two main Kurdish factions in northern Iraq (the PUK and the KDP) so that they could effectively join in the coming fight. Obtaining the weapons was not a problem for us, but getting them there was another matter. The Turks refused to allow the weapons to transit their country.

CIA then chartered several large transport aircraft but kept getting turned down when we requested overflight rights from all the neighboring countries. The Kurds were exasperated at the delay. "Where are the weapons you promised us?" they asked over and over again. We had no satisfactory answer. Finally, in

February 2003, about a month before the start of the war, Tom S., the head of our NILE team in Suleimaniya, was told by the local PUK representative, "Never mind." He was stunned to watch as trucks rolled up to a warehouse only fifty feet from his base and tons of weapons were delivered to the Kurds by the Iranian Revolutionary Guard Corps (IRGC).

The NILE team stayed in close contact with Langley, passing back to headquarters hundreds of intelligence reports. In turn, they would be kept apprised of what was going on in Washington. In one conversation, operations officers in Washington told the field of a major development back home. The Starbucks at CIA headquarters had just switched over to a twenty-four-hour-a-day operation. Agency officers in the field speculated that this move signaled an imminent start to the war, and they were right.

Operation Iraqi Freedom began a little bit earlier than we anticipated because of a tip-off from one of the NILE's best sources on the possible whereabouts of Public Enemy Number One: Saddam Hussein. Some of the group's members were involved in providing communications for top Iraqi officials, including Saddam. A status board in the regime's communications headquarters showed green lights when Iraqi networks were functioning correctly and red lights when they were not. Generally, the lights were green. Our source noticed that Saddam's security forces always cut off communications in the areas where he was about to travel—presumably to prevent disloyal military personnel from revealing his whereabouts to enemies.

Temporarily cutting off communication, however, caused red lights to go on near Saddam's intended destination. Over time our source was able to confirm his suspicions. The red lights would go on, and he would later learn that Saddam had been at that location. Saddam would leave and the green lights would return. Thanks to this glitch in the system, the status board was basically broadcasting Saddam's whereabouts.

Two days before a U.S. deadline for Iraqi compliance was to

expire, our source got wind of a possible meeting that evening at Dora Farms, an estate owned by Saddam's wife. Although it was unclear who was going to be present, indications were that Saddam's sons and perhaps the entire family might be planning a meeting there, presumably to discuss what might happen should the United States invade.

Our source relayed this news to the NILE team in northern Iraq. They immediately flashed the news back to headquarters and to the CIA liaison with Gen. Tommy Franks, in Doha, Qatar.

The next morning, March 19, the CIA officer briefed Franks on the previous evening's intelligence. Later that evening, our source rang in again. The red lights on the status board were once more showcasing Dora Farms. Odds were that the Iraqi leader would be going there again that evening. Other human sources involved in providing security near Dora Farms had also heard that a major meeting of Saddam's family might happen at the farm that night.

At that point, we ordered U.S. overhead reconnaissance to examine the site closely. What we saw was a large contingent of security vehicles, precisely the kinds that would typically precede and accompany Saddam's movements, hidden under trees at the farm.

It just seemed too good a scenario to pass up, so I called Don Rumsfeld and asked if we could come brief him right away on something potentially significant. He said yes, by all means. I gathered John McLaughlin and the head of our Iraq Operations Group and we made our way to the garage. En route we ran into Steve Kappes, the number-two in our Directorate of Operations. "Come with us," I shouted, and we dragged him into the elevator. We whisked Steve into our armored SUV and roared off the CIA compound before he could find out where we were going or why.

When we got to the Pentagon, we were immediately ushered into Rumsfeld's spacious office. We quickly laid out the facts for him. He understood the importance in an instant and said, "We've got to take this downtown." Seconds later he was on the phone and had arranged for all of us to see the president right away. Back in the SUV, we sped off to the White House, but Rumsfeld's limo and accompanying security vehicles handily beat us there.

We got right in to see the president. The vice president, Andy Card, and Condi Rice were already there, and before long the president asked that Colin Powell join us. The chairman of the Joint Chiefs, Dick Myers, had come along with Secretary Rumsfeld.

In the private dining room just off of the Oval Office, we rolled out some maps and briefed the president on the intelligence we had. We were honest with him about the limits of our knowledge. We thought the information was pretty good—very good, as these things go—but we could not guarantee that the information was not wrong. Nor could we swear that this wasn't a trick, or prove that Saddam hadn't moved an orphanage onto the site to set us up for a PR disaster. Ultimately, deciding to strike would not be an easy call.

We told the president that we were unlikely to get any additional information to help him make the decision. Then, moments later, we got more. A source providing security on the scene had gotten another call out. He said that there were rumors among his colleagues that Saddam himself might show up between 3:00 and 3:30 A.M., Baghdad time.

Soon another report came in. Real-time intelligence reports arriving in the middle of a crisis happen all the time in Hollywood, but this is highly unusual in reality. The chief of our Iraq Operations Group was called away from the Oval Office to take a secure telephone call at the desk of the president's scheduler. The latest information said that whoever was going to be there would be in a *malja*—an Arabic word that could mean "basement" or "place of refuge." If it were a bunker, cruise missiles would not be able to penetrate it. That meant that manned bombers would be required as well.

Clearly, we were on the brink of a momentous decision. What wasn't clear was whether it would be a good or bad one. F-117 bombers would have to be employed before Iraq's air defenses could be neutralized. The air crews would have to rely on stealth and surprise to survive such a mission. This upped the stakes.

The president took all the information in and polled those of us present for our thoughts. You could see him moving from the information-gathering mode to the decision-making mode. Then he moved behind his Oval Office desk and ordered that the strike go ahead. Karen Hughes and Dan Bartlett started crafting remarks for a presidential announcement a few hours hence, announcing the strike and the fact that the war had commenced.

Back in Doha, Central Command was putting together a strike package. Cruise missiles had to be launched hours ahead of the desired time of impact. Meanwhile, targets were being passed to F-117s already aloft and carrying bunker-busting bombs. Dora Farms was a large complex with a number of buildings. Tommy Franks made a decision to take off the targeting list the villa that was associated with Saddam's wife. He was concerned that the building would be full of women and children and he didn't want to increase the likelihood of unintentional collateral damage. We anxiously waited for the results of the attack, hoping that through some miracle the war might be concluded with a minimum loss of life or destruction.

Several hours later, some forty cruise missiles and a number of bombs from the F-117s smashed into the facility. Before long, the first intelligence reports from the scene started coming in. One of our sources was killed in the attack, and two others escaped and deserted their military units. (Their wives were later reportedly tortured by Saddam's henchmen.) As daylight broke in Baghdad, another of our sources reported to us that he had spotted someone who looked to be Saddam being pulled from the rubble, looking blue. That person, he said, was loaded into an ambulance and spirited away. For several hours we had reason to hope that our goal of regime change might have happened in the first seconds of the war.

Unfortunately, it was not to be. The next morning we brought to the Oval Office overhead imagery of Dora Farms. It was clear that a large villa on the compound was still intact. Had Saddam and his sons escaped death in the one building scratched from the target list? We were told after the fact that there had been a meeting of senior Ba'ath Party officials at Dora Farms that evening, but apparently, despite the red lights on the status board, Saddam was not among those who attended. We were confident that the technical source was telling us what he believed to be true. The second source who reported having seen Saddam being pulled from the rubble, however, was probably embellishing his story. When Agency officers were able to reach Dora Farms a few weeks later, they determined that the source could not have seen what he reported from his vantage point.

Given the same information and the same circumstances, I still would have recommended to the president that he authorize the strike. As to how history might have changed had we been able to remove Saddam on the first night of the war, all we have to go on are questions. How many lives might have been saved? How much damage would have been averted? Without Saddam lingering in the shadows, would the conditions that spawned an insurgency have flourished? We will never know. We do know that many Iraqi military members told us that they would never work with us as long as Saddam was alive because they feared his coming back to power more than they feared the United States.

Since the long-shot "regime decapitation" failed, the invasion of Iraq proceeded as planned. Inside CENTCOM headquarters and at CIA, plasma screens called "Blue Force Trackers" showed the positions not only of U.S. and allied military units, but also of CIA officers in the field and of the Iraqi sources who were feeding real-time intelligence to the war fighters. Constantly updated, these screens helped prevent attacking U.S. military forces from accidentally targeting our own forward-deployed personnel.

One of our prewar objectives in the south had been to get two Iraqi divisions opposing us out of the fight. Up to 90 percent of these divisions were populated with Shia. One Iraqi Shia whom we had recruited to conduct sabotage operations was a veteran of the first Gulf War and had many contacts in these Iraqi divisions. Through smuggling networks, we sent in money and phones for him to reach out to relatives and members of his tribe. The military gave us permission to tell these divisions that the United States would provide an unmistakable sign that hostilities were about to commence. When they saw it, they were instructed to change out of their uniforms and go home.

The sign was indeed unmistakable. Napalm and artillery were fired on top of Mount Jebel Sinam in southern Iraq. As U.S. forces drove through the foxholes and pillboxes of Iraqi divisions, they found weapons, equipment, and uniforms left behind. Any resistance encountered in Nasiriyah came from the Fedayeen Saddam, a group of Ba'athist thugs loyal to Saddam. We had not counted on the Fedayeen being as strong as it was. Our own Iraqi sources and contacts had dismissed them as an ineffective fighting force.

The invasion was a huge initial success. Iraqi military resistance melted, the regime dispersed, oil fields stayed largely undamaged. But as U.S. and allied forces streaked toward and into Baghdad, a giant sucking sound could be heard in their wake. Clearly, the Coalition lacked adequate troop strength to secure the flanks of the attacking forces. The hope had been that the speed of the advance and the "shock and awe" of the strike would render enemy forces docile, and that, freed from the yoke of oppression, the Iraqis would allow peace and stability to break out. The reality was somewhat different.

Some of our intelligence networks—scores of human assets in key locations—were reporting to us that the war was not having that much of an impact on the average Iraqi. In some ways, U.S. military precision was too good. Air strikes were so carefully targeted that Iraqi citizens took to referring to it as the "Disney war"—a lot of noise and lights but nothing that was having a significant impact. Indeed, until U.S. troops showed up in Baghdad, many Iraqis did not believe a full-scale invasion was actually under way.

An old axiom holds that no military plan survives its first contact with the enemy. Parts of this U.S. plan, though, unraveled long before that. Many months in advance of the start of the war, a U.S. Army colonel visited CIA headquarters and told our Iraq Operations Group staff that he had been charged with putting together a fighting force of Iraqi exiles—something he called the Iraqi Freedom Force. The plan, this colonel said, was to train and equip a full division, about fifteen thousand men. Some of our more seasoned Iraqi hands told him that this was fantasy, that he would be lucky if he could get a thousand men. No, we were assured, a force of twelve to fifteen thousand was entirely doable if the United States focused on it, and for that the colonel offered no less an authority than Ahmed Chalabi.

One of the most controversial characters in the Iraq drama, Chalabi was an émigré whose family had left Iraq in 1958, when he was just a boy. He grew up in Great Britain and the United States. Chalabi had almost no following in Iraq but quite a large one among some circles in the U.S. government. An extremely bright man with a Ph.D. in mathematics from the University of Chicago, Chalabi is slick, charming, and talks a great game. In the late 1980s he was tried and convicted in absentia of bank fraud in Jordan. Following the first Gulf War, he was instrumental in creating, with CIA assistance, the Iraqi National Congress. But in the ensuing years, CIA found him to be a most unreliable partner. Although CIA came to take everything we heard from Chalabi with a healthy dose of skepticism, others, such as the vice president, Paul Wolfowitz, and Doug Feith, welcomed his views.

Agency officers again suggested caution to the colonel. Many people will tell you they will sign up for such an adventure, but when it comes down to leaving their comfortable homes in Europe, elsewhere in the Middle East, and in the United States, the reality will be quite different. The colonel, however, would not be dissuaded, and so the INC reportedly started distributing in the mosques of Europe applications for joining the Iraqi Freedom Force. The response was even worse than we had predicted; only a handful of people signed up. By the fall of 2002, Agency officers suggested to DOD that they scrap the idea of a fighting force of Iraqi exiles and focus instead on identifying a reasonable number of people—perhaps twenty-five—who could do something useful, such as serve as translators or interpreters. We were scoffed at once again. By the time the war started, what had once been envisioned as a division amounted to seventy-seven poorly trained individuals.

I thought we had heard the end of them, but we had not. On Friday, April 5, 2003, I was stunned to learn that the U.S. military had airlifted into southern Iraq hundreds of members of the Iraqi Freedom Force, led by Ahmed Chalabi. I was attending an NSC Principals Committee meeting when someone simply informed us that Chalabi had landed in Nasiriya, 230 miles south of Baghdad. If there had been any discussion of the wisdom of introducing Chalabi and his contingent into the ongoing fight, it had not been conducted within my earshot or that of any of my senior personnel. Long after I left office I heard that Chalabi had been lobbying senior Central Command generals to transport him and his supporters into the war zone so that they could legitimize themselves. Senior CENTCOM officials turned down this request on the night of April 4. When they woke up April 5, they found that their orders had apparently been countermanded by Paul Wolfowitz at the Pentagon.

Just as mysterious as how Chalabi had gotten there was the question of where the troops had come from. According to the press accounts, Chalabi's meager band of seventy-seven would-be warriors suddenly numbered "hundreds" of fighters. We later learned that he had paid many former Badr Corps members to swell his ranks. (The Badr Corps was created by former Iraqi Shia military men who had defected during the

Iran-Iraq war in the 1980s and had been operating as a militia in Iran with the support and backing of Tehran.) As a fighting force, the IFF proved to be totally feckless. Some of its members, however, evolved into a private militia for Chalabi, and set about commandeering property, vehicles, and wealth for the use of his Iraqi National Congress. We weren't the only ones bewildered by the arrival of Chalabi's small private army. At the time, one Iraqi asked a senior CIA official a pertinent question: "I thought Chalabi ran a political party? In the United States do your political parties have their own militias?"

Despite such distractions, the plan to take Baghdad was executed with precision. The men and women of the U.S. military, their allies, and our intelligence officers deserve huge credit for their skill, courage, compassion, and restraint. CIA teams entered Baghdad by April 7. On the eighth, Saddam's government essentially ceased to exist. On a scale of one to ten, the plan to capture the country scored at least an eight. Unfortunately the plan for "the day after" charitably was a two. The war, in short, went great, but peace was hell.

CHAPTER 22

The Hunt for WMD

Sometime around the end of May, shortly after declaring an end to major combat operations in Iraq, I was with President Bush in the Oval Office when he described a meeting he had recently had with Jerry Bremer and Tommy Franks. The president said he had asked them who was in charge of the hunt for weapons of mass destruction. "They went . . ." the president said, then took his two index fingers and pointed left and right, suggesting that both Bremer and Franks pointed at each other. Not a good sign. The president looked at me and said, "As a result, you are now in charge, George."

The Pentagon was still calling the shots in Iraq—that hadn't changed—but it already had enough to do on the ground and was more than happy to see CIA shoulder the responsibility of the WMD hunt. Logistically, this was a little tricky. Military personnel would have to do the lion's share of the actual searching and provide almost all of the physical security for those engaging in the mission. To get around that hurdle, we carefully negotiated a memorandum of understanding with DOD, spelling out how a senior advisor appointed by me would work with, but not command, what was called the Iraq Survey Group (ISG), which would stay technically under the command of a two-star general reporting to the secretary of defense. For those who haven't lived and worked inside the Beltway, such issues might seem minor or arcane, but sorting out lines of authority and chains of command can be some of the most difficult tasks to handle inside a bureaucracy.

The size of the WMD-hunt would prove mammoth. Iraq had

130 known ammunition depot sites, two of them roughly equal to the square mileage of Manhattan. As many as 1,400 people were attached to the ISG at any single time, mostly Americans but also Brits and Australians. People often cite that number with disapproval—that so many people were dedicated to this mission in a war zone. In truth, the actual commitment was much smaller. The size of the ISG varied considerably over the months, and most of its personnel were engaged in support activities—logistics, security, and admin—for the between 100 and 200 core specialists trained to collect and analyze information related to WMD.

A lot of time had been lost. The major fighting in Iraq had been over for two months, and we were only now really getting organized to look for the WMD that the U.S. government had cited as a primary justification for having gone to war. In that time, Iraqis had been deliberately destroying records, other potential evidence was being carted off by looters, and still more Iraqi government files were being seized by the truckload by groups such as the Iraqi National Congress (INC)—raising questions about the validity of any information that might later be discovered in those documents.

As we were grappling with how to organize and conduct such a search, and with finding someone to lead it, David Kay visited CIA headquarters to read a paper and consult with someone on the National Intelligence Council. At the time, his appearance seemed a gift from heaven, but appearances can sometimes be deceiving.

Kay, a former UN weapons inspector, had just returned from Iraq, where he had served as a consultant for NBC. While he was there, a trailer was found near Mosul in northern Iraq in late April that looked remarkably like the mobile biological weapons facilities featured in Colin Powell's UN speech and in our NIE. Kay was interviewed on *NBC Nightly News* on May 11, 2003. Crawling around in the trailer and explaining to the reporter how it supposedly worked, Kay said that after personally examining the vehicle, he was sure "there could be no other use" for it other than to produce biological weapons. He expressed this view again on June 8 on CNN, saying that alternative theories "did not pass the laugh test," including the idea that the trailers might have been designed to produce hydrogen for meteorological balloons (ironically, the use judged most likely by Kay's successor, Charles Duelfer, a year or so later). Kay could not have appeared more certain, and his confidence seemed to recommend him as an expert who could sort through all of this.

Several days later, John McLaughlin and I met with Kay in my office. He shared with us his impressions of the environment in Iraq and the likelihood that we would eventually find the WMD that all of us expected to be there. I realize now that Kay's public statements and testimony before the war had actually been more confident than even the most assertive statements in our NIE, but back then, all I was certain of was that (a) he talked a good game and had previous experience in Iraq, and (b) we needed to move quickly. Kay was appointed my senior advisor on June 11, and headed out to the region a few weeks later, after getting briefed in Washington.

Our instructions to Kay were simple. Find the truth. We promised him the resources he needed and an absence of interference from the home front. I am confident that we delivered both.

Kay apparently had the impression that coming to a resolution on the presence of WMD was not going to be as difficult a task as it turned out to be. But Saddam had been playing cat-and-mouse with his weapons programs for more than a decade. That should have been warning enough. Worse, the deteriorating security conditions in Iraq made searching for anything almost a life-and-death struggle. On arriving in Iraq, Kay set up shop inside the heavily protected Green Zone, in central Baghdad. The majority of his troops, meanwhile, were based on the outskirts of town, at the far more combustible Baghdad International Airport.

One of the first things we did when Kay signed on was to streamline Washington's role in managing the process. While the hunt was still in DOD hands, there had been multiple meetings, phone calls, and video conferences on the issue. We cut this back to one weekly secure video conference with Kay and his team in Baghdad and occasional e-mail exchanges. We wanted to get out of the way and let the experts do their jobs. I attended many of these weekly video gatherings but let John McLaughlin preside most of the time. Kay and his team would report on their activities and needs, and we would do our best to provide what they needed or to sort out problems on the Washington end.

Three months after arriving in Iraq, Kay returned to the United States to deliver an interim report to Congress. He prepared this report entirely on his own, and John McLaughlin stressed that Kay was to have the final word on everything in it. We protected Kay's independence fiercely. Of course, the White House was intensely interested in what Kay would say. But McLaughlin did not let anyone there see Kay's report until the morning it was delivered—not because we feared the White House would try to change it; we simply wanted to be able to say unequivocally that no policy official had even had the *opportunity* to tinker with it.

In Kay's October 2 testimony before Congress, he described how Iraq had intentionally misled United Nations inspectors prior to the war. He stated that the ISG had discovered evidence of Saddam Hussein's intent to develop WMD and of his having retained some capacity to do so. Kay told reporters that it might take an additional "six to nine months" of searching to reach more definitive conclusions.

Discovered were dozens of activities related to a WMD program as well as significant amounts of equipment. He also talked about finding a clandestine network of laboratories and safe houses run by the old Iraqi intelligence service. These facilities contained equipment for continuing research into chemical and biological warfare. Strains of organisms were found concealed in a scientist's home, at least one of which could have been used to produce biological weapons.

On the nuclear front, documents and equipment useful in

resuming uranium enrichment by centrifuge and electromagnetic isotope separation had been found buried outside scientists' homes and elsewhere. Just as alarming, the ISG had found plans and advanced design work for new long-range missiles with ranges up to at least 1,000 kilometers—well beyond the 150-kilometer range limit imposed by the UN. Missiles of that range would have allowed Iraq to threaten targets throughout the Middle East, including Ankara, Cairo, and Abu Dhabi. The ISG had also uncovered evidence of clandestine Iraqi attempts between late 1999 and 2002 to obtain prohibited North Korean ballistic missile technology.

Collectively, Kay's interim testimony was a damning portrait of deception and dissembling by a man capable of horrendous acts. Yet in the resulting headlines, the press stressed only what Kay had *not* found—stockpiles of WMD. I recall Kay expressing frustration at this—he thought that any of the things he *had* found would have been headlines had they been known before the war.

None of it, however, was the "smoking gun" that would justify our NIE estimates and validate the allegations in Powell's UN speech. Much of the media focus on Kay's testimony regarded his still being unable to come to any final resolution of the purpose of that mobile biological weapons trailer—the one he had already told *NBC Nightly News* and CNN had no possible function other than biological weapons production.

Shortly thereafter, Kay returned to Baghdad to resume the weapons search. In his absence, the already shaky security situation there had deteriorated considerably. On October 9, a suicide bomber drove his car into a group of Baghdad policemen, killing nine and injuring forty-five. Three days later, a bombing outside a Baghdad hotel used by senior Coalition officials killed at least eight. On October 27, the first day of Ramadan, four coordinated suicide attacks against three other Baghdad police stations and the Islamic Red Crescent killed forty-three people and wounded more than two hundred. Six days later, on November 2, sixteen

U.S. soldiers were killed and twenty-one injured when a helicopter was shot down. November would go on to be the bloodiest month up to that point for U.S. military personnel, with seventy-five dead.

Central Command generals were scrambling to try to find out where the attacks were coming from and to figure out ways to stop them. Not surprisingly, they looked to the Iraq Survey Group as a resource for analysts who might help stop the bleeding. The request wasn't large—Central Command was seeking the temporary loan of a handful of area experts, in the single digits—but Kay objected.

A senior military officer later told me of a conversation he had with Kay. The official was "flabbergasted," he said, when Kay refused to lend some of the ISG's experienced intelligence analysts to help him find insurgents *"that are killing us."* Kay said he could not afford to do that because it would "destroy his operation." He didn't want any assets pulled away from the weapons hunt, despite the fact that the insurgency was making the mission of his troops nearly impossible to complete.

After weighing many competing demands, John McLaughlin managed to identify a few other intelligence community personnel who could be sent to Iraq to replace anyone diverted from the ISG staff. Still, Kay could not be placated. On our periodic video conferences with him, he became obstreperous, claiming that he was not getting the support he needed to do his job. In one call with McLaughlin, Kay said that he would not "stake his name and reputation" on this mission unless he got everything he wanted.

Had he been a regular CIA officer, I would have relieved Kay of his command and ordered him home. American servicemen and women were dying; Gen. John Abizaid needed help. Instead, McLaughlin made a visit to Iraq in November and met with Kay and the ISG leadership. McLaughlin expected at least small thanks for the difficult choices that were being made to divert people to Kay's mission, but instead he found Kay quite brusque, and insistent that his needs be met.

On November 19, a month and a half after Kay had told the media his mission would require six to nine more months, I learned from the rumor mill he was planning on quitting that very day. I called him, and he confirmed the rumor. He didn't provide any rationale for wanting to abandon his post other than expressing a general unhappiness that anyone in the ISG might be asked to help control the increasingly deadly insurgency. His threat to resign would occur on the day President Bush was arriving in London for a state visit to the United Kingdom. "No," I said in reply. "I won't allow you to embarrass the president in that way." I reminded Kay, without much effect, of the extraordinary support and assets he had been provided and the importance of coming to some final resolution on WMD. "Look David," I said, "why don't you come home for the holidays, take some time off to think about continuing the job?" He agreed to do so, but when he left Baghdad, his colleagues couldn't help but notice that he cleaned out the trailer he was living in and took home all his personal effects.

While Kay spent much of December decompressing, Maj. Gen. Keith Dayton, the senior military officer who headed the Iraq Survey Group, led the search for WMD. Then, sometime around Christmas, Kay informed us that his mind was made up and that he was not going back. I asked that he withhold any public announcement of his departure until we could identify a suitable replacement for Kay.

John McLaughlin undertook the effort to find a replacement for Kay. We developed a list of about five candidates and began checking them out. As McLaughlin gathered recommendations from various proliferation experts, one name kept coming up—Charles Duelfer, a former UN weapons inspector with a wealth of experience on the ground in Iraq and inside the UN, where he had served as the deputy chief of the weapons inspection effort.

Duelfer had the reputation for being iron-willed and dogged about his work. One person McLaughlin talked to cautioned us that Duelfer had a strong independent streak and was no slave to bureaucracy. But that was exactly what we were looking for. Although the security situation had continued to slide downhill, we felt certain Duelfer wouldn't be intimidated. After all, the guy does free-fall skydiving for fun. And from his past lengthy time on the ground in Iraq, he knew the country, its culture, and, most important, many of its leaders—a depth of knowledge that would prove invaluable in the months ahead.

We prepared a press announcement of Kay's departure and Duelfer's hiring. As is the form in such matters, I said some nice things about the individual departing. No matter what my personal feelings about him, the man had given up six months of his life to live in Baghdad, and he deserved our thanks. Our press office coordinated the statement with Kay, including a quote from him about there being many unresolved issues for the ISG to pursue.

In a final meeting in my office, with John McLaughlin present, Kay said that he was going to leave "quietly and like a gentleman." We invited him to stay on for the swearing in of Duelfer later that morning, but Kay said he had to go. Within forty-five minutes of leaving the CIA headquarters compound, Kay was being quoted by Reuters intelligence correspondent Toby Zakaria as saying that he concluded that "there were no Iraqi (WMD) stockpiles to be found." Although this later proved to be correct, it was quite a change from his comments just weeks before that it would take another six to nine months to know for sure. As for his promise to go quietly and allow his successor to finish the job, I can only say that I greatly regret the manner of Kay's departure.

Five days later, on January 28, 2004, Kay testified before the Senate Armed Services Committee, carried live on all the cable networks. He started by saying that "we were almost all wrong." Kay also inserted the familiar theme that the ISG needed more resources, ones that would be devoted entirely to the WMD hunt.

Why we would need more resources to hunt for weapons that he had concluded were not present went unexplained.

Kay proceeded to describe the global significance of what he hadn't found in Iraq and even added some commentary about how wrong the United States had been about Libya and North Korea, two accounts about which he was not briefed and about which he was spectacularly misinformed. Kay ended his opening statement by saying,

> *And let me just conclude by my own personal tribute, both to the president and to George Tenet, for having the courage to select me to do this, and my successor, Charlie Duelfer, as well. Both of us are known for probably, at times, a regrettable streak of independence. I came not from within the administration, and it was clear—and clear in our discussions, and no one asked otherwise—that I would lead this the way I thought best, and I would speak the truth as we found it. I have had absolutely no pressure, prior, during the course of the work at the ISG or after I left, to do anything otherwise.*

I mention the above quote not for the supposed "tribute" he gave me, but because several months later Kay would start telling people that he had concluded Iraq had no WMD before he left his post but had not been allowed to say what he thought. Apparently, this was part of an ongoing revision of his own recent performance because after he returned to the private sector, Kay stopped giving me any tributes altogether and became instead my long-distance psychoanalyst.

In a widely reported interview taped for PBS's *Frontline*, Kay said that "George Tenet wanted to be a player . . . and if you didn't give the policy makers what they wanted . . . your views wouldn't be taken and you wouldn't be invited into the closed meetings." He concluded that I had "traded integrity for access, and that's a bad bargain any time in life. It's particularly a bad bargain if you're running an intelligence agency."

Ringing allegations. Great TV drama. And as wrong as any words can be. Never did I give policy makers information that I knew to be bad. We said what we said about WMD because we believed it.

In October 2004, I ran into David at a conference hosted by Ted Forstmann in Aspen, Colorado. Sir Richard Dearlove and I had appeared on a panel together moderated by Charlie Rose. One of the topics we discussed was how our respective intelligence communities had reached their judgments regarding Iraq and WMD. David approached me afterward and said, "You know, we are not much in disagreement on the substance." I looked at him and said, "There is one big difference: you have made this personal." Appearing on PBS, he had talked about my meetings and interactions with senior policy makers that he had never attended. He did not have a shred of evidence to back up his allegation.

Although Kay expressed the view that the WMD job was almost over, nobody in Baghdad believed it. He didn't deliver the evidence needed to make that case persuasively and in a definitive way that would put the issue to rest. It was never enough merely to cite Kay's *opinion* that there were no WMD and that the job was done. Why? Because to close this chapter of history in a responsible way, we needed hard data, lots of it, organized and presented in a manner that would give future policy makers and historians confidence that we had gone about this thoroughly and professionally. We also wanted our own analysts to have the data necessary to understand what went wrong and what lessons should be drawn from it.

That is what Charles Duelfer delivered. When he arrived in Baghdad early in 2004, Duelfer installed himself at the ISG's airport headquarters, not downtown in the relative safety of the Green Zone, and then set about putting his own stamp on the WMD search. A number of the analysts had been working on draft chapters for a possible next report, but Duelfer put that effort on hold. He told the staff he didn't want to buy into any further interim conclusions unless he personally had had an opportunity to understand the underlying information. In particular, he wasn't going to make incremental decisions on important issues such as mobile biological weapons trailers.

We resumed with Duelfer the weekly secure video conferences that we had held with Kay. Duelfer would bring a varying set of participants to the meetings on the Baghdad end, and he kept us carefully apprised of what he was and was not finding. It was apparent even via a long-distance video hookup that he was exercising hands-on leadership and restoring momentum to the effort.

As it turned out, Duelfer arrived in Baghdad at about the same time that I was making my second visit to that country, in February 2004. Shortly after arriving, I asked for an all-hands meeting of the ISG at the airport. I used the occasion to tell Duelfer's troops that although he was obviously more than a little crazy for jumping out of perfectly good airplanes, he was going to be a great leader.

I also wanted to let them know we appreciated their heroic work in what had become a very, very tough environment. I gave them a pep talk about the importance of their mission and how much they were appreciated. My remarks seemed well received at the time, but a couple of years later some of the foreigners present complained anonymously to the media that by ending my remarks with something like "Now, go out and find WMD," I was subtly suggesting that there was only one permissible result of their mission. That, of course, is nonsense. My guidance to Duelfer—just like my guidance to Kay—and to everyone in the ISG was simply to go out and find the truth.

Duelfer turned out to be a remarkably good choice for the job. He had a wealth of experience in Iraq and knew senior bureaucrats in almost every one of Saddam's key government ministries. In a large room at ISG headquarters they turned one entire wall, about twenty feet long, to a time line plotting anything to do with Iraq and WMD. The timeline covered the period from 1980 to 2003. At any point in that time span, they could draw a line down and say, this was Saddam's worldview at this point. The time line also gave context to the data and the interviews that the analysts were accumulating. On another wall a second time line plotted when Iraq made funds available for weapons programs. The ISG was thus able to track the relationship between funding and WMD activity. Duelfer was convinced that the answer to the questions "Did Saddam have WMD, and if not, why not?" would come not from documents or scavenger hunts but from talking to the right people.

Getting to the right people was hard to do. The security situation in Iraq made the ISG's job nearly impossible. Large parts of the country were simply inaccessible for search without a huge military contingent to provide protection.

Because of the increasingly dangerous environment in Baghdad, to protect our personnel we purchased armored sedans wherever we could find them on the open market. One day an ISG team en route to a suspect site found themselves riding in an armored BMW that we had just had flown into Iraq. Originally intended for some European industrialist, the BMW came equipped with a DVD player in the backseat. One of the team members accidentally hit the DVD's Play button, not knowing that there was a copy of the movie *Saving Private Ryan* already in the machine, and the volume was on high. Seconds later the sound of gunfire and explosions came blasting through the car's speakers. It was the opening scene of the movie. For a few seconds, the vehicle's driver and security team thought the gunfire was live. While that might have been a humorous incident, most of the travel around Iraq was no joking matter. The threats were real and considerable.

On April 26, 2004, the ISG conducted a well-planned and well-rehearsed inspection of an area of Baghdad known as the "Chemical Souk," looking for people and materials that might have been involved in chemical weapons production. Teams of armored military vehicles with .50-caliber gun turrets escorted the ISG team to the scene. Overhead a UAV provided surveillance video. The inspectors, wearing full body armor, which adds up to forty pounds to a person, inspected a building full of leaking barrels of mysterious chemicals. Suddenly, a huge explosion erupted, nearly trapping in the basement an Australian scientist from the team. She narrowly escaped as the building collapsed above her. The fireball blew outward from the building to the rear, where soldiers were providing perimeter security. Two sergeants were killed, and five other soldiers were very badly burned.

On November 8, 2004, Charles Duelfer was traveling along the airport highway toward downtown Baghdad with three or four security vehicles. A civilian car loaded with explosives, known as a mobile improvised explosive device (IED), tried to insert itself in the middle of the convoy. Before it could get close enough, one of the security vehicles cut it off. The car detonated, killing two soldiers from the Kansas National Guard and seriously wounding another. Duelfer's car was severely damaged but he was unhurt. After he returned to the United States, Duelfer traveled to visit the families of the soldiers to thank them personally for their sacrifice. Throughout its existence, the ISG worked heroically to find the truth.

Duelfer told me much later that when he watched the Powell UN speech, he had the gut feeling that half of the information in it was wrong. "I just didn't know which half," he said. "With the Iraqis there was often some wacky, implausible, but true explanation for the way things seemed," he said.

From his subsequent conversations with Iraqis, Duelfer said that they were convinced that no matter what they did in the period prior to the war, it was not going to be good enough to satisfy us. Therefore, why try? Given our deep suspicions of Iraq, their track record of deception, and Saddam's desire to restart his weapons programs as soon as possible, Duelfer's contacts might have been right.

A number of the people Duelfer had known during his previous visits to Saddam-controlled Iraq were now in detention. So he spent a lot of time talking to these officials, trying to get to ground truth. He explained that any ability he had to influence the treatment the detainees received would go away when sovereignty was turned over to a new Iraqi government, around June 30. If they had useful information to share, now was the time to share it. Among those he talked to was Saddam himself.

According to Duelfer, "Saddam Husayn so dominated the Iraqi regime that its strategic intent was his alone. He wanted to end sanctions while preserving the capability to reconstitute his weapons of mass destruction (WMD) when sanctions were lifted." Duelfer wrote that Saddam wanted WMD to deter Iran, in his view Iraq's principal enemy. The belief that he had such weapons would also, Saddam thought, deter hostile groups inside Iraq. Maintaining a calculated position of ambiguity on whether he had WMD was, in Saddam's view, essential to deterring these external and internal threats. The Oil-for-Food (OFF) program (a UN program that allowed Iraq to sell oil on the world market and use the proceeds for food and medicines but not to rebuild its military), and the associated corruption, had terminally undermined the effect of sanctions on Iraq. Saddam believed he could simply wait out the sanctions and then begin re-creating Iraq's WMD capabilities.

In April of 2004, Duelfer met me in Amman, Jordan, where he asked me to support the release of an entirely declassified final report. I quickly agreed, viewing the report as a way to renew some faith in the intelligence community. I knew Charlie Duelfer would be thorough and fair and that he wouldn't pull any punches or spare anyone's feelings, including CIA's. In the end, that's just what happened. I had been gone from office for three months before Duelfer delivered his roughly thousand-page report to the new DCI, Porter Goss. As with Kay, Duelfer was given complete independence in putting the report together. He had the final word on what it said, but CIA did give a last-minute heads-up to a few key policy makers on what Duelfer had discovered about corruption in the UN Oil-for-Food program, because the documents he uncovered would prove embarrassing to several of their foreign counterparts.

In Duelfer's report, the ISG noted, as had just about everyone else, that Saddam had cheated consistently on United Nations sanctions, but on the critical issue that had been used as justification for the war, the report concluded that Saddam did not possess stockpiles of biological, chemical, and nuclear weapons at the time of the U.S.-led invasion of Iraq in March 2003, and that he had no active program to produce them. Asked in testimony by Senator Edward Kennedy what the chances were that WMD might still be uncovered, Duelfer replied, "The chance of finding a significant stockpile is less than five percent." That still sounds right to me.

Throughout this process, CIA and the intelligence community were committed to finding the truth and learning lessons from it. This is quite remarkable, and not very typical of what normally goes on in Washington. We oversaw a process that independently and unflinchingly drew unflattering conclusions about our work. Duelfer's report, produced solely under our guidance, was then used as the basis for many of the harsh judgments of the intelligence community rendered by the Silberman-Robb Commission. This willingness to look at itself critically is one of the strengths of the community I was privileged to lead and one of the few points of pride to come out of the whole WMD episode.

CHAPTER 23

Mission Not Accomplished

I first flew into Iraq just about the time Jerry Bremer took over as head of the Coalition Provisional Authority, or CPA, during the third week of May 2003. I took a helicopter ride with Jerry right over Baghdad. It was daylight. The helicopter door was wide open, and I was looking out as we flew. I remember thinking, as we scudded along, how precise the U.S. military action had been. There had been no massive carpet bombing; whatever they intended to get, they'd hit.

On the ground, the environment was strikingly permissive, considering that a foreign army had just invaded the capital and deposed the country's long-term dictator. People were going out, eating in restaurants. You half expected to see double-decker buses rolling down the main streets, with curious tourists gaping out the windows.

That same sense of optimism pervaded our station in Baghdad. Half the people there were young men and women who had just finished up their training. Mixed in with them were seasoned older pros and retired guys who had come back to work as contractors. I knew a lot of the veterans from odd spots all around the globe. Now they were in Baghdad, to help finish up the job of launching a new and democratic nation.

When I returned to Iraq in February 2004, the environment had changed dramatically. We flew into Baghdad at night, because you couldn't come in during the day. The C-17 bringing us there made a full-combat landing—a steep dive, quick on the ground. I was seated far forward, wearing flak jacket and helmet. There was no sightseeing this time. We flew into the Green Zone at treetop level and landed in the dark, on an unlit tarmac. I never felt in anything other than competent hands, but when you are flying black and wearing Kevlar, the pucker factor is hard to ignore.

By this time, CIA's presence in Iraq had grown quite large. Many of our officers showed up for a get-together that our senior man in Baghdad had arranged. Just about everyone arrived in body armor. I'd never seen so many stressed-out young people in one place in my life. I stayed three or four hours, talking with them. Then I was off again. I had to be somewhere else the next day, and in Baghdad in early 2004, you could fly out only at night.

In those intervening ten months, Iraq had become a very different place, but not at all in the way that the U.S. government had intended. How did it get that way? Through a series of decisions that, in retrospect, look like a slow-motion car crash.

In fact, the problems started well before the war. There was little planning before the invasion concerning the physical reconstruction that would follow. But regarding the political reconstruction of Iraq—how the country was to be administered and what role, if any, Iraqis would play in determining their political future—there was a great deal of spirited interagency discussion, often at the highest levels. Condi Rice and the vice president took an intense interest and often participated directly. The usual deputy- and undersecretary-level officials represented their respective agencies. John McLaughlin and Bob Grenier, a senior CIA operations officer who was our "mission manager" for Iraq, split the duty from our side.

The debates generally broke down along familiar lines: State, CIA, and NSC favored a more inclusive and transparent approach, in which Iraqis representing the many tribes, sects, and interest groups in the country would be brought together to consult and put together some sort of rough constituent assembly that might then select an advisory council and a group of ministers to govern the country. No one advocated immediate introduction of

Jeffersonian democracy, but many believed that the Iraqis should be encouraged to participate in a process that would quickly help identify—and legitimize—genuine leaders of a future democratic Iraq.

The vice president and Pentagon civilians, however, advocated a very different approach. Rather than risking an open-ended political process that Americans could influence but not control, they wanted to be able to limit the Iraqis' power and handpick those Iraqis who would participate. In practice, that meant Ahmed Chalabi and a handful of other well-known, longtime exiled oppositionists, along with the leaders of the essentially autonomous Kurdish areas. The differences in approach were clear and starkly articulated. The vice president himself summed up the dilemma: The choice, he said, was between "control and legitimacy." Doug Feith clearly stated his belief that it would not be necessary for the Iraqi exiles to legitimize themselves: "We can legitimize them," he said, through our economic assistance and the good governance the U.S. would provide. They never understood that, fundamentally, political control depends on the consent of the governed.

No consensus was ever reached, and no clear plan ever devised. In early January 2003, however, President Bush signed National Security Presidential Directive Number 24, giving the Department of Defense total and complete ownership of postwar Iraq. We didn't fully realize it at the time, but in the end, NSPD 24 would determine who made the final decisions on these momentous questions, and set the direction of the postwar reconstruction.

Hovering over this entire process was the figure—seldom acknowledged, almost never mentioned—of Ahmed Chalabi. Time and again, during the months leading up to the invasion and for months thereafter, the representatives of the vice president and Pentagon officials would introduce ideas that were thinly veiled efforts to put Chalabi in charge of post-invasion Iraq. Immediately before the invasion, the effort took the form of a proposal, put forward insistently and repeatedly, to form an Iraqi "government in exile," comprised of the exiles and the Kurdish leaders. These exiles would then be installed as a new government once Baghdad fell. My CIA colleagues were aghast. As Grenier later recalled, it was as though Defense and the vice president's staff wanted to invite comparison with the Soviet invasion of Afghanistan, when Russian troops deposed the existing government and installed Babrak Karmal, whom they had brought with them from Moscow.

At an NSC meeting about three months before the war got under way, President Bush asked Gen. Tommy Franks what he was going to do about security and law and order in the rear areas. Franks told the president, "It's all taken care of, sir. I have an American officer who will be lord mayor of every city, town and hamlet." That simply did not turn out to be the case. Whether that was part of CENTCOM's planning early on or not, I cannot say. In practice, though, the U.S. troop strength was sufficient to defeat the Iraqi army, but woefully inadequate to maintain the peace—just as Gen. Rick Shinseki, the former army chief of staff, had predicted.

Before the Iraq war began, an NSC staffer prepared an estimate of the troop strength necessary to stabilize postwar Iraq. The answer: 139,000 if the model was Afghanistan; more than 360,000 if the model was Bosnia; and a little shy of 500,000 if it was Kosovo. Which one was Iraq? Well, the war strategists erred on the side of Afghanistan when they went into Iraq, and we've been paying ever since.

The Pentagon's first man in charge of "post–major conflict" Iraq was retired Lt. Gen. Jay Garner. Named to his position some months before the invasion, Garner was then sent forward to Kuwait to assemble and prepare his team. When he and his team arrived in Iraq on April 18 to take responsibility for the newly created Office of Reconstruction and Humanitarian Assistance (ORHA), it quickly became apparent that the task before Garner was monumental and the advance planning woefully insufficient. ORHA was set up in one of Saddam's abandoned palaces, but found itself without adequate communications, short of sufficient Arabic speakers, and lacking in contacts and understanding of the Iraqi people. Garner was a good man with an impossible mission. He had responsibility without authority, and a bad situation immediately got worse.

The CIA tried to help. They set up meetings with a cross section of important Iraqi technocrats—people who could help make the country work—and brought them together to meet with senior U.S. military. Right off the bat, however, they ran into difficulty. Did the groups we were assembling include members of the Ba'ath Party? they were asked. Of course they did. You couldn't advance in Saddam's Iraq without joining the Ba'ath Party. Just as the governments in newly democratic Eastern Europe would inevitably include former members of the Communist Party, any group of skilled bureaucrats in Baghdad would have to include people who once held Ba'ath Party membership. Nobody questioned this initially, but the understanding that was obvious to us was less so to the new ORHA, a portent of far more serious problems to follow.

Similar problems arose when the United States started looking for candidates to populate an Iraqi provisional government. U.S. officials kept searching for, as one Agency officer put it, "Mohammed Jefferson," to launch Jeffersonian democracy in Iraq. The problem was that anyone who neatly fit that description would have long before been killed off by Saddam.

In the spring of 2003, Jay Garner, with NSC senior director Zal Khalilizad's assistance, began the process of holding regional conferences in Iraq in the hope of recognizing and taking advantage of different centers of power. According to CIA officers with him, Khalilizad believed that it was essential that Iraqis legitimize themselves. There were inherent risks in this. You can guide such a process, but you cannot control it. This was, after all, the essence of the democracy we had been preaching. It was important for the future stability of the country that Iraqis see people whom they recognized as having specific gravity being involved in the political process. This did not happen. The messy process of Iraqis legitimizing themselves came to a screeching halt. And Zal and Garner were out.

The assumption the U.S. government was working under was that this was going to be like the occupation of Germany, a supine country at our feet that we could remake in essentially whatever way we chose. The United States was going to completely demolish the Ba'ath Party. In the view of Paul Wolfowitz and others, you could replace "Ba'athist" with the word "Nazi." It soon became clear to us and very clear to the Iraqis that the purpose of the U.S. invasion was fundamentally to remake their society.

In early May 2003, I got a call from Colin Powell asking what I knew about Jerry Bremer. "I don't really know him," I said. From what I'd heard, Bremer was a tough-minded former ambassador who for a while had been head of the State Department's office of counterterrorism. "I certainly haven't heard anything bad about him."

Colin went on to say that the administration was considering Bremer as a replacement for Jay Garner. A few days later, on May 6, the White House made it official: Bremer had been selected to lead the effort to rebuild Iraq's infrastructure and help set up a new government. Although he was a presidential envoy, Bremer would report directly to the secretary of defense. His organization was given the title Coalition Provisional Authority. Once CPA had been established, Condi Rice ordered the interagency committee that had been constituted to deal with postwar planning issues to fold its tent. It was only a short while later, however, that, as one White House official told me, "The shit hit the fan and we had to rely on the British to tell us what was going on because we were getting no political reporting out of CPA." Rice then ordered the NSC process to start up again. But by then, fundamental decisions on disbanding the army and de-Ba'athification had already been made. The early returns filtering back to me on CPA indicated that it was not running smoothly.

The news was disquieting. It was just as worrisome that CPA was not being staffed with people with the requisite skills to enable our success. Many possessed the right political credentials but were unschooled in the complicated ways of the Middle East. What Iraq needed were Arabists and Foreign Service officers who understood the country's tribal allegiances, or who at least knew a Sunni from a Shia. What CPA seemed to be getting were people anxious to set up a Baghdad stock exchange, try out a flat-tax system, and impose other elements of a lab-school democratic-capitalist social structure. One of my officers returned from a trip to Iraq a month or two after CPA had taken over and told me, "Boss, that place runs like a graduate school seminar, none of them speaks Arabic, almost nobody's ever been to an Arab country, and no one makes a decision but Bremer."

The State Department had earlier assembled a team of experts to plan for a postwar Iraq, and Rich Armitage had 737s all lined up to fly them and their computers and some eighty Arabic linguists with regional knowledge out to Baghdad to begin setting up an embassy-in-waiting. The Pentagon, though, had other plans, and they certainly didn't include the Department of State, which many in Rumsfeld's circle thought had performed poorly in Afghanistan. Time and again, Marc Grossman, the undersecretary of state for political affairs, would raise the matter with Doug Feith, and time and again, Feith would say he was going to look into it. Before long it became apparent that, from the Pentagon's point of view, the State Department team of experts could sit on the runway at Dulles or Andrews Air Force Base, waiting for a lift to Baghdad, until hell froze over.

The security situation in Iraq started heading south remarkably soon after Saddam's statues fell. A reasonable question is: Did the U.S. intelligence community fail to predict the possibility of civil strife? Did we buy into the notion that Americans would be "greeted as liberators"? The answer, as so often is the case, is not black or white.

Although CIA was not among those who confidently expected Coalition forces to be greeted as liberators, we did expect the Shia in the south, long oppressed by Saddam, to open their arms to anyone who removed him. And, initially, Coalition troops were well received in the south.

Our expectation, though, wasn't open-ended, and it wasn't blind to other possibilities. Simultaneously, we produced a document that we titled, prophetically as things turned out, "The Consequences of Catastrophic Success." Our analysis said that there would be a feeling of relief among the Iraqi people that Saddam was gone but that this would last for only a short time before old rivalries and ancient ethnic tensions resurfaced. During this critical period, we needed to demonstrate an ability to provide the services that a country demands—food, water, electricity, jobs—while creating also a sense of safety and security that was absent under Saddam.

That, to me, is where plans went awry. Our analysis assumed there was a plan for ensuring the peace. In fact, there was no strategy for when U.S. forces hit the ground. This playbook wasn't written until long after kickoff.

In a January 2003 CIA paper, we said:

Iraq would be unlikely to split apart, but a post-Saddam authority would face a deeply divided society with a significant chance that domestic groups would engage in violent conflict with each other unless an occupying force prevented them from doing so. Rogue ex-regime elements could forge an alliance with existing terrorist organizations or act independently to wage guerilla warfare against the new government. In the early months after the forceful ouster of Saddam, stability in Iraq would depend partly on the perspectives of Iraqis towards whatever interim authority, military or civilian, foreign or indigenous was in control, as well as the ability of the authority to perform the administrative and security tasks of governing the country. The top priorities of most Iraqis would be to obtain peace, order, stability and such basic needs as food and shelter. . . . US-led defeat and occupation of Arab Iraq probably would boost proponents of political Islam. Calls by Islamists for the people of the region to unite probably would resonate widely. Fear of US domination and a widespread belief probably would attract many angry young recruits to extremists' ranks.

The same paper said, "Iraq's history of foreign occupation, first the Ottomans then the British, has left Iraqis with a deep dislike of occupiers. An indefinite military occupation with ultimate power in the hands of a non-Iraqi officer would be widely unacceptable. Iraqi military officers who oppose Saddam find the idea of a Western power conquering and governing Iraq anathema and a motivation to fight with Saddam where they otherwise would not."

In another paper we cautioned that the demobilization process would be full of pitfalls and suggested that "Baghdad's immediate post-war security needs may require that demobilization be delayed until Iraq [is] ready to begin building the armed forces."

We warned that, "Regardless of US postwar policy for Iraq, Iraqis would become alienated if not persuaded that their national and religious sensitivities, particularly their desire for self governance were part of the foundation for reconstruction. Iraqis would likely resort to obstruction, resistance and armed opposition if they perceive attempts to keep them dependent on the US and the West."

A National Intelligence Council paper in January 2003 titled "Can Iraq Ever Become a Democracy?" said that "Iraqi political culture is so imbued with norms alien to the democratic experience . . . that it may resist the most vigorous and prolonged democratic treatments."

In March 2003 we warned that "Iraqi patience with an extended US presence after an overwhelming victory would be short," and said that "humanitarian conditions in many parts of Iraq could rapidly deteriorate in a matter of days, and many Iraqis would probably not understand that the Coalition wartime logistic pipeline would require time to reorient its mission to humanitarian aid."

Our prewar analysis of postwar Iraq was prescient. The challenge for CIA analysts was not so much in predicting what the Iraqis would do. Where we ran into trouble was in our inability to foresee some of the actions of our *own* government. If you don't know the game plan, it is tough to do good analysis. As a result, did we exactly predict everything that would unfold? No.

Bremer would later write that three days after the White House announced his appointment, and shortly before going to Baghdad, he met with Doug Feith in the Pentagon. Feith, he says, urged him to issue an order as soon as possible upon arriving in Iraq that would prevent former Ba'ath Party members from having a role in the new government. Bremer did just that, on May 16, just four days after landing in Iraq. That morning's *New York Times* carried a hint of what was to come: "Shortly I will issue an order on measures to extirpate Baathists and Baathism in Iraq forever," Bremer was quoted as saying. "We have and will aggressively move to seek to identify these people and remove them from office."

Just a few weeks before the war started, senior U.S. officials were saying publicly that the conflict might be avoided if Saddam and a few dozen of his top henchmen simply left. This concept was never embedded in our war goals. Now, the war having been waged, the United States apparently was saying that thousands of officials around the country would be aggressively removed.

Bremer writes in his memoir that the intelligence community estimated that this order would affect only about 1 percent of the Iraqi population. That could be taken to imply that we supported the move and thought it was a good idea, but that was definitely not the case. In fact, we knew nothing about it until de-Ba'athification was a fait accompli. Clearly, this was a critical policy decision, yet there was no NSC Principals meeting to debate the move. As for the 1 percent number Bremer cites, he didn't ask for that estimate until the day after he issued the order, and once he got it he ignored the twofold context: first, that many of those Ba'athists were technocrats of exactly the sort Iraq would soon need if it were to again resume responsibility for its governance, and, second, that every Ba'athist "extirpated" from Iraq, to use Bremer's word, had brothers and sisters and aunts, uncles, and cousins with whom to share his anger.

Privately, in fact, the senior CIA officer in Iraq and others strongly advised against this step when they were finally informed of it, and they continued to argue after the decision was made. A senior NSC staffer told me that when he briefed the president on de-Ba'athification, the staffer talked about South Africa's Truth and Reconciliation program. Just as South Africans had done, Iraqis themselves should determine who had too much blood on his or her hands to be permitted to take part in a new government. Bremer's plan put the process in the hands of an Iraqi all right. Ahmed Chalabi was named to head the de-Ba'athification Council, and as a result the implementation of the order was even more draconian.

We soon began hearing stories about how Iraqis could not send their kids to school because all the teachers had been dismissed for being members of the Ba'ath Party. In the context of a country armed to the teeth, this was not a good thing. If the kids and teachers were not in school, they were on the streets. I went to see Condi Rice and complained that the indiscriminate nature of the de-Ba'athification order had swept away not just Saddam's thugs but also, for example, something like forty thousand schoolteachers, who had joined the Ba'ath Party simply to keep their jobs. This order wasn't protecting Iraqis; it was destroying what little institutional foundations were left in the country. The net effect was to persuade many ex-Ba'athists to join the insurgency. Condi said she was very frustrated by the situation, but nothing ever happened. Several months later, with a full-blown insurgency under way, an interagency group headed by Deputy National Security Advisor Bob Blackwill desperately looked for ways to reach out to dissident Sunni Arabs. We again raised the subject of rolling back the de-Ba'athification order. Doug Feith retorted that doing so would "undermine the entire moral justification for the war."

Bremer's de-Ba'athification order became known as CPA Proclamation Number One. As bad as that was, CPA Proclamation Number Two was worse. Again, without any formal discussion or debate back in Washington—at least any that included me or my top deputies—Bremer, on May 23, ordered the dissolution of the Iraqi army.

To be sure, elements of the Iraqi army, especially the Special Republican Guards (SRG) and the Special Security Organization (SSO), did have much blood on their hands. However, we viewed many Iraqi military officers as professionals, driven by national Iraqi values rather than loyalty to Saddam, who could form the core of a new Iraqi military, but the order struck a broad blow at the Sunnis, who comprise 20 percent of the national population and who occupied virtually all of the top ranks in the army. Granted, they were never going to be completely satisfied, short of having Iraq handed back to their control, but along with the de-Ba'athification order, this second order had effectively alienated one fifth of the population and much of the center of the country.

NSC officials were expecting Proclamation Number Two to include some language about how Iraqi military members below the rank of lieutenant colonel could apply for reinstatement. After all, the majority of army members were conscripts just trying to feed their families. CPA Proclamation Number Two appeared to be punishing them—and even the Shia who made up the bottom rung of the military—equally with those who had ruled the roost. When the pronouncement was issued, however, that provision was not mentioned. So, as far as the rank-and-file members were concerned, Bremer had just announced that they were all unemployed.

Jay Garner, who was still in Iraq at the time, went to see Bremer along with our senior CIA officer in the country. They both told him that the demobilization order was madness. Garner had been counting on using some of the former Iraqi military for stabilization and security. Our officer told Bremer that the action would only "give oxygen to the rejectionists."

The argument from some supporters of CPA Proclamation Two was that the army had essentially dissolved itself anyway, so what was the big deal? Our officer on the ground at the time, however, estimated that the majority of the army could have been recalled within a two-week period and put to useful work.

Bremer was unmoved. He reportedly told Garner that he could raise the issue with the secretary of defense if he wanted to, but that this was a done deal and a decision made at a level "above Rumsfeld's pay grade."

Whoever had made the decision, the reaction from former Iraqi army members was swift. A *New York Times* report on a May 25 demonstration in Basra by dismissed Iraqi soldiers quoted one former Iraqi tank driver as saying, "The U.S. planes dropped the papers telling us to stay in our homes . . . They said our families would be fine," he said. More ominously, a lieutenant colonel told the reporter, "We have guns at home. If they don't pay us, if they make our children suffer, they'll hear from us."

Eventually some army members were paid and allowed to apply to rejoin the new Iraqi army, but all officers with ranks of lieutenant colonel and above were permanently banned—despite the fact that, like many non-Western armies, Iraq had a disproportionate number of army members with high ranks. A typical Iraqi lieutenant colonel did not have the same level of authority or influence wielded by his U.S. Army counterpart.

At meetings in the White House and in Baghdad after the two proclamations were issued, we argued that the orders were having unintended negative consequences. The actions had taken large numbers of common Iraqis and given them few prospects beyond being paupers, criminals, or insurgents. One of our senior officers tallied the numbers, including affected family members and the like, and came up with a pool of a hundred thousand Iraqis who had been driven toward the brink by the de-Ba'athification order alone. In the end, too many of them chose insurgency.

For some officials in the Pentagon, the accelerating violence simply proved the wisdom of excluding these Ba'athists and ex–army members from the future of Iraq. As late as the spring of 2004, at a meeting in the White House, one of our officers was asked for "out-of-the-box" ideas to stem the violence. He suggested rescinding CPA Proclamation Two and mounting an aggressive campaign to round up former army members and enlist them to help secure Iraq's borders and maintain internal security. As later described to me, a U.S. Army colonel present, who had been DIA's liaison to Ahmed Chalabi and the Iraqi National Congress, said, "I agree. We should round them all up and shoot them."

The moves the U.S. government was making were driving a wedge between the various factions in Iraq. Charles Duelfer was told by an Iraqi friend that, in the past, Iraqis were not accustomed to thinking of themselves primarily as Shia or Sunni. But the way we implemented democracy had led people to believe that they deserved a piece of the pie based on their membership in a certain group. So the whole dynamic was to pull away from the center. The decisions we made tended to fracture Iraq, not to bring it together.

On one of his trips to Iraq, Wolfowitz told our senior man there, "You don't understand the policy of the U.S. government, and if you don't understand the policy, you are hardly in a position to collect the intelligence to help that policy succeed." It was an arrogant statement that masked a larger reality. In many cases we were *not* aware of what our own government was trying to do. The one thing we were certain of was that our warnings were falling on deaf ears.

In the midst of all this, we started pushing for the establishment of a new Iraqi intelligence service. Any government intent on protecting people needs an organization to acquire information regarding internal security and external threats. That much seems obvious, but we ran into strong and immediate resistance to our suggestions on building such a service.

John McLaughlin tried to get authorization through the Deputies Committee to help set up such a capability, only to be thwarted. In all the years that I have known John, I don't think I have ever seen him more exasperated. "The only country in the world where the U.S. intelligence community doesn't have a counterpart is Iraq," he remembers saying at one of the deputies meetings. "The best way to get a handle on who is causing the violence in Iraq is to have Iraqis figure it out." That message, too, never seemed to get heard.

On another occasion, Steve Kappes, the then second-ranking operations officer at CIA, was pushing the same theme at a meeting where Condi Rice was present. "How do I know you guys aren't going to create another KGB?" Condi asked. "We didn't create the first one," Steve reminded her. Condi's comment was emblematic of the mind-set we were up against. Policy makers didn't seem to want us dealing with anyone who wasn't "politically acceptable" to them on some firm but unannounced scale. Our point was that Americans were dying, jihadists were running all over the country, and it was time to figure out how to vet Iraqis who had the capabilities to do something about it.

We'd been through this before. When the Soviet Union fell and the West inherited Eastern Europe, we set about building intelligence services there out of what was already on hand to work with. Was there a high probability that Soviet agents still peopled those services? Sure. Is there a high probability that over the course of time, they'll be weeded out? Sure, again. The point was, you have to take some risk if you want to make the government work.

After many months lost, months during which insurgents and dissidents gained a valuable foothold, we began the process of setting up an Iraqi intelligence service.

Gen. Mohammed Shawani, the hero of the Iran-Iraq war, was finally selected to head it up and build a service drawn from across the country's ethnic, religious, and tribal groupings. He spoke frankly to the Bush administration in the months after the liberation of Iraq, highlighting his concerns to the president and vice president about the developing insurgency. He was the first senior Iraqi official to identify and speak of Iran's hand in destabilizing his country. (He continued to serve as the director of Iraq's National Intelligence Service as of early 2007, although Iran and elements of the Iraqi Shia groupings were working to have him removed because of his anti-Iranian stance.) It may be fair to say that our analysis before the war never precisely predicted the dire circumstances that would unfold on the ground in Iraq after the initiation of hostilities. What is absolutely clear, however, is that the intelligence generated by our officers on the ground after the war told the story, and the reasons for a deteriorating situation, with great clarity.

How does an insurgency grow in a place like Iraq? It happens when you are late securing your lines. Or when you create a vacuum to be filled by opportunists like al-Qa'ida. It occurs, in essence, when you disenfranchise many of those most able to help you. Also when you refuse to avail yourself of indigenous resources that could provide you with intelligence on insurgent activity. And, finally, when you blind yourself to the evidence that is steadily mounting in front of your eyes.

As the situation turned for the worse, the senior CIA officer on the scene would send in field appraisals. These cables are known internally as "Aardwolves." (The Agency has called such assessments this for many years—although the origin of the name is obscure. One theory is that in the early days of CIA, someone opened his dictionary to page one, looking for an apt code word, and "Aardwolf" just leapt out.) The common thread running through all the Aardwolves during this period was the threat of the rising insurgency.

On July 8, 2003, a report from CIA's senior officer in Baghdad noted that while normalcy seemed to be gradually returning for "average Iraqis," security for Coalition forces was crumbling. "Among the factors contributing to hostility toward allied forces is a general sense of disappointment at the slow progress in rebuilding Iraq and producing tangible evidence that life will be better . . . than it was under the former regime." The report went on to mention the demoralizing effect of widespread looting in the aftermath of Saddam's fall, the rise of opportunistic terrorist groups, and the lack of an "effective internal security service."

The report also stated, "In the current environment of confusion, uncertainty and dissatisfaction, the risk exists for violence to quickly become acceptable and justified in the minds of broader sectors of the population."

Six weeks later, on August 20, another Aardwolf noted that "the insurgency is the most pressing security issue the CPA faces in Iraq today. . . . Success against the insurgents and terrorists requires an immediate and enhanced effort on the part of the coalition. The liberation of Iraq has sparked a revolution among the Shia community. This revolution . . . will only begin to gather momentum. We will face violence and instability in the Shia heartland as soon as this sorts itself out."

To this assessment, Bremer added his own comment that read in part: "It is not clear to me that at its current level, or even if it picks up, this low intensity conflict could erase our gains. The insurgency could certainly challenge parts of the reconstruction program, and it has. But on balance, reconstruction has gone forward . . . even in the face of this low level conflict." Some journalists have written that we hesitated to pass our negative reports up the line for fear they would spark an unpleasant reaction. This is absolute nonsense. They all went straight to the top policy makers.

We held nothing back. Reports from the field during my tenure were remarkably prescient, and some were leaked to the media at warp speed by various recipients.

The fact that these often gloomy assessments found their way to the press led some in the administration to believe that CIA was trying to undermine the administration's efforts in Iraq. That was not the case. Although Aardwolves were originally very closely held documents, in recent years they have gotten a much wider dissemination. Typically, they are now read at senior levels in the departments of Defense and State, and at the NSC. I have no idea where the leaks came from, but I have no reason to believe that they originated within CIA.

Whoever leaked the Aardwolves could have been motivated by the notion that CIA's assessments were important and deserved public airing, but he or she could have been equally motivated by the sentiment, shared widely in some parts of the government, that these guys from CIA "don't quite get it and aren't with the program." Leaks, after all, are the improvised explosive devices of inside-the-Beltway warfare.

I remember hearing, after some of the first Aardwolves seeped out, that NSC officials were calling our senior officer in Iraq a "defeatist." That shoot-the-messenger theme came up time and again. He was, of course, being nothing more than a realist, and we did everything we could to see that he got heard on the home front. In addition to disseminating his written report, I brought the senior officer to the Oval Office, when he was back in Washington in November 2003, to give the president his frank assessment of the situation on the ground. Yet as late as April 2004, when it was plain to see that the situation had unraveled, Jerry Bremer was still complaining that one of our senior officer's reports was "over-the-top pessimistic." The newest report, Bremer wrote, "begins to smell like classic CYA."

Our senior officer in Baghdad wasn't a lone voice in the wilderness. Bob Grenier sent me a report on Iraq on November 3, 2003, saying that "Security conditions in the center of the country are going from bad to worse." And that attacks on Coalition forces, if allowed to proceed unchecked, threatened the "de facto political dismemberment of the country." In another report to me, Grenier wrote, "It is important to stress that the Sunni Arab insurgency is primarily a political problem, rather than a military one. . . . We cannot find and kill all those who oppose us, particularly if their members' numbers can grow over time."

Braced and deeply concerned by the consistent, troubling messages I was getting from my team, I felt an obligation to make sure that policy makers got the clear, unvarnished truth as we saw it. We held a series of senior-level briefings in my conference room, where the recipients were removed from their phones, aides, and BlackBerrys. The first was for Condi Rice, Steve Hadley, and several of their key deputies on the NSC staff. We spent about three hours in briefings and discussion. Hadley, in particular, seemed to get our message—that unless we could reassure significant elements of the Sunni Arab community and bring them into a political process, the insurgency would continue to grow and ultimately split the country. He asked us to prepare an integrated plan for how all elements of U.S. power could be harnessed to arrest this slide. I asked Bob Grenier to prepare it, and he set about with several others to put it together.

An important message delivered was about the magnitude of the challenge we would face in Iraq. The analyst giving the briefing had covered jihads for over a decade. She noted that Iraq would represent roughly the nineteenth in a long series of jihads since the Soviet invasion of Afghanistan. Many Iraqi factional leaders were primed for the greatest jihad yet, against Americans in the Arab heartland. She noted that al-Qa'ida had always been nothing more than an exploiter of jihads, and this one would come exactly at a time when the organization was on the ropes and would allow al-Qa'ida to keep itself alive and to make a comeback.

Apparently the word spread, because we quickly received a request from the vice president for a similar briefing. He, his chief of staff, Scooter Libby, and several of their close aides spent several hours with us, listening carefully and asking thoughtful questions.

The Sunni Arab insurgency that we began to clearly identify in the summer and fall of 2003 was primarily in our view a political problem rather than a military one. While military operations were important, they could be effective only as part of an Iraqi-driven political process, coupled with an economic program that recognized the obvious. Iraq's governates were racked with unemployment, making large numbers of unemployed young men susceptible to recruitment by insurgents. We worked with the military to reach out to Iraqi tribal leaders, moderate clerics, businessmen, and professionals, seeking to provide them with the financial basis to expand their influence and gain a constructive political following. From our perspective there were three critical enablers in reaching out to the Sunni community without which the chances of success would be remote—a shift in de-Ba'athification, a restoration of at least part of the army, and economic assistance to quickly put money in the hands of Iraqis.

Our military units enjoyed considerable success with the modest reconstruction funds at their disposal. Yet the funds made available were insufficient and could not be sustained in a meaningful manner to allow us to get traction. The majority of the billions of dollars at U.S. disposal in Iraq were tied up in major long-term projects targeted at structural reform and long-term economic development, which, while valuable on paper, were divorced from the needs on the ground. And as a result, we ended up ceding much of the political space to the insurgents.

The continued sense of isolation in the Sunni heartland, the complete dissolution of the Iraqi army, rigid de-Ba'athification, and the lack of economic opportunity or political direction provided fuel for the insurgency. In fairness, we cannot say whether some combination of these enablers would have made our efforts with the Sunnis more successful, but none of them was implemented.

CIA was not alone in sending out a dire message. On November 10, 2003, Colin Powell weighed in from the State Department with an assessment every bit as dark as the ones we were providing. "Given mounting popular discontent with occupation," he wrote, "we cannot sustain the current CPA arrangement long enough to allow completion of the complicated process of drafting a Constitution and holding full-fledged elections. . . . A credible political process leading to an early transition of power is critical to subduing the growing insurgency that coalition forces face."

That same day, a new Aardwolf came in warning that growing numbers of Iraqis were becoming convinced that the U.S.-led Coalition could be driven from the country and were joining the insurgency. The combination of this Aardwolf and Colin's message sparked the White House to act: the next day, November 11, the president called a quick meeting in the White House Situation Room to hear from CIA what was now becoming a very polished brief. It was Veteran's Day, a federal holiday, and I had to track down some of our top Iraq analysts, who were enjoying a rare day off, and drag them in for the meeting.

Despite the short notice, the president had assembled quite a crowd. As I recall, he was joined by the vice president, the secretaries of state and defense, Condi Rice, Steve Hadley, Rich Armitage, Paul Wolfowitz, and, in a surprise to us, Jerry Bremer, who was back in town. I brought with me John McLaughlin; one of our most senior operations officers, Rob Richer; Grenier; and three of our analysts. The president said he wanted to find out what the current situation was in Iraq. Don Rumsfeld quickly deferred to CIA. Rich H., one of our lead Iraqi military analysts, started to give a briefing—influenced in large part by the Aardwolf that had come in just the day before. Early in the briefing he mentioned the ongoing "insurgency" in Iraq.

Rumsfeld immediately interrupted and pointedly asked, "Why do you call it an insurgency?"

"Sir," Rich said, "the Department of Defense's definition of insurgency is . . ." and then he proceeded to list the three necessary conditions that DOD required before the term "insurgency" could be used. All three conditions had obviously been met in Iraq.

The message out of the Oval Office that day was, "No one in this administration will make any reference to an insurgency." Apparently, that message did not filter down, because a few days later, much to the dismay of some at NSC, Gen. John Abizaid, by then head of the U.S. Central Command, described the current uprising—quite accurately—as an insurgency.

At the same briefing, another CIA analyst described how Iraq was the latest in a long series of jihads for Islamic fundamentalists. "Iraq," she said, "came along at exactly the right time for al-Qa'ida." It allowed them to tap into deep wells of support and to inspire a permanent jihadist movement and lure Iraqis into the fight. They were being aided and abetted by experienced facilitators whom we had encountered previously—in Afghanistan, in Bosnia, in Chechnya, and elsewhere.

We ended the presidential briefing with a plea, again, for measures that would address the Sunnis' concerns, and set the conditions that would enable our people on the ground to organize an indigenous opposition to those who were attacking U.S. troops and Iraqi security personnel. We hadn't counted on having Jerry Bremer in the room to hear such a direct attack on the policies he had implemented, but as soon as we finished, the president abruptly turned his gaze on Jerry: "What do you say, Bremer?"

With an air of resignation, Bremer recounted how he, too, had attempted to identify responsible and capable Sunni Arab leaders. There were none, he said. The Iraqi army, moreover, had dissolved itself, and would not be coming back. And as for de-Ba'athification, as strongly as the Sunnis might feel about it, the Shia leaders with whom he dealt were every bit as passionate, and would never accept a rollback. The message: there's nothing to be done but to continue on the current line of march.

By mid-November 2003, it was clear in the minds of many that something was going to have to change in Iraq. Condi Rice asked Ambassador Robert Blackwill of the NSC staff to go to Baghdad just before Thanksgiving. Blackwill asked Grenier to accompany him. On the way out, Grenier asked him, "What is your mandate?" Blackwill said that Rice had charged him with trying to bring about some changes and that he was going to have a "Socratic dialogue" with Bremer. Nobody wanted to give Bremer specific marching orders. According to Blackwill, Rice felt she could not *order* changes, but she wanted Blackwill to lead Bremer in the direction they thought they needed to go. A major component of that was to be an integrated program of Sunni outreach, including something on de-Ba'athification and a more effective reconstruction of the Iraqi army. In the process, Blackwill met with all the senior British and American officials in CPA, with a number of the provincial coordinators, and with senior U.S. military officials in the field.

On the way back, Blackwill and Grenier agreed that CPA was essentially hopeless; as currently constituted, it would be neither willing nor capable of doing what was necessary. Blackwill summed up his feelings to Grenier: "The only hope we have is you, CIA, and the deployed military. So it is over to you guys, to figure this thing out and do what you can." According to Grenier, Blackwill came back and wrote a trip report for Rice that was quite stark.

Equally futile, or so it seemed, were our efforts to form a credible and durable Iraqi governing body. In Afghanistan, we had started from the ground up, allowing the various political groups to legitimize themselves, then building toward a central, representational government. In Iraq, the process couldn't have been more different. We never had a conference comparable to the Afghan Loya Jirga that produced a leader, Hamid Karzai, around whom the country could coalesce. Rather, we essentially determined that we would legitimize the Iraqis. We had won the war; we had the guns, the tanks, the soldiers, and the air power. We were in charge, and by God, we knew what was best. Alas, what too many people in the U.S. government were convinced would be best was an Iraqi government headed up by Ahmed Chalabi.

At another meeting in May 2003, one of our officers said he thought it was unwise for the United States to try to anoint Chalabi or anyone as the new Iraqi leader. Condi Rice asked why. "Iraq has no water, no electricity; employment is in the pits," our officer said. "Anyone we try to install will be seen as responsible for all that and will fail." Steve Hadley reached over and patted the officer on the knee. "I once thought that, too," he said, "but I've come to know differently. It just doesn't work that way."

Sometimes Chalabi's name would be strangely absent from the discussion, although he was obviously on everyone's mind. We would sit around these White House meetings expressing the hope that a strong, unifying Iraqi leader would emerge, and while you could tell that one name was on the minds of many in the room, no one would utter it. You had the impression that some Office of the Vice President and DOD reps were writing Chalabi's name over and over again in their notes, like schoolgirls with their first crush. At other times, so persistent was the cheerleading for Chalabi, and so consistent was our own opposition to imposing him on Iraq, that I finally had to tell our people to lay off the subject. "They all know what we think about him," I can remember saying at one senior-level staff meeting. "He's now in Iraq. He's either going to succeed or not, but Iraqis are going to have to make the decision for themselves."

My view was that Chalabi was not going to fare very well, and I ended up being right. In the parliamentary elections, once they were finally held, his party got practically no votes, no seats. By then, though, we had gotten pretty much accustomed to political controversy in Iraq.

The Coalition struggled to get the new Iraqi government functioning, and CIA tried to help. In prewar discussions about postwar authorities, we sought permission to assist in identifying nascent Iraqi political figures who could create a new democratic government. Playing a role the Agency had played in many other countries over the years, we asked for authorities to work with Iraqi tribes, to get them to engage in the political process. This time, though, there was a reluctance to allow us to play that role. The reasons are not entirely clear to me, but some elements of the administration were obviously concerned that long-standing animus between the Agency and the INC would stand in the way of the political advancement of Chalabi.

As relayed to me, CPA meetings with Iraqi leaders tended to have an imperious and condescending tone, more in the manner of lectures than discussions. As the security situation continued to spin out of control, potential future leaders among the Iraqis were reluctant to come forward.

Efforts to rebuild an Iraqi army and security force were going badly, but CPA officials kept trying to put a smiley face on that, too, as if wishing would make things so. At one point, when Armitage's boss, Colin Powell, came out to the region to receive briefings, our CIA senior rep pulled him aside and said that the information being presented about new Iraqi army equipment sets and deployable units was being exaggerated. "I can see that, son," Colin told him. "Believe me, I know a brigade when I see one."

CIA also tried to help out on the political front—and met opposition at almost every turn. We set up a program with some of the Sunni chieftains, exchanging humanitarian assistance for their cooperation, but Bremer refused to support it. "You are dancing with CIA's old pals," he told one person, referring to the tribal chieftans. On another occasion CIA set up a meeting in the Green Zone with a number of Sunni leaders to try to get them to buy into a new government. One of my officers later told me Bremer walked into the conference room where they were meeting, delivered a twenty-minute diatribe, and walked out again. The Sunnis were furious. We lost contact with half of them in the aftermath.

On yet another occasion, our senior officer on the ground arranged a meeting with fifty-seven former Iraqi generals. The intention was for them to open a dialogue with Lt. Gen. Rick Sanchez, commander of U.S. Army troops in Iraq. The meeting was supposed to be a possible first step toward an interim government, even if none of the ex-generals could serve in it. At the last minute, Bremer told Sanchez not to go. "We will not engage with the enemy," he said.

In May 2004, the CPA was trying to persuade Dr. Iyad Allawi, a prominent Iraqi neurosurgeon and head of the Iraqi National Accord, to agree to take on the position of defense minister in the new provisional government. A Shia, Allawi had once been a Ba'ath Party member but had broken ranks with Saddam. In 1978, while living in London, he and his wife were attacked in their home by one of Saddam's assassins wielding an axe. Allawi was left for dead. In the mid-1990s, he had been active in the abortive efforts to overthrow Saddam.

I'd met Allawi a number of times before, in Washington and London. We didn't know each other well, but as DCI, I was a beneficiary of all the trust and goodwill that the CIA had built up over the years with him and the INA. For that reason, I was asked to go see Allawi and urge him to accept the offer to become defense minister.

We met in a hotel room in Amman, Jordan, just the two of us. My marching orders were to talk tough with him, to make him understand that he had to do this, but I knew Allawi better than that. I knew what he had suffered and what he had placed at risk, and I knew that I wasn't going to be able tell him what to do or how to do it. That's not the way to approach a meeting like this anyway. Instead, I went intent on letting him talk and listening as he voiced his frustrations; and that's what he did.

Allawi, it turned out, had little regard for the CPA. He had been approached to be defense minister, he said, but no one would tell him just what that meant. The bottom line was that he was very uncertain whether he wanted to participate in anything like this, because he understood there was a high probability that the provisional government simply wouldn't work.

I waited until he was through venting before chiming in. "Iyad," I said, "I can't tell you that you must take this job, but I need to tell you that you must carefully consider it. If good men like you will not put themselves forward for important positions, there is no hope for Iraq."

"George," he responded, "I can't get anyone at CPA to tell me what the duties of this defense minister would be—what his authorities would be, what his limitations are. How can I accept a job that no one will describe?"

I promised that I would ask someone to provide him with details. When the meeting was over, I picked up the phone and called Steve Hadley back in Washington. "Steve," I said, "this is a proud man. No one has given him a clue about what is expected of him. You have to get people to reach out to him and explain the process—don't just try to tell him what to do. Consult with him. Ask him how we can get to where we need to go. Uncle Sam ordering guys like this around ain't going to work."

I must have gotten through, because when Allawi returned to Iraq, some of the information he was looking for about the CPA's vision started to flow to him. Very quickly, he was interested enough that he met with a number of other Iraqi leaders and debated next steps. And then, the next thing I knew, Allawi was sending word to Bremer that he was not interested in the defense minister post. However, he *was* willing to accept the position of interim prime minister in the provisional government. Allawi, it turned out, had managed to assemble a large number of other Iraqi leaders who fully supported him.

My first reaction when I heard the news was: Great! Although I wasn't sure Allawi was the right man for the top job—whether a former Ba'athist, Shia expatriate could effectively lead a coalition—the bigger point to me was that at last Iraqis were emerging on their own to legitimize their future government. But instead of looking on this as a godsend—finally, some home-brewed unity and leadership!—many in Washington viewed Allawi's emergence as a CIA plot. Almost immediately, Bremer ordered our senior officer in Baghdad to stay away from Allawi, a man whom days earlier we had been asked to meet with and urge into greater involvement in the political process.

Iyad Allawi was and remains far too independent to be anyone's puppet. He knew his country, he knew the challenges, and he had perhaps the best chance of bringing order out of the chaos that had become Iraq. In the end he had to fight various sectarian opponents to achieve success. The fight proved too hard. To my mind, it was a loss. But that could be said about Iraq in general, too.

Perhaps the greatest disappointment of postwar Iraq was trying to create an Iraqi army. By the time Allawi took over as prime minister of the interim Iraqi government in June 2004, it was clear that the training effort was going badly. Although battalion-strength units were being turned out, their discipline was poor, and they would often dissolve in battle. Senior U.S. military officers began to mutter darkly that the problem was not U.S. training but Iraqi leadership. This came as no surprise to some. For months, General Shawani had been complaining loudly, including to senior White House officials, that the U.S. training effort was deeply flawed. Armies, he said, are built from the top down. You've got to begin with a respected general who can put together a competent divisional staff. The brigade and battalion staffs and their subordinate units can then be built out in turn. The traditional Iraqi army had always been based on those highly personal ties of loyalty and trust. The United States, Shawani said, was not building an army; it was training a series of militias, with no indigenous logistics or support, no respected leadership above the battalion level, and no Iraqi command and control. As an antidote, Shawani proposed that a number of respected senior Iraqi generals, whom he and others could identify and vet, be called back to reconstitute the five traditional "territorial" divisions of the Iraqi army. They would be allowed to form their own staffs, and then incorporate U.S.-trained units into a coherent division-level command structure. In this way, a provisional Iraqi government could rebuild a unifying national institution in service of a unified state.

Word was that this was what Prime Minister Allawi intended to do. Immediately after his accession, however, a DOD delegation led by Paul Wolfowitz traveled to Baghdad to meet with Allawi. When he explained his plan to them, they listened politely and then inquired how he intended to pay for it. It was clear that DOD would not; they would continue to train battalions completely dependent upon American support.

There were many bizarre twists to the Iraq story, none more so than the continuing theatrics of Ahmed Chalabi. During President Bush's State of the Union speech on January 20, 2004, Chalabi was given a seat of honor in the gallery near the First Lady. Just a few weeks later, he was quoted in the British newspaper the *Daily Telegraph* saying that he and his INC were "heroes in error" and that he had no qualms about information he had passed to the U.S. government, since his organization had been "entirely successful" in achieving what they wanted, the removal of Saddam Hussein. In March he appeared on CBS's *60 Minutes* blaming U.S. intelligence for not doing a good enough job checking out the flawed information *his organization* was peddling.

"What the hell is going on with Chalabi?" the president asked me at a White House meeting that spring. "Is he working for you?" Rob Richer, who was with me at the meeting, piped up, "No sir, I believe he is working for DOD." All eyes shifted to Don Rumsfeld. "I'll have to check what his status is," Rumsfeld said. His undersecretary for intelligence, Steve Cambone, sat there mute. "I don't think he ought to be working for us," the president dryly observed.

A few weeks later the president again raised the issue. "What's up with Chalabi?" he asked. Paul Wolfowitz said, "Chalabi has a relationship with DIA and is providing information that is saving American lives. CIA can confirm that." The president turned to us. "I know of no such information, Mr. President," Richer said. The president looked to Condi Rice and said, "I want Chalabi off the payroll."

At a subsequent meeting, chaired by Condi Rice, DIA confirmed that they were paying the INC $350,000 a month for its services in Baghdad. We knew that the INC's armed militia had seized tens of thousands of Saddam regime documents and was slowly doling them out to the U.S. government. Beyond that it was unclear to me what the Pentagon was getting for its money. Somehow the president's direction to pull the plug on the arrangement continued to be ignored.

It was about this time that we received reliable information that Chalabi was passing highly sensitive classified information to the Iranians. This should have been the final straw—but nothing is ever final with Chalabi. The CPA ordered a raid on his offices. Chalabi later claimed that CIA was behind a plot to undermine him. In truth we didn't even know about the raid until after it had taken place. Finally, in May 2004, the INC's services contract with DIA was terminated. While Chalabi was accused of all manner of malfeasance, nothing ever came of the charges. In the December 2005 elections, Chalabi's party garnered about 0.5 percent of the vote and won not a single seat in Parliament.

The true tragedy of Iraq is that it didn't have to be this way. I can't begin to say with absolute clarity how things might have worked out, but I have to believe that if we had been more adept at not alienating entire sectors of the Iraqi population and elites; if we had been smarter at the front end; if we had thought about reconstruction from the perspective of how much money we could put in people's hands so that they would know they had a steady stream of income; if we had figured out a way to let Iraqis know that they actually did have a role in their future that went beyond words, a role they could see being implemented in practice on the ground—we would be far better off today.

To be certain, we were never going to return to the Iraq of old. Sunnis would never occupy the privileged positions they once enjoyed. We backed an increase in Shia power and we did not allow any sort of equivalent Sunni alternative to form.

Whenever you decide to take the country to war, you have to know not only that you can defeat the enemy militarily but that you have a very clear game plan that will allow you to keep the peace. There was never any doubt that we would defeat the Iraqi military. What we did not have was an integrated and open process in Washington that was organized to keep the peace, nor did we have unity of purpose and resources on the ground. Quite simply, the NSC did not do its job.

As early as the fall of 2003, it was becoming clear that our political and economic strategy was not working. The data were available, the trends were clear. Those in charge of U.S. policy operated within a closed loop. Bad news was ignored. Our own subsequent reporting—reporting that eventually would prove spot-on in its predictions of what came to pass on the ground—was dismissed. Yet little was done to make the adjustments necessary to avoid being overwhelmed by a growing domestic insurgency. Too big a burden was placed on the military to deal with problems that at their root could not be solved simply by using more force. We could never subdue an entire country, because we were not meant to stay.

Despite the consequences of decisions regarding de-Ba'athification or disbanding of the army, and the inability to use the billions of dollars at our disposal to implement a political strategy that might have succeeded, not much was done to change course. In the way of Washington, it is too easy to blame Jerry Bremer, who gave up a year of his life to serve in difficult circumstances, and who worked in a chain of command. In many ways, he was set up for failure.

The president was not served well, because the NSC became too deferential to a postwar strategy that was not working. This was no time for a subtle "Socratic dialogue" with Jerry Bremer. The National Security Council was created in 1947 to force important policy decisions to be fully discussed, developed, and decided on. In this case, however, the NSC did not fulfill its role. The NSC avoided slamming on the brakes to force the discussions with the Pentagon and everyone else that was required in the face of a deteriorating situation. By sending Bob Blackwill out to chat with Bremer, NSC substituted a time-tested process for one almost guaranteed to fail.

The critical missing element was an Iraqi government that could have helped us. We decided instead to have Americans administer Iraq. It may have worked in World War II, after the entire world fought against Nazi Germany for many years. But in the context of the Middle East, it was not going to work any more than the French occupation of Algeria. To Arabs it looked as though this was all about occupation as opposed to liberation. We were dismissive about the capacity of Iraqis to control their own future. We have struggled ever since.

CHAPTER 24

Sixteen Words

"Condi, we have a problem."

The national security advisor hated it when I would tell her that, but not as much as I hated saying it. Unfortunately, my job sometimes required that I use those words.

Now, in mid-June of 2003, I was obliged to use them again. I had called to tell her that it was time—past time, actually—that we all admit that some language in the president's State of the Union speech six months prior should not have been there. The words at issue were: "The British government has learned that Saddam Hussein recently sought significant quantities of uranium from Africa." Those words would later create a firestorm, but at the time of the State of the Union speech, they were barely noticed.

This story begins on Saturday, October 5, 2002. I was at work in my office when several members of my staff came to say they were having trouble getting the White House to remove some language from a speech the president was preparing to deliver in Cincinnati. The sixth-draft speech asserted that Saddam's regime had "been caught attempting to purchase up to 500 metric tons of uranium oxide from sources in Africa—an essential ingredient in the enrichment process." Analytically, the staff said, we could not support such a statement. Having testified to Congress only the day before on the matter, I was well familiar with the controversy. I picked up the phone and called Steve Hadley. Our conversation was short and direct. "Steve, take it out," I said, telling him that he did not want the president to be a "fact witness" on this issue. The facts, I told him, were too much in doubt.

My executive assistant followed up with a memo to the speechwriter and Hadley to confirm our concerns. It said in part: "Remove the sentence [regarding Saddam's attempt to purchase uranium oxide] because the amount is in dispute and it is debatable whether [uranium oxide] can be acquired from the source. We told Congress that the Brits have exaggerated this issue. Finally, the Iraqis already have 550 metric tons of uranium oxide in their inventory."

The White House removed the language, but the next day, Sunday, one of our senior analysts sent yet another memo to 1600 Pennsylvania Avenue, further driving home the reasons why CIA thought the offending words should not be uttered by the president. That memo said in part:

> *More on why we recommend removing the sentence about [Saddam's] procuring uranium oxide from Africa: Three points (1) The evidence is weak. One of the two mines cited by the source as the location of the uranium oxide is flooded. The other mine cited by the source is under the control of French authorities. (2) The procurement is not particularly significant to Iraq's nuclear ambitions because the Iraqis already have a large stock of uranium oxide in their inventory. And (3) we have shared points one and two with Congress, telling them that the Africa story is overblown and telling them this was one of two issues where we differed with the British.*

The memo has a handwritten note on the bottom from Mike Morell: "This has been sent to the White House (Rice, Hadley, Gerson)." (Mike Gerson was then the White House chief speechwriter.) Despite all that, the African yellowcake story would unhappily reemerge three months later, in the president's 2003 State of the Union address.

Thanks to some stories in the press, the ill-advised inclusion of those words in the State of the Union had become a flap. I picked up the handset on my "MLP"—a bulky white "secure" telephone over which one can discuss highly classified information without fear of the call being intercepted. If I pressed one button, I could call the president; if I pressed another, I would have the secretary of defense on the line, the secretary of state, or, as I did this day, the national security advisor.

I was calling from my office on the seventh floor of CIA's headquarters. Except for the addition of technological advances like the MLP, the office hadn't changed much in the forty years since the building opened: wood paneled on three sides, with a long expanse of floor-to-ceiling windows looking out over trees along the Potomac and toward Maryland and the District of Columbia.

Saddam and his search for African uranium had been based on questionable intelligence. In truth, the case suggesting that Saddam was reconstituting his nuclear weapons program was much weaker than the evidence suggesting that he was working on chemical and biological weapons. But the vision of a despot like Saddam getting his hands on nuclear weapons was galvanizing. The notion provided an irresistible image for speechwriters, spokesmen, and politicians to seize on.

Our NIE had said that Saddam was unlikely to have a nuclear weapon before the end of the decade. But it had also said that if someone gave him fissile material, he could have a weapon much sooner. If Saddam were smuggling uranium, it would mean he was going to the trouble to enrich his own.

The issue was not trivial—even if this bit of intelligence, his supposed attempts to obtain uranium suitable for enriching, known as "yellowcake," was far from solid information. The allegation was worthy of investigation. Based on what we found, however, it was not worthy of inclusion in a presidential speech.

When President Bush addressed the Joint Session of Congress on January 28, 2003, the handful of words toward the end of the lengthy address received very little attention from most people. But at that moment, they got absolutely no attention from me. I was at home, in bed, asleep. You won't find many Washington officials who will admit to not watching the most important political speech of the year, but I was exhausted from fifteen months of nonstop work and worry since the tragedy of 9/11. Frankly, too, I was relieved that, unlike in the Clinton administration, where my job had Cabinet status, I no longer was obliged to attend ritual events like the State of the Union.

In addition to dealing with the usual array of difficult counterterrorism decisions over the previous few weeks, I had also been handling some political infighting over the planned Terrorist Threat Integration Center (TTIC), whose creation the president planned to announce in his speech. TTIC, which later evolved into the National Counterterrorism Center (NCTC), was very controversial within the intelligence community. The president's plan called for CIA, FBI, and the Department of Homeland Security to have parts of their organizations stripped away to create this new entity. It wasn't clear who would be in charge of TTIC, who would select its leadership, or what functions the various agencies might lose. (If you want to stir up a hornet's nest in Washington, try taking responsibility away from proud agencies.)

The planning for the move was being held in strict secrecy, so that the announcement in the State of the Union speech would make news. The underlying secrecy made the bureaucratic players even more paranoid. I had to calm jangled nerves of several of my senior deputies, who feared that the loss of people to TTIC would render their own organizations ineffective.

Six weeks later there was a brief flurry of interest when the International Atomic Energy Administration (IAEA) determined that some documents they had been given by the United States relating to charges of Iraqi interest in Niger's uranium were forgeries. But the report came out just days before the start of the Iraq war, and the issue was lost in the noise. By that time, the die was already cast, and there was not much debate going on about bits and pieces of the underlying intelligence.

A second minor squall blew up in May when *New York Times* columnist Nicholas Kristof wrote that a U.S. envoy had been sent to Niger and reported back to the CIA and State Department, debunking the Niger uranium story. But again the story did not have, at least in Washington, what they call "legs." The column appeared just days after the president declared an end to major combat in Iraq while standing beneath a banner that read, "Mission Accomplished."

The story came back to life again in June when Walter Pincus, a veteran intelligence reporter for the *Washington Post*, started asking questions around town about a former U.S. ambassador who, he said, had been dispatched by CIA in response to questions from the vice president about the Niger uranium allegations. When Pincus first called us, the press office needed a day or two just to figure out what he was talking about. The ambassador's trip sixteen months earlier had been authorized at a low level within CPD, the Counterproliferation Division of the Directorate of Operations at CIA, and had produced such inconclusive results that the press office had trouble finding people who remembered the details of the trip. Eventually, our spokesman was able to figure out the story behind Pincus's inquiry. Yes, they told Pincus, there was such a trip but, no, the mission had not been undertaken at the vice president's behest, and the vice president was never briefed on the trip's less-than-compelling results.

What they didn't know at the time, of course, was that Pincus had learned about the Niger mission from Ambassador Joseph Wilson, the man CPD had asked to undertake the trip.

How did the trip happen? Several of our briefers had received questions not only from the vice president but also from the State Department and DOD about a February 2002 Defense Intelligence Agency report that first raised the possibility of Iraq having sought uranium from Niger. "What more do you know about this?" they were asked. "Hardly anything," was the answer. Midlevel officials in CPD decided on their own initiative to see if they could learn more. Someone had the idea that Joe Wilson might be a good candidate to look into the matter. He'd helped them on a project once before, and he'd be easy to contact because his wife worked in CPD. Wilson agreed and undertook the assignment without compensation. Only his expenses were reimbursed.

Critics have subsequently suggested that our selecting Wilson demonstrated that the Agency had it in for the administration. After all, wasn't he a supporter of the Democrats? I would argue that his selection illustrates that Agency officers often don't give a second thought to U.S. domestic politics. The report that Saddam might be getting yellowcake from Niger was not an issue of left or right—it was either right or wrong.

Not surprisingly, local officials in Niger denied illegally selling uranium to Iraq. Wilson didn't even write up a report; he gave an oral briefing to two CIA analysts at his home one evening over Chinese takeout food. Their summary of his remarks said that the officials denied selling yellowcake to Iraq but that one official admitted Iraq had been seeking expanded trade relations with Niger. The presumption was that the only thing Niger had worth trading was yellowcake.

This unremarkable report was disseminated, but because it produced no solid answers, there wasn't any urgency to brief its results to senior officials such as the vice president. Had the vice president been in Washington at the time, his personal PDB briefer might have mentioned it, but as it happened, Cheney was on a ten-day overseas trip when the report came out. By the time he returned to Washington, there were undoubtedly more pressing things to brief him on. As far as we could tell, the Wilson summary was never delivered to Cheney. In fact, I have no recollection myself of hearing about Wilson's trip at the time.

Pincus's story, which ran in the *Washington Post* on June 12, revived interest in the State of the Union address and yellowcake, and for several days thereafter the rest of the news media chased the issue, trying to sort out who had said what to whom—and how those sixteen words had gotten into the speech. Several follow-on stories by Pincus cited sources close to the vice president complaining that CIA had "failed" to keep them informed. It was pretty clear that some anonymous staffers in the vice president's office were trying to make sure that if there were any fallout over the issue, CIA would solely be held at fault. This became a familiar theme for us.

Then the issue seemed to have died again. And for me, the matter was nowhere near the top of my list of things to worry about that spring. Yes, the temperature was rising, but at any given time dozens of such issues are bubbling in Washington. Try as you might, you never know which mini-crises will subside and which will boil over. When I called Condi that June to express my concern over the matter, I was troubled by the weak intelligence that underlay the phrase, not with Joe Wilson. I have to admit, I did not see trouble looming when I first learned that Wilson's wife, Valerie, was a CIA employee. I did not view that as a big deal or a political vulnerability, or much of anything, for that matter. Condi called several days after my call to say the White House would not be issuing any statements saying that the Niger material should not have been used. Condi made it clear to me that this was not her decision.

Sunday morning, July 6, dawned a typical Washington summer's day. I tried not to go into work on Sundays so I could spend as much time as possible with my family. But work always came to me. My ever-present security detail delivered a stack of overnight cable traffic, intelligence analysis on critical issues, and a thick package of clippings from the morning's newspapers, called the Media Highlights, with stories relating to intelligence. Prominently displayed in the Media Highlights was a column written by Ambassador Wilson that appeared in the morning's *New York Times*. Apparently, he had decided that feeding anonymous stories to Kristof and Pincus had not achieved his goals, so this time he outed himself in an Op-Ed piece titled "What I Didn't Find in Africa."

While the earlier Kristof and Pincus articles had touched off brush fires, the Wilson Op-Ed and subsequent TV appearances ignited a firestorm. I'd been around Washington long enough to know that when you attach a name to an allegation, the story has much greater traction. If there were any doubt, it was removed when I tuned in to NBC's *Meet the Press* that morning and saw the guest host, Andrea Mitchell, interview Joe Wilson on his allegations that the administration had ignored his findings and hyped the Niger information even though, in his estimation, they "knew" the claim not to be true.

By Monday morning virtually every major news organization was chasing the story. Ari Fleisher, the soon-to-depart White House spokesman, was swamped with questions at his early morning press "gaggle," an on-the-record but off-camera media briefing. Ari told reporters that there was "zero, nada, nothing" new in the weekend's coverage other than the fact that Wilson's name was now attached to the allegations. He was pressed on whether the White House still stood by the words in the "SOTU"—Washington-speak for the State of the Union address.

Fleisher danced around that, but later that day—after the president, the White House staff, and the traveling press corps departed on a trip to Africa—Ari's staff finally released a brief statement acknowledging that the uranium language should not have been included in the speech. The White House had finally gotten around to saying the obvious—saying, indeed, what I had said to Condi Rice a few weeks before. I know of no meeting that was convened to come to this decision. The White House staff simply read the tea leaves after Joe Wilson's weekend media appearances and decided to commit truth.

That should have ended the matter. The White House essentially admitted that "mistakes were made," "we're sorry," and "let's move on." Each day brought fresh stories quoting anonymous officials pointing fingers at each other's organizations. Pundits began opining that the White House had deliberately misled the American people. The word "lied" was bandied about by administration critics.

Presidential overseas trips are especially likely times for self-inflicted crises to spring up. A huge press contingent and many staffers travel with the president—so many people that two 747s,

Air Force One and its twin, are not big enough to handle them all. The White House staff spends too much time cooped up together, and they tend to get one another spun up with the latest tales of what they are hearing from back home. Meanwhile, the press contingent is hungry for any tidbits or insider bickering to report. In the hothouse environment of Air Force One an "us against them" attitude often leads to badly thought through reactions.

As it turned out, I had some traveling to do of my own: a long-standing speaking engagement in Sun Valley, Idaho. This was at an event sponsored by Herbert Allen, whose investment banking company specializes in working with major figures in the entertainment, communications, and technology fields.

I made an hour-long off-the-cuff presentation in the Sun Valley Lodge's conference room about the state of the world as I saw it. It was my second appearance in front of this crowd; I enjoyed the informal banter in the subsequent question-and-answer period with the eclectic group of participants.

At one point, one of the attendees, NBC anchorman Tom Brokaw, suggested that I should be making this same kind of presentation on national TV. In front of the assembled group, which included some of his competitors, he offered me an opportunity to do so on NBC.

"Yeah, Tom," I said, "it's always been my dream to be grilled by you on national TV."

"Well, George," he replied, "you know we are in Sun Valley, and they call this 'the place where dreams come true.'"

Tom earned a big laugh—but no interview.

In addition to offering a chance to speak to an influential group of people, the trip also provided me an opportunity to take a day or two off in a beautiful setting. After checking with our ethics attorneys and agreeing to pay her expenses, I was able to bring Stephanie along for what I hoped would be a relaxing couple of days. But there was to be no relaxation. Almost from the moment I arrived in Idaho, I'd been bombarded with calls from Washington about growing concerns regarding the State of the Union controversy. Now, rather than enjoying the mountain trails and streams, I found myself fielding a never-ending flow of phone calls from headquarters telling me of the latest sniping going on across the Potomac and now across the Atlantic.

Stephanie and I were staying in a room in the main lodge that was said to have once been occupied by Ernest Hemingway. Unlike "Papa" Hemingway, though, we had to take over the adjoining room as well for a "command post." That was standard procedure. Whenever I traveled, even to a garden spot like Sun Valley, a team of communicators would arrive ahead of me and set up an office with sophisticated satellite communications equipment, allowing me to be in touch with national command authorities and to receive highly classified voice and data transmissions. The team would work in shifts to ensure that someone was always in touch with our headquarters back home. When taking trips with multiple stops, communications teams would have to leapfrog ahead of me, moving hundreds of pounds of equipment that would permit encrypted communications as soon as I stepped off the airplane at the next destination.

The communications this time were virtually nonstop. The classified fax machine kept humming, spitting out news stories, briefing transcripts, and editorials—a barrage that made it crystal clear this story wasn't going away soon.

Finally, at one point I decided that I had had enough. I called Steve Hadley at the White House. "We need to put an end to this," I told him. As I had explained to Condi in my call to her a few weeks before, including the uranium language in the State of the Union speech had been a mistake. Now, I said, I had decided that I would issue a statement accepting responsibility for the Agency's shortcomings in allowing the uranium language to make it into the speech. I would stand up and take the hit. Obviously, the process for vetting the speech at the Agency had broken down. We had warned the White House about the lack of reliability of the assertion when we had gotten them to remove similar language from the president's October Cincinnati speech, and we should have gotten that language out of the SOTU as well. It was because of my failure to fully study the speech myself that I took responsibility. We owed it to the commander in chief, and we had failed him, and now, I told Hadley, was the time to own up to that.

Hadley candidly responded that the process had not worked well at the White House, either—and that they would stand up with us. "It will be shared responsibility, George," he told me. For that reason, I fully expected Condi Rice to publicly state that she joined me in accepting responsibility.

I wasn't just being magnanimous. Part of the fault truly was mine. The day before the State of the Union, I was at a Principals' meeting in the White House Situation Room, a place where it seemed I spent more time than in my own home in recent years. As the meeting broke up, several of us were handed copies of a draft of the forthcoming speech. I remember going back to headquarters and giving the draft to one of my special assistants, unread, and asking that it be put "into the system for review."

I gave it no further thought. As always, other crises were banging on the door, but I fully expected that if there were any problems with the State of the Union draft, someone would have come and alerted me. That's exactly what had happened with the Cincinnati speech the previous fall. On another occasion, involving the 2002 State of the Union speech, my chief of staff, John Moseman, and spokesman, Bill Harlow, intervened at the last moment to stop the president's speechwriters from including language about the number of terrorists believed to have been trained in Bin Ladin's camps in Afghanistan, a number that was tens of thousands beyond what we thought true. Moseman called the NSC staff and said, "Look, if the president goes out and says that and tomorrow media call us and ask if we agree with the number, Harlow is going to have to say no. The number was corrected at the last minute—so late that an advance text copy of the speech put out at a background briefing at the White House that evening still contained the unsupportable tally.

In early 2003, though, the same system and same people that had rescued the president from incorrect assertions in previous speeches failed to catch the troublesome language in the State of the Union. Later, in trying to find out why alarm bells hadn't gone off, I was told that Alan Foley, head of WINPAC, had focused on clearing the speech for "sources and methods," rather than for substance. In other words, as long as the language didn't give away any secrets about how the intelligence was collected, they didn't worry about whether we believed the assertions in the speech were accurate. That was a terrible mistake. Our job was never to clear solely for sources and methods, but also for substance. And the last time I looked, as good as the British intelligence service is—and it is very good—it does not work for the president of the United States.

On the morning after I talked with Steve Hadley, I called Washington, pulled Bill Harlow out of the morning staff meeting, and told him that I had decided to issue a statement taking our share of the blame for the mix-up. I gave him a sense of how I wanted the statement to go, and read him a few opening paragraphs I had scribbled on a yellow legal pad overnight, since I hadn't been able to sleep.

My instructions were clear: "I want this statement scrubbed carefully. It has to be as accurate as we can make it. Factual, clear, and no whining allowed." But more than just saying "we screwed up and we're sorry," I wanted to lay out to the extent possible what had happened. The statement also needed to be a roadmap and to convey the clear impression that we never believed the Niger story. Most important, I wanted to say that we regretted having let the president down and that I took personal responsibility.

My deputy, John McLaughlin, and Bill Harlow labored long and hard trying to construct a statement that would accomplish what we wanted and stand up to scrutiny. It was a painful process.

They wrote version after version trying to get the language right, faxing drafts back and forth to me in Idaho and checking with all the appropriate players at CIA headquarters. Among the people they needed to consult was Alan Foley, a senior CIA official who had discussed and eventually cleared the language for the State of the Union with Bob Joseph, a senior NSC official. John and Bill wanted to ensure that they understood Foley's actions and position, but as it turned out, he was on an official trip to Australia. So there I was in Idaho, coordinating a statement with my staff in Washington while they were reaching out to a key player in Australia, and we were all looking for more incoming flak from the traveling White House in Africa.

Early in the process, I decided that I wanted to inject some perspective. Yes, it was a bad thing that some of the language drafted for the president's remarks didn't rise to the level of certainty that one would expect, but after all, we were talking about a tiny fraction of his speech. That's when Bill counted and found that we were talking about only "sixteen words"—a phrase that would take on a life of its own. Later some would allege that this handful of words was critical to the decision that led the nation to war. Contemporaneous evidence doesn't support that, but just try convincing people of that today.

A better case could be made that the "sixteen words" started an unintended war between the White House and CIA. That was certainly not our intention. If there was such a war, it was largely one-sided. Neither I nor my senior leadership ever considered ourselves at war with the vice president or anybody else.

At one point, Steve Hadley asked me to call Scooter Libby, the vice president's chief of staff, to discuss my forthcoming statement. I refused to do so. The statement was to be mine and no one else's. I've subsequently seen reports that Libby and Karl Rove debated what they would like to see in my statement. Perhaps so, but I was unaware of their views at the time.

Sometime between drafts one and seventeen of my "mea somewhat culpa," Bill Harlow was interrupted by a call from syndicated columnist Bob Novak. Novak said that two administration sources had told him that the real story on the Joe Wilson trip was that Wilson's wife worked for the Agency and was responsible for sending her husband. Bill struggled to convince Novak that he had been misinformed—and that it would be unwise to report Mrs. Wilson's name. He couldn't tell Novak that Valerie Wilson was undercover. Saying so over an open phone line itself would have been a security breach. Bill danced around the subject and asked Novak not to include her in the story. Several years and many court dates later, we know that the message apparently didn't get through, but Novak never told Bill that he was going to ignore his advice to leave Valerie's name out of his article.

I was amused to hear Novak subsequently say that he is confident that I must have been aware of his call at the time and that if I had only phoned him to tell him not to run the item, he would have complied. I was not aware of Novak's call. I was consumed with the "sixteen words" flap, wondering if in the next few days I would need to resign or would perhaps be fired. About two weeks after Novak's column appeared, CIA lawyers sent to the Justice Department a formal notification that classified information may have been inappropriately leaked to the media. CIA lawyers had to make that kind of notification about once a week on average. I was informed after the fact that a "crimes report" had been submitted. I supported the action but had nothing to do with the decision. It's been suggested I ordered the action to get back at the White House for some reason. This is absurd. At the time we had no idea where the leak had come from but were obligated by law to report it to the proper authorities. I was angered that someone, whether intentionally or not, blew the cover of one or our officers and that they appeared to be implying that some desk-bound analyst at Langley was sending her husband on a boondoggle. This was never the case. Nor can we have outsiders determine who is legitimately undercover, ever—because it suits the politics of the moment. To do so is irresponsible and dangerous.

Even as we were drafting our statement taking responsibility, we were hearing from reporters that the sniping at us aboard Air Force One was intensifying. I told my staff to stay calm and not be taken in by one of the oldest reporter's tricks in the book: "Did you hear what they said about you?" Still, it was maddening that we were seeing no signs of that "shared responsibility" that I had been promised by Hadley. Reporters kept calling our press office with accounts from "senior administration officials" on Air Force One who continued to insist that the CIA's share of the fault was 100 percent.

Late on Thursday, July 10, I asked John McLaughlin to send a copy of the draft of my statement to Hadley. "Make clear to them, John," I instructed, "that we are sending the draft over for their information only. We were not soliciting their concurrence, and we are for damn sure are not seeking their edits."

Around 2:00 A.M., Mountain Time, I was rousted from my bed by my special assistant, Scott Hopkins, to take a call from Condi Rice, who was somewhere in Africa.

Condi might have been responding to the draft of our statement I had sent to Hadley. He, no doubt, had forwarded it to Air Force One. Or maybe she was reacting to a *CBS Evening News* report by Pentagon correspondent David Martin. According to sources, Martin said, CIA officials had warned the White House that the Niger reporting was "unreliable," but the White House had gone ahead with it anyway. Martin had the story only partly right. We had warned the White House against using the Niger uranium reports previously but had not done so with the State of the Union; still, a story like that was bound to spike the blood pressure on Air Force One. CIA seemed to be deflecting blame. Here was a perfect storm, with all the key players in different time zones and continents.

Early Friday morning, the CIA press office was suddenly inundated with calls from reporters looking for a reaction to a press briefing that had just taken place on board Air Force One about the Niger issue. En route to Entebbe, Uganda, Condi Rice had conducted an on-the-record press briefing of nearly an hour, during which time she was peppered with questions, mostly about that single sentence in the State of the Union speech. Soon wire stories began appearing quoting Condi as saying, "If the CIA, the Director of Central Intelligence, had said, take this out of the speech, it would have been gone, without question." The Reuters wire service carried a story headlined "White House Points at CIA over Iraq Uranium Charge."

In response to questions, Condi denied that she was blaming CIA and she stressed that the president still had confidence in me and the Agency. She was sure I would not "knowingly" have put false information in the speech, even though the line somehow got in there. That was hardly a ringing endorsement, but the question itself was just as worrisome. When reporters start asking if the president still has confidence in you, you know you are in a world of trouble.

Later that morning, McLaughlin received a call from Hadley, who, despite our admonishments, had a few suggestions to offer to "improve" our draft. The opening paragraph of the draft, for example, was not as strong as I had wanted it to be with regard to our taking responsibility. I knew that Condi and Hadley would press us on taking the blame more directly. They did not disappoint us. I conceded a few points and strengthened that part, and was pleased that the administration was not focusing much on the latter portion of the statement, which for anyone who read it carefully, laid out a roadmap for arriving at the complete story. That portion was a neon sign that pointed to the fact that we were especially unhappy at having allowed the sixteen words to get into this speech, since we had previously expressed serious doubts about the reliability of the information and did not think that it was a reason to believe that Saddam was reconstituting his nuclear weapons program.

I guess we struck a nerve. Although I didn't know it at the time, it was revealed in Scooter Libby's trial in February 2007 that the draft of my statement was being passed around the White House. Someone, whose handwriting reportedly resembled the vice president's or perhaps Steve Hadley's, wrote "unsatisfactory" on the draft. Also penciled in was a proposed change that we did not accept that would have rendered the press release factually incorrect. They wanted us to say that Niger was "just one" of the factors we relied on to make the nuclear reconstitution case. In fact, we said it was "not one" of the factors.

Despite what some White House officials have subsequently said, I was anxious to get the statement out. The story had taken on a life of its own, and I didn't want to go through another weekend with more media speculation as to who said what to whom. I also didn't want to issue the statement late on a summer Friday night, a technique usually reserved in Washington for statements that officials want to bury. That was not the case with this statement. The only reason for getting it out there was so it *would* get attention.

Even as we were preparing to release the statement, we began to hear from other precincts. Senator Pat Roberts, chairman of the Senate Intelligence Committee and a close confidant of the vice president's, told reporters that he was "disturbed by what appears to be extremely sloppy handling of the issue from the outset by the CIA." Roberts reportedly said that he was most concerned about "a campaign of press leaks by the CIA in an effort to discredit the president." To top things off, he accused me of failing to warn the president about any doubts at the Agency regarding the Niger information. The chairman convicted us of trying to discredit the president and of sloppy work—without ever once bothering to ask us the facts. I wondered at the time, "Where is he getting his information?"

All this sniping was going on while we were working to finalize the text of a statement in which we would take our "share" of the responsibility. Meanwhile Hadley called wanting to set up a conference call with Condi Rice to talk about the draft. Reluctantly, I agreed. There were four of us on the line—me in Sun Valley, Condi in Uganda, Hadley from the West Wing, and John McLaughlin holding down the fort at Langley.

I could tell by the tone of her voice that Condi was furious. I resisted asking why she had felt it necessary to hold an airborne press conference hours earlier hanging all the responsibility on me. I had an equal right to be angry, but that wasn't going to help get a statement out.

Finally, I told them that I was comfortable with my statement, and I asked John McLaughlin to tell Bill Harlow to send it out. As the call was wrapping up, someone expressed the hope that we could get this issue behind us. I still hadn't heard any sign of "shared responsibility" from the administration. "What are you going to do about Cincinnati?" I asked Condi. There was dead silence on the line. I reminded her that I had intervened to get similar language out of the Cincinnati speech, and yet it had found its way back into the State of the Union address. The conversation ended uncomfortably.

I felt a certain sense of relief once the decision was made to release the press statement. "We're finally free to take in some of this Idaho scenery," I told Stephanie when I got off the conference call. Soon thereafter, we got in an SUV driven by my security detail and headed through the mountains to a nearby lake for some much-needed relaxation—or as close as you can get to relaxation when you're DCI.

My staff used to joke about how I would claim, when going off on a rare vacation, that I wasn't going to give work a moment's thought, and then, before my car had left the Agency compound, I'd call in on my cell phone to see how things were going. Here in Idaho, it was no different. I was anxious to learn what the reaction was to the release of my statement. Unfortunately, though, none of our sophisticated cell phones seemed to work in the mountains of Idaho. My communications team was still in Sun Valley, so we decided to stop at a rustic roadside store in search of a pay phone—a place called the Smiley Creek Lodge, in Sawtooth City. Not exactly a major metropolis. It turned out the place had only one working pay phone, and four people waiting in line to use it.

One of my security team asked if I wanted him to tell those

waiting that it was a national emergency so we could jump ahead of the queue. "That's all I need," I thought, "some guy flashing a badge to get me head-of-the-line privileges." I opted to wait for the folks ahead of us to complete their calls, although I did allow one of my security detail to take my place in line while I got a milkshake and fries. (I highly recommend both the next time you are in Sawtooth City.) When my turn for the phone came, I learned that the Agency press staff was swamped with incoming calls, but it was too soon to gauge how the story was playing.

When we finally got to the lake, Stephanie and I got in a two-person kayak and paddled around, taking in the majestic beauty of the nearby mountains. It was peaceful, quiet, and quite romantic—just Stephanie, me, and the other canoes with my security detail. Some of the beefier members of my security team almost swamped their kayaks.

On the way back to Sun Valley, we stopped at the Smiley Creek Lodge again to use the pay phone. By now, the predictable uproar was in full swing. All three network news programs had led with stories about my taking the blame for the now-famous sixteen words. Every major newspaper was covering this as well, and many speculated that my days as DCI were numbered as a result.

Early the next morning, a Saturday, I was awakened in Sun Valley, this time not by Condi Rice but by a call from our then-sixteen-year-old son, John Michael, who had stayed behind at our home in Washington's Maryland suburbs. He was quite upset. "Dad," I can remember his saying, "there are a bunch of television camera crews out in front of our house. They are just standing on the neighbor's lawn with their cameras pointed at our house. What should I do?"

I tried to explain to him that this is what happens when you find yourself all over the front page. ("CIA Director Takes the Blame," the *New York Times* headline shouted that morning, I would later find out.) But my son thought a bunch of strangers "staking out" our house was a bit too much.

"Dad, I'm going to go out there with my baseball bat and slug one of them," he said, full of a sixteen-year-old's bravado.

I was glad his mother was not on the line.

"No, John Michael, those cameramen are just doing their jobs." I reminded him that one of our closest family friends, George Romilly, whom he called "Uncle George," was a cameraman for *ABC News.* Had he been on duty that morning, Uncle George might have had to stake me out just like the others.

I called the security officer on duty in the basement of our house and told him to slip our son out the rear door, across the yard of our back-door neighbor's, and have him wait on a nearby street, where Stephanie's brother, Nick, would pick him up. In the meantime, I asked CIA's very able deputy spokesman, Mark Mansfield, to race over to my house and chat up the TV crews.

"You guys are welcome to stay out here and stare at that house," Mark told them once he arrived. "But I thought you ought to know that Director Tenet is out of town. You could be in for a long, long wait."

"When will he be back?" they asked.

"Can't say—we never discuss his movements, for security reasons," Mark told them with a smile, as he wiped his brow to emphasize that the temperature was ninety degrees and certain to climb higher. "You guys could be in for a lot of overtime." Mark left, and shortly thereafter, the TV crews did, too. I returned, as previously scheduled, late that evening.

That same day, the White House sent around draft talking points for administration officials who would be interviewed on the Sunday talk shows the next day. My chief of staff, John Moseman, was stunned to see that the talking points still tried to justify their including the "sixteen words" in the State of the Union speech. John called the NSC staff and told them they were nuts to keep beating that dead horse. He suggested they just take my statement from the day before and stick with it. Those words should never have been in the president's speech. Period.

I took some comfort in a small article buried in the *New York*

Times on the day after I returned to Washington from Idaho. The article reported that, at CIA's behest, the White House had removed any mention of African uranium from the Cincinnati speech in 2002. I was especially pleased that the reporter attributed this fact to "Administration officials involved in drafting the speech." This had to have come from the White House. Perhaps they were about to step up and admit some error, too.

On Sunday, July 13, I got a call from Secretary of State Colin Powell asking me to come over to his home. I was just back from Sun Valley; Colin was just back from the African trip along with Condi, the president, and others. Together, we drank lemonade on his back patio. Colin, it turned out, had been asked by the president to deliver a message to me.

"Keep your building quiet," he said. Washington is the only place in the world where buildings are believed to speak. What he meant was that I was somehow supposed to get the thousands of Agency employees to quit responding when officials in the administration took rhetorical shots at them, deserved or not.

Colin also wanted to give me some of the atmospherics from Air Force One. There had been a lively debate among staffers on the aircraft and back in Washington, he said, about whether to continue to support me. In the end, the president said yes, and said so publicly. But Colin let me know that other officials, particularly the vice president, had quite another view.

Reactions to my "mea culpa" continued to pour in, and not just from the media. My old boss and mentor, Senator David Boren, now president of the University of Oklahoma, was livid. He sent word that he was very disappointed that I had not consulted with him personally before issuing the statement. Had I done so, he insisted, he never would have agreed with the wisdom of my accepting blame for the incident. He had been after me to resign from the Agency for some time. If I left now, however, everyone would believe I had been fired. "You're stuck," he said.

While my staff continued researching what had gone wrong with the State of the Union process and what had gone right with the Cincinnati speech, the "who screwed up?" stories percolated day after day, fed by a White House spin machine that kept trying to find ways to turn the issue to its favor.

During the middle of the week, NSC officials called asking us to declassify just a couple of paragraphs from page twenty-four of the NIE dealing with uranium from Africa. The person responsible for handling the request at the Agency refused to do it. "It's misleading," he explained to John Moseman. "Put out those two paragraphs and you imply that the Niger stuff was a major part of our thinking. It wasn't. We did not even cite the reports as among the reasons we thought Saddam was reconstituting his nuclear weapons program."

Moseman told the NSC we wouldn't do it. On July 17, a written request came in asking that we declassify the reasons why we thought Saddam was pursuing nuclear weapons. That was followed the next day by another written request that we declassify the NIE's "Key Judgments" and the paragraphs concerning yellowcake from page twenty-four. Both requests were signed by Condi Rice. Although less than an ideal solution, it was better than declassifying the Niger stuff alone. We complied.

In fact, it was a few years later that we learned through court papers and the media that, much earlier, the White House had apparently declassified parts of the NIE without telling us. Special Counsel Patrick Fitzgerald said in a court filing on April 5, 2006, that "[Libby] testified (before the Grand Jury) that the Vice President later advised him that the President had authorized [Libby] to disclose the relevant portions of the NIE." From the court documents, it is clear that these briefings occurred on or before July 12, 2003.

I now believe that one of the reasons some people in the White House were unhappy with my "mea culpa" statement was that the details in it might lead some of the journalists who received background briefings on the NIE—without our knowledge—to discover that they had been misled regarding the importance we attached to intelligence reports alleging that Iraq had vigorously pursued yellowcake in Niger. My statement made clear that we put little stock in that reporting and we did not rely on it for our judgment regarding whether Iraq was reconstituting its nuclear weapons program.

On the afternoon of Friday, July 18, two senior White House officials held a lengthy background briefing during which they discussed the situation with the media. At the start of the briefing they released to the press the Key Judgments and the Niger paragraphs from the NIE, both of which we had declassified that morning. Their intent was obvious: they wanted to demonstrate that the intelligence community had given the administration and Congress every reason to believe that Saddam had a robust WMD program that was growing in seriousness every day.

The briefers were questioned about press accounts saying that the White House had taken references to Niger out of the Cincinnati speech at CIA's request. Why then, did they insert them again in the State of the Union? The senior officials said that the material that was taken out of the first speech was quite different from the material the president used before Congress. That simply wasn't so. It was not clear to me then, nor is it clear now, whether they even understood the facts, but it was clear that the entire briefing was intended to convince the press corps that the White House staff was an innocent victim of bad work by the intelligence community. Here, again, was the familiar mantra: the intelligence community made us do it. Apparently, I was expected to go along with the notion that *only* we had screwed up. In any event, instead of spiking the sixteen-words story, the briefing just gave it more life. More stories about "what the White House knew and when they knew it" kept rolling out all weekend long.

Just before six o'clock on Monday morning, July 22, the MLP secure telephone rang in the command post in the basement of my home. One of the security officers on duty buzzed me on the intercom. Condi Rice wanted to talk to me. I wearily dragged myself downstairs to take the call. I had the impression that Condi was already at work. She told me that the administration had decided that this was the day that the White House would accept their share of responsibility. Finally.

"Don't do anything until I can talk to you," I said. "I want to make sure you've seen all the same material I have."

Later that morning I went to the White House as usual for the president's daily intelligence briefing. I brought with me two memos that my staff had recently dug up—memos we had sent the White House in October 2002 explaining in detail why the president should not cite the yellowcake information in his Cincinnati speech. Condi had told me earlier that she wouldn't be available that morning—she was traveling—so I went to see Steve Hadley before the briefing and handed him copies of the memos. As he read them, I could see his face go ashen.

We didn't have time for a lengthy discussion about the memos' content—the briefing was about to begin—but I had brought along a second set of the same memos to show the president's chief of staff, Andy Card, someone whom I admired and respected greatly. Just before the PDB got under way, I asked Andy if I could see him privately in his office once we were finished. "Sure," he said, "go down there and wait; there are a few things I need to discuss with the boss first." As I recall, the vice president and Hadley also stayed behind when the briefing was over.

Afterward, I waited in Andy's office for what seemed like an hour, a highly unusual circumstance. When he finally showed up, I handed him copies of the two memos.

"Andy," I told him, "some folks here still don't get it. Not only did I personally call Steve Hadley last October and demand he remove the yellowcake stuff from the Cincinnati speech, but my staff sent two, count 'em, two follow-up memos to make sure the NSC got the point."

Apparently, Andy had already been acquainted with the memos while I was cooling my heels waiting for him in his office. He told me that he just learned that Hadley, Rice, and the chief speechwriter, Michael Gerson, had read the memos when they were received in October. All three must have known from the memos that our objections to the Niger information were much broader than was alluded to in the background briefing at the White House the previous Friday.

"Why are you giving me these memos only now?" Andy asked. He looked stunned.

"I wanted to double-check on my end to make sure that not only did we write the memos, but that they were received as well. I had my staff confirm with the folks who keep the secure fax machine logs that the memos were in fact sent and received," I said.

Just to remove any doubt, I passed Andy a slip of paper indicating the precise times each memo had arrived at the White House Situation Room.

"Besides," I said, "I presumed you were doing the same thing around here—looking for the facts. If I have the memos, surely your staff gave them to you, too, didn't they?"

Andy shook his head and simply said, "I haven't been told the truth."

Days later my staff was still digging through our files, trying to come up with a better understanding of the history of CIA's involvement with attempts to get the yellowcake information out of presidential speeches. That's when my executive assistant found a copy of draft remarks for a September 2002 speech, dated several weeks before the Cincinnati speech brouhaha. The White House staff had sent us some comments planned for use by the president in a Rose Garden event scheduled for September 26, 2002, following a meeting with congressional leaders. In the draft were these words:

We also have intelligence that Iraq has sought large amounts of uranium and uranium oxide, known as yellowcake, from Africa. Yellowcake is an essential ingredient of the process to enrich uranium for nuclear weapons. With fissile material, we believe Iraq could build a nuclear bomb within one year.

A footnote in the draft, typed in by the White House speechwriters, noted that the NSC and CIA were debating these three sentences. Apparently, we had earlier raised our concerns and were trying to persuade them to drop that segment of the speech. One of my assistants later marked the three sentences for deletion and penned in a note that read:

> *9-24-02 (8 PM)*
> *Rice proposed simply removing the bracketed text. Jami concurred.*

I don't believe that this earliest attempt to get the yellowcake information in the president's mouth has ever been publicly mentioned before. Why do so here? What's the significance of this nonevent? Either people overwhelmed with data and meetings had simply forgotten, or, for the White House speechwriters, the third time was the charm.

On the afternoon of July 22, the same day I gave Andy Card copies of the memos regarding the Cincinnati speech, Steve Hadley and Dan Bartlett were again in the White House press room. This time they were "on the record." The single-subject briefing lasted for one hour and twenty-three minutes. Hadley admitted having been reminded just that morning of our two October memos, which described weakness in the Niger uranium evidence and the fact that Iraq's effort to procure the yellowcake was not particularly significant to its nuclear ambitions because the Iraqis already had in inventory a large stock, 550 tons, of uranium oxide. Hadley said that "the memorandum also stated that the CIA had been telling Congress that the Africa story was one of two issues where we differed with the British intelligence." He said that the memo was received by the Situation Room and sent to both Dr. Rice and himself. One reporter asked Bartlett if they were saying that the mess was not George Tenet's fault as had been said the week before. Bartlett ducked the question. That,

I suppose, is what the White House meant when it promised to "share" the blame.

Only sixteen days had elapsed since Joe Wilson's Op-Ed piece about the sixteen words appeared in the *New York Times*. In that brief period, my relationship with the administration was forever changed.

CHAPTER 25

Going

At some point in a job like mine, you just give out. You've been going on adrenaline for so long. The relentless pressure and middle-of-the-night phone calls take their toll. The work matters enormously, and it's never over. But the family time lost, the high school lacrosse games missed, the vacations cut short or not taken—they all add up. And then something comes along, some essential trigger, and that's it. You know you've hit the wall.

I had just about reached that point when the sixteen-words flap broke out. Internecine warfare and finger-pointing are inside-the-Beltway intramural sports, but this time the pushing, shoving, and back-biting seemed to have been taken to an Olympic level.

A few months earlier, in May 2003, Senator David Boren had asked me to come to the University of Oklahoma to deliver the commencement address. That afternoon, following the graduation, David and his wife, Molly, took Stephanie and me out to the site of the new house they were preparing to build. There, on a hill in the middle of a field, David once again argued vehemently that it was time for me to resign. I had put in my time, served under two presidents, and weathered 9/11, David said. No one could ask for more from a DCI. It was best to go out on a high note. I know of no more astute observer of the ebb and flow of politics than David, and I listened carefully to him. Back in Washington after that trip, I told Andy Card that I was considering stepping down, but I hadn't fully made up my mind.

During this time, I learned that the administration was talking with Jim Langdon, chairman of the president's Foreign Intelligence Advisory Board, about taking my job. Whether that was as a result of my conversation with Andy Card or an independent initiative, I have no idea. But beyond that, I heard very little until that September, when the president asked me to come in early one morning in advance of the daily briefing.

Alone in the Oval Office, George Bush looked at me and said, "I really need you to stay." It wasn't a long conversation, and under the circumstances, with a war still going on in Iraq and the fight against terrorism still raging in Afghanistan and around the world, it would have been hard to say no to the president.

At a personal level, yes, I was probably ready to go. The most important reason to leave was my son, then a sophomore in high school. The job was toughest on him, and the public pounding I was taking did not help. I was worn out, but the CIA had men and women committed on many fronts. Leaving them or the rest of the Agency workforce in the middle of that would have been difficult. We'd worked too hard together, put in too many long hours, and accomplished too much. I felt an enormous obligation to them; they had become family to me. Nobody is indispensable, yet I also knew there was so much more to be done. And in truth, while catching Khalid Sheikh Mohammed had been big, I wanted to be at the helm when Usama bin Ladin was brought to justice.

Almost as important in my mind were the 9/11 Commission hearings looming on the horizon. You didn't need a pitch-perfect ear to know that they would be contentious and politically charged. I was going to be called to testify whether I was still DCI or not, but the Agency was sure to be thrown into turmoil by the hearings. I couldn't in good conscience leave that mess waiting for whoever my successor might be.

So I settled back into the DCI's chair, continued to put in the long hours, and did everything I could to keep morale high at an Agency that was being stretched perilously thin by Afghanistan, Iraq, and the global war on terror. As I had been doing for years,

I also worried day and night about what al-Qa'ida and other like-minded groups might next have in store for us.

On February 5, 2004, I delivered a major speech at Georgetown University, laying out the Agency's record on Iraqi WMD and affirming our professional commitment to call them as we saw them. Seven weeks later, on March 24, and again on April 14, I testified publicly before the 9/11 Commission. Both commission appearances were grueling experiences. In the end, though, I tried to represent the Agency well. Then, on April 17, three days after my last appearance before the 9/11 Commission, I picked up my *Washington Post*, saw the front-page story touting Bob Woodward's new book on the run-up to the Iraq war, and read the following in the second paragraph of the story:

"The intensive war planning throughout 2002 created its own momentum, according to *Plan of Attack* by Bob Woodward, fueled in part by the CIA's conclusion that Saddam Hussein could not be removed from power except through a war and CIA Director George J. Tenet's assurance to the President that it was a 'slam dunk' case that Iraq possessed weapons of mass destruction."

That's when I pretty much knew the wheels had come off the train.

As I wrote earlier, I'd had some advance notice of this. Woodward called just before *Plan of Attack* came out and, in an awkward way, raised the "slam dunk" issue. I guess he was trying to warn me it was going to be controversial, but my first reaction was almost a total blank. I remembered no such seminal moment. Now seeing the words in the *Post*, I felt as if I were reading about someone else in a parallel universe. Within days, though, Woodward's book had ignited a media bonfire, and I was the guy being burned at the stake.

This controversy was the last thing I needed. I took off for a few days and went up to the New Jersey shore, by myself. I wanted to get my thoughts together, and the beach, to me, is about the most serene place on earth. This, however, was not a peaceful time. Yes, we at CIA had been wrong in believing that Saddam had weapons of mass destruction. In the National Intelligence Estimate, in testimony on the Hill, in briefings to almost every member of Congress, I, John McLaughlin, and others had delivered the same message: our analysis showed that Iraq had chemical and biological weapons and was working on a nuclear capability, though they were years away from achieving it. There was no secret about it. Now, thanks to White House spin, our long, complex record on a difficult subject had been reduced to some ridiculous scene out of a comic opera. It was like I was Tom Cruise jumping on Oprah Winfrey's couch.

It was obvious to me that this whole Oval Office arm-waving, jumping-off-the-sofa, slam-dunk scene had been fed deliberately to Woodward to shift the blame from the White House to CIA for what had proved to be a failed rationale for the war in Iraq. Woodward's books, dependent as they are on insider access, have long been used in just this way—to deflect blame and set up fall guys. Now it had happened to me.

I remember sitting there at the beach contemplating all we had accomplished in my seven years on the hot seat—the rebuilding of a broken Agency, the restoration of morale, the successes in Afghanistan and the larger war on terrorism, the takedown of A. Q. Khan and the neutralizing of WMD development in Libya, our role in the Middle East peace process, my own role as personal envoy to the Crown Prince of Saudi Arabia and Pervez Musharraf, and so much else—and thinking, my God, none of that really matters to this administration. What I couldn't stop wondering was, had the president been convinced by some of his advisors that the blame should be shifted onto me? In the end, I will never know the answer to that question.

I like the president, plain and simple. We had been bound together after 9/11 by a national trauma and a common purpose. All of us at the storm's center believed we were doing the right thing, and every one of us, the president included, had given it his or her absolute best effort. His staff, though, had different priorities. For them, preserving the president's reputation—particularly with an election coming up and a war plan coming apart—was job one. Perhaps I was just collateral damage.

Maybe my second day at the shore, I phoned Andy Card at the White House and laid it on the line for him. "Andy," I remember saying, "I'm calling to tell you that I'm really angry. Yes, we wrote a National Intelligence Estimate, we expressed our confidence levels, John McLaughlin and I briefed almost every member of Congress; we were fairly strident about the fact that we believed Saddam had weapons of mass destruction. But what you guys have gone and done is made me look stupid, and I just want to tell you how furious I am about it. For someone in the administration to now hang this around my neck is about the most despicable thing I have ever seen in my life."

Andy is one of the most honorable, decent people I've ever worked with. What's more, he was always very good to me. But he is also extremely disciplined about what he says when he talks to you, and this time he said nothing in response. There was only quiet from the other end. Yet in that silence, I understood that there had been a fundamental breakdown of trust between the White House and me. In short, it finally, absolutely was time to go. I couldn't quit immediately over something that had appeared in a book, but I didn't see any way I could or should stay on much longer.

Over the next six weeks, I tried to think through the resignation process with Stephanie, my brother, Bill, John McLaughlin, John Moseman, and Bill Harlow. I also talked to David Boren about it in this period, as well as to my old friend Ken Levit, who went back years with me in the Senate and had served as my special counsel at CIA. That Memorial Day weekend, back at the beach, I had several long conversations with my brother. He was adamantly against resignation, because he felt that if I stepped down, the administration would dump on me whatever else they wanted to. "They've already done that," I pointed out to him, "and I'm still in the job!" Stephanie was also opposed to my resignation, because she did not want me to leave while the country was at war and our men and women were at risk on the ground in Afghanistan and Iraq. As for me, I already knew the answer. Unfortunately, to the outside world, my credibility had been undermined. My staying on would only hurt CIA. And then, as if by magic, someone appeared to confirm my decision to go.

That Sunday evening we were cooking hamburgers, but we didn't have any buns. I volunteered to run over to the A&P to get them. I've always found food stores and food shopping to be very therapeutic, probably as a function of growing up in my family's diner. So there I was at the A&P, pushing my cart down aisle seven, and unbeknownst to me, Louis Freeh, a dear friend who three years earlier had stepped down as FBI director and was a fellow devotee of the Jersey shore, was simultaneously pushing his cart down aisle eight. At the end of the aisle, Louis made a left turn, I made a right turn, and—yes, it's true—our carts smacked right into each other.

I looked up and said, "Well, Louis, how are you?" He said fine, and asked after me, and since we knew each other well and had gone through some of the same battles, I told him how upset I really was. We were both in shorts and T-shirts; my security detail was waiting outside. I explained my thinking, and we discussed my dilemma standing there in the middle of the A&P, our carts blocking the aisle. Louis first tried to talk me out of resigning. I looked at him and said, "I can't stay. Trust has been broken." Louis finally said to me, "You're right. It is time to leave. Now, here's how you do it."

To begin with, Louis said, you pick the date; no one else does.

"Fine," I told him. "Thursday." Four days hence.

"Okay, Thursday. You go in to see the president late Wednesday night. You ask to see him alone. You tell him that it is your intention to resign and to issue a public statement the following morning, and you ask him to keep this between the two of you until that occurs. Then, once he announces you're leaving, you announce it to your workforce. The key thing is to allow no more than ten to twelve hours to separate your conversation with the president and your announcement to your own people. That's why you don't see him earlier in the day or in the middle of the day. You see him as late as you can possibly see him because you want to keep this buttoned up. The worst thing that could happen is that word reaches your people before you tell them. You don't want to be in that position."

I shook hands with Louis when he was through and went back to the house feeling terrific. "God has spoken," I said when I walked into the house. "Louis has told me how to do it, and that's what I'm going to do." After I explained what that was about, Stephanie felt better about it, too, but she wasn't yet convinced this was the right course. As for me, I slept great that night, better than I had slept in months, maybe in years.

Before Louis and I parted at the A&P, we had agreed to meet the next morning with our families, under the American flag, for the terrific local Memorial Day parade. Louis, his wife, Marilyn, and Stephanie and I chatted away about the situation. For Stephanie, I know it was an important moment. She got a great deal of comfort from Louis and from the way he reinforced that this was the right thing to do, and by the time we set off that afternoon to drive back to Washington, she had come into camp, too. Louis Freeh swore me into office in 1997, and now he was telling me how to quit. Life had come full circle.

Wednesday morning, I set out to put the Freeh Plan into motion. The president and Andy Card were traveling that day, so I placed a call to Andy's office and he called me back from the road. "I want to see the president tonight," I told him. Andy didn't ask why. He told me when they were due back early that evening and said he would try to fit me in around eight o'clock. "That's fine," I said. "I'll see him then."

That evening I drove down to the White House and entered the grounds by the southwest gate. I'm not sure what the security people thought was happening, but John Moseman, Bill Harlow, and Dottie Hanson, my executive secretary, were the only CIA people who knew for sure what I was doing. The three of them waited back at Langley to learn how it had gone.

Inside the West Wing, I stopped briefly by Andy Card's office. "It's time to go," I told him. "I want to tell the president myself." Andy was considerate as always. As always, too, he didn't tip his hand as to whether he was surprised by my announcement.

In short order, Andy led me upstairs to the residence, where President Bush greeted me in the library, and the three of us sat down together. "It's time for me to go," I repeated. "I've been doing this a long time. I have a boy who needs me, a family that needs me. I've done all I can do. This is a good time for me to go, and I feel very strongly about it."

"When do you want to announce it?" the president asked.

"Tomorrow morning," I told him. I think that caught him off guard a bit, but it also raised a logistical problem. John Howard, the Australian prime minister, was coming early the next day, and he and the president were scheduled to have a joint morning press conference. Howard had been one of our closest allies. Not only had he deployed troops to Iraq, but he'd also had the enormous political courage to say that he'd gone to war in Iraq not because of what the intelligence said but because he'd believed it was the right thing to do. The president didn't want to do anything to step on Howard's visit. Nor did I. Instead of launching the daily news cycle with my resignation, the president decided that he would hold off on that news until after the Howard press conference, and then make the announcement as he was heading to the helicopter for yet another overseas trip. In the meantime, we would enforce silence.

"We tell nobody," the president told Andy Card. "We don't tell Rice. We don't tell anybody about this until tomorrow morning."

I thanked him for making that extra effort, and he told me how much he appreciated what I had done, but unlike our previous conversation, in September 2003, there was no attempt to talk me out of resigning.

Afterward, I walked back out the gate I had entered and found

Stephanie, who had come down to the White House with me, waiting at the base of the monument honoring the 1st Infantry Division, a magnificent sixty-foot column topped by a fifteen-foot gilt representation of Winged Victory.

"You look twenty years younger," she told me.

"I feel great," I said.

The two of us then sat together by the monument for what must have been fifteen or twenty minutes. I told her about my meeting, and Stephanie said that while I was inside, a dark cloud had suddenly appeared, accompanied by a hard downpour. One of the security guys, Bob Woods, had come running over with an umbrella, and he and Stephanie raced back to the car. They were just about to get in it when the sky cleared and the setting sun reappeared in a brilliant show of color, and just at that moment, she said, I walked out of the White House grounds.

I learned later that while Stephanie and I were talking, the contingent back at headquarters had migrated from my office down to the "cage," where the security detail operated, and were frantically radioing Bob Woods and others for a heads-up on just what the DCI and "Daphne," Stephanie's code name, were droning on about. I think they were as relieved as I was when I finally got back to Langley that evening. I filled them in on what the president had said, and assured one and all that the show really was over.

Thursday morning, still sticking to Louis's script as well as I could, I assembled our top people in my conference room, roughly fifteen minutes before I knew the Howard press conference was going to go off. I told them that I had submitted my resignation the night before and that the president would soon be making the announcement. I did not let anyone leave the conference room until the president had finished and was headed to the helicopter to take him to Andrews Air Force Base.

A happy side effect of my departure plan was that by the time the president announced it, most of his staff was already airborne, en route to a summit meeting in Europe, so their ability to spin the reasons behind my departure was mercifully constrained—at least for a few hours.

Maybe an hour later, I went to the Bubble. The rest of the people at Langley, at outlying buildings, and at many locations overseas could watch on closed-circuit TV. Stephanie and John Michael were waiting in the front of the audience when I walked in. By then, I'm sure, the surprise was gone—we are an intelligence agency, after all—but I told everyone that I was leaving all the same, and how proud I was to have worked by their side. Stephanie tells me that I was far from the only person in the auditorium getting choked up. By then, even our always-cool security detail was getting a bit misty-eyed. Near the end, I looked at John Michael and said, "You've been a great son, and I now am going to be a great dad." That's when I lost it, completely.

A footnote to the story: A few days earlier, when Stephanie and I discussed my resignation with John Michael, I told him that he was the main reason I was stepping down. I'd missed too many good times with him. That wasn't going to happen any longer. As much as John Michael appreciated that, he also expressed the fear that the president would be mad at him for causing my departure. I told the president that story when we met Wednesday evening. Thursday afternoon, after my resignation, the president called John Michael from Air Force One to assure him that, no, he wasn't mad at him and to tell him that his father had done an outstanding job.

That wasn't the first time George Bush had gone the extra mile for my son. He knew, from firsthand experience as the son of a former DCI, what it was like to see your dad get chewed up in the press, and he always asked about John Michael and how he was bearing up. Back in February 2004, three months before I left for good, I had told the president that John Michael was having an especially rough time watching me get pummeled, and the president invited him down to the White House for a chat. John Michael never told us about their conversation, but he came home feeling a lot better about life.

I set my resignation date for July 11, in part so I would have time to hand matters over to my successor in some reasonable shape, but also for sentimental reasons. I had been sworn in on July 11, 1997, exactly seven years earlier. Four days before my last day, Stephanie and I flew out to Sun Valley, Idaho, to attend the annual conference sponsored by Herbert Allen and to see the hundreds of wonderful people who had made us feel so welcome over the years. I even got back to the Smiley Creek Lodge for a milkshake and fries.

We returned home on the eleventh. Late that afternoon I decided to go back to my office one last time. It happened to be a Sunday. The headquarters was all but deserted as I went up to my office on the seventh floor. As I entered, I walked up to the charred American flag at the far wall that had been pulled out of the rubble of the World Trade Center shortly after 9/11. I sat at my desk for a while, thinking about what an amazing nine years it had been since I had come over to CIA as John Deutch's deputy. I was rolling events through my mind when I remembered that I had stored away a great Cuban cigar that King Abdullah of Jordan had sent me. I found it and lit up, and then I walked alone around the CIA compound—my own way of saying good-bye to a place I loved.

AFTERWORD

My journey as DCI, which began along the Towpath on the C&O Canal, had more twists and turns than I could ever have imagined. My relinquishing the helm after seven years, in July 2004, did not lead to the calm that usually follows a storm. In fact, the performance of the intelligence community became a debating point in the 2004 presidential campaign. The political arguments generated much heat but little light. Each party tried to bludgeon the other, using American intelligence as a cudgel. The debate also led to a rush to reorganization—an effort destined to provide only a false sense of progress and security.

Somehow the country survived, and shortly after the 2004 election, Brett Kavanaugh, the president's staff secretary, surprised me with a call saying that the president wanted to present me with the Medal of Freedom, the nation's highest civilian commendation. Kavanaugh explained that I was to be honored jointly with Tommy Franks and Jerry Bremer. I was not at all sure I wanted to accept. We had not found weapons of mass destruction and postwar Iraq hadn't been the cakewalk that some had suggested it would be.

I asked Kavanaugh why the president wanted to honor me, and to read me the proposed citation. It was all about CIA's work against terrorism, not Iraq. Fair enough, I thought. Perhaps I could accept a medal on that basis, not for me so much as for the Agency. But I was a long way from convinced. "I'll get back to you," I told him.

I understood the politics clearly, but I weighed that against what I believed the medal would mean to the many heroic men and women of CIA and U.S. intelligence who had performed superbly in responding to the attacks of 9/11. In the end, I said yes for that reason. I also hoped the ceremony might bring a kind of closure to my tenure as DCI and help ease the pain of those last months for my family. Family often gets forgotten in these difficult times; but believe me, they feel the sting of criticism every bit as much as the principals do.

On December 14, 2004, in the East Room, the president showered praise on us. The part I recall best were the words meant not for me but for the Agency I had led: "In these years of challenge for our country," he said, "the men and women of the CIA have been on the front lines of an urgent cause, and the whole nation owes them our gratitude." What meant the most, though, was the look on my son's face as the ceremony proceeded. I don't think I've ever seen him so happy, so proud, so much at peace.

The ceremony turned out to be only a momentary interlude. Time has passed, and controversy has continued to swirl. But in that time, I have given considerable thought not just to the lessons learned in my seven years as DCI, but also to what lies ahead for the country and the intelligence community.

First and foremost, it must be said that intelligence is not the sole answer to any complicated problem. Often, at best, only 60 percent of the facts regarding any national security issue are knowable. Intelligence tries to paint a realistic picture of a given situation based on expert interpretation and analysis of collected information. The results are generally impressionistic—rarely displayed in sharp relief.

Being able to obtain these impressions, however, is critical. To do so, a nation must devote constant attention and resources to its intelligence capabilities—not just in times of crisis but always. Years of neglect cannot be overcome quickly, no matter how intense or well intentioned the recovery effort. The investments made today—in developing intelligence collectors and analysts and in nurturing relations with foreign partners—may not pay dividends for decades to come. But ignore those requirements now, and the cost in terms of lives and treasure will be exponentially higher.

No matter how conclusive intelligence assessments may be, policy makers must engage and ask tough questions. Intelligence alone should never drive the formulation of policy. Good intelligence is no substitute for common sense or curiosity on the part of policy makers in thinking through the consequences of their actions.

Terrorism and Iraq were the two most pressing issues of my tenure, but as critical as they are, we should not be blind to other issues in that troubled region. The Middle East is less stable today than at any time in the past quarter century. The security of Israel is at greater peril than at any time I can remember. The United States entered into the war in Iraq and acted as if our actions there had no relationship to the Middle East peace process, events in Lebanon or Syria, or to the broader struggle against Sunni Islamic terrorism. In fact, these issues are intertwined and now require a strategy that sees them as inextricably linked.

Take the ill-fated Palestinian-Israeli peace process. Had we seriously tried to rejuvenate discussions several years ago, we might have mitigated the agitation in the Sunni world and created an environment more conducive to regional peace and security and less hospitable to the forces of Islamic extremism that we see today.

In the mid- to late 1990s, security cooperation between the Palestinians and Israelis was made possible by a political process dedicated to Palestinian and Israeli states existing side by side in peace. So long as a political process was alive, extremists had little base of support on the Palestinian street for terrorism, and Palestinian security forces could work against extremists and not be seen as collaborators.

True, Arafat's flawed policies and tactics, and his reliance on violence, were major obstacles to peace, but we failed to seize the initiative, upon his death in 2004, to create a political process that offered real hope to the Palestinian people. As a result, they were driven toward extremists who offered them false hope through violence. Security deteriorated, and without a partner, the Israelis properly took measures to protect themselves. In the Middle East, the window of opportunity opens only for brief moments. Sadly, when the window presented itself upon Arafat's death, we did not reassert ourselves as honest brokers seeking to bring a solution to the issue.

When the Bush administration pushed for elections in the Palestinian territories, those elections only served to deliver power to Hamas, which is now ascendant. Hamas's victory was disastrous for the peace process. An Israeli friend asked me, "Why did you Americans insist on elections?" Both the Palestinian Authority and the Israeli government, he said, had requested a delay. The implication of the elections' going forward was that that "the United States was on the side of Hamas." My friend's comments illustrate the fundamental contradiction, in this region, between stability and democracy, especially when democracy is equated only with elections. Was insistence on elections worth Hamas's accession to power? No.

We need to understand that people in the Middle East need a foundation that will allow them to migrate to more representative forms of government in their own way and at their own pace. Simply shouting "democracy" without the existence of a vibrant civil society, and without paving the way for the educational, economic, and institutional transformations required as a foundation for that democracy, may well take us backward and empower the very extremists whose strength we are trying to diminish. Once these extremists gain power, they are unlikely to let it go. Their concept of democracy is "One man, one vote... one time." I believe that if we insist on trying to remake the world in our image, we will fail. Still, we must engage relentlessly to foster a solution to these problems, because the region that served as the cradle of civilization also holds the potential to be its grave.

Unfortunately, the task ahead is made more difficult as a result of the United States' current low standing in the Middle East. Commentators have talked about American arrogance and incompetence as the cause for this. Whatever the reason, we should stop acting as if it were irreversible. A bold new framework for security, stability, and the growth of reform in the Middle East is required, with the people of the region leading the effort and the United States serving as their most ardent and forceful supporter.

Overlaying the very general problem of instability in the Middle East is the very specific challenge of the war in Iraq. The wisdom of our entering that war will be debated for years to come. No doubt, the uncertain road to war was paved, in part, by flawed performance from the U.S. intelligence community, which I led. The core of our judgments on Iraq's WMD programs turned out to be wrong, wrong for a hundred different reasons that go to the heart of what we call our "tradecraft"—the best practices of intelligence collection and analysis. It is no comfort to know that other intelligence services made the same misjudgments. In the case of Iraq, we fell short of our own high standards.

Even if the invading coalition forces had discovered stockpiles of WMD in Iraq after Saddam's ouster, the current situation on the ground would be the same. The same U.S. post-invasion policies would have produced the same disastrous results. While we got it wrong on much of our WMD analysis, we correctly anticipated what might ensue during an extended occupation. What I did not know at the time was how badly our government would mishandle the invasion's aftermath and the effort to win the peace. Once on the ground, CIA provided clear warning of a growing insurgency. The problem was that our warnings were not heeded. For too long our government was either unable or unwilling to look at new facts and transform its policy. As a consequence, a domestic insurgency in Iraq worsened daily and the political and military situation spiraled out of control. We followed a policy built on hope rather than fact.

Perhaps I should have pounded the table harder. But let me be clear: I am not among those who, with twenty-twenty hindsight, now say, "If only they had listened to me, we never would have gotten into this mess." I did not oppose the president's decision to invade Iraq. Such decisions properly belong to the policy makers, not to intelligence officials.

The lessons derived from our national nightmare in Iraq are many and have been painfully learned. To start, I would say that even the world's lone superpower must see that there are some mountains too high for it to climb, and that military might alone cannot solve the endemic political and social problems of other nations. We should enter into wars of choice only with the greatest reluctance, and then only after being completely honest with ourselves and the world about our rationale for undertaking such missions. It is not enough to know how to win wars; equally important is having the knowledge, and the will, to secure the peace. Going into Iraq, the United States let the desire to bring down Saddam's regime overwhelm the recognition that we were unprepared to create the conditions that would put a workable model in its place.

In Iraq we removed a Sunni-dominated and tribally based cult of personality and backed an increase in Shia power without allowing any Sunni alternative to develop. We did this without a broader political strategy that contemplated an outcome whereby Iran would be deterred and contained, and without a strategy to pull Syria away from the Iranian orbit of influence. In effect, we kept Syria and Iran in the same orbit, shunning them and refusing to talk to them about important issues in the region. Over time both countries became determined to resist us. Rather than seeking to create a broad regional consensus for our goals in Iraq, we have isolated Iraq within the region and, more important, isolated the United States.

The administration did not understand that in the volatile Middle East it is often imperative to fight and talk at the same time. We need to talk to the Arab world about issues they care about, not simply issues of concern to us.

The problems in post-Saddam Iraq grew in large part out of an erroneous belief that we could impose our vision of the future on a diverse set of people with very different motivations and expectations. Some in our government felt that the United States could dictate to the Iraqi people our view of their sovereign will, that we could provide legitimacy for their new political leaders simply through our military strength. They were sadly mistaken.

I don't know whether putting more American troops on the ground in Iraq in the middle of a sectarian conflict is going to work. At this writing, such a new strategy is being implemented by Gen. David Petraeus. It may have worked more than three years ago—before a country that believed it had a national identity reverted to the politics of religious and ethnic identification—but whether it will work now only time will tell. My fear is that sectarian violence in Iraq has taken on a life of its own and that U.S. forces are becoming more and more irrelevant to the management of that violence.

In the end it will not matter how many troops the United States puts on the ground. Only Iraqis can determine what kind of country they want and whether they want to pursue a national reconciliation that allows them to remain unified. They can no longer use the U.S. presence as an excuse for failing to make fundamental decisions about their future as a nation.

Any surge in U.S. forces must continue to be accompanied by the ongoing diplomatic effort to bring to the table all the regional stakeholders. This must include the Iranians and Syrians. This is not a question of sanctioning Iranian behavior that leads to the killing of our troops in Iraq; this behavior is unacceptable and it must be addressed on the ground there. Nor is it a question of fearing that Iran will want to discuss their nuclear program. This subject should properly be dealt with separately.

But Iran is not a monolith. It has serious internal problems, including rising unemployment and a very young population that believes that Khomeini's revolution has failed the Iranian people. Chaos and civil war in Iraq may be a threat to the Iranian regime, too. Is it possible that there is a convergence of interest between us and the Iranians? We will know only after we talk to them in front of their Sunni counterparts in the region. Should the Iranians resist such a dialogue, what would be lost?

What we don't want is Sunni countries stoking the flames of the Sunni insurgency, which would increase the likelihood of a broader Sunni-Shia conflagration that could spill over Iraq's borders and further endanger the region. A Shia political revival is occurring across the Middle East. It needs to be understood and considered in any plans for broad political reform in countries throughout the region. Only in this way can Iranian attempts to gain greater leverage and cause more mischief be restrained.

All of this requires carefully managed and staged discussions, sometimes with the United States, and sometimes without us. But we can never be far away from the process.

As difficult as the problems in Iraq, Iran, and the Middle East might seem, they pale in comparison to the global challenge of terrorism. Our highest priority must be to continue to fight terrorists around the world. The campaign against terrorism will consume the next generation of Americans the way the cold war dominated the lives of their parents and grandparents. It will require an intensity of focus unmatched by any other challenge. Let down our guard for a moment, and the consequences could be devastating. When your enemy wants to kill you, is not afraid to die himself, and actively looks forward to the prospect, then you have a daunting challenge.

Few understand the palpable sense of uncertainty and fear that gripped those in the storm's center in the immediate aftermath of 9/11. Before 9/11, our country was without any systemic program of homeland defense. We allowed ourselves to exist on the home front without the capability to prevent the onslaught of a determined enemy. Moving quickly to compensate for what we did not know—the potential al-Qa'ida cells that I believed were likely already in our country planning another round of attacks—we implemented a surveillance program that critics said was an abuse of our rights as Americans.

This was never so. I sat in on every briefing given to the leadership of the House and Senate Intelligence Committees where then NSA director Gen. Mike Hayden methodically walked through the surveillance program, how it was being implemented, and the care NSA was taking to ensure that its sole focus remained on providing us with the speed and agility we needed to protect the country.

As for the treatment of detainees, the senior leadership at CIA understood clearly that the capture, detention, and interrogation of senior al-Qa'ida members was new ground—morally and legally. We understood the tension between protecting Americans and how we might be perceived years after the trauma of 9/11 had faded from the nation's memory. History had taught us that decisions made to protect the public from another more devastating al-Qa'ida attack might be viewed later as our sanctioning torture or abuse, thus jeopardizing the CIA and public trust in it. None of this was taken lightly. The risks were understood.

By speaking out about the use of certain interrogation techniques, Senator John McCain engaged the country in an important moral debate about who we are as a people and what we should stand for, even when up against an enemy so full of hate they would murder thousands of our children without a thought. We at CIA engaged in such a debate from the beginning, struggling to determine what was required to protect a just society at so much risk. But from where we sat, in the late summer of 2003, preventing the death of American citizens was paramount. It is easy to second-guess us today, but difficult to understand the intensity of our concerns when we made certain decisions and the urgency we felt to protect the country.

Leaders of our country must find a way to build a broad political consensus on the lengths American citizens will expect intelligence, law enforcement, and military personnel to go to protect the United States. To find such consensus, there must be a sound foundation of consultation and understanding.

After 9/11, gripped by the same emotion and fears, Congress exhorted the intelligence community to take more risks to protect the country. But if the elected representatives of the American people do not want an NSA surveillance program, no matter how rigorous the oversight, then the program should be shut down. If they believe that certain actions taken during an interrogation process put us in a difficult place morally—even if we believe those actions to be disciplined and focused, in compliance with the law, and invaluable for saving American and foreign lives—then we should not employ those actions. Our role as intelligence professionals is to inform policy makers of both the hazards and the value of such programs. We should say what we think but the final decision belongs to the political leadership of the country. It is they who must engage the American people.

In all these programs, we believed we were doing what was right for the country; we calibrated the risks and discussed the tensions. But the debate must be broadened, the guidance made clear, and the consequences of either taking or not taking an action clearly understood.

But I ask that we all remember those decisions when the next terrorist attack occurs. We must understand collectively that if we decide not to empower our intelligence-collection activities, we have to be willing to take the risk and pay the price. If we do not have that debate now, the pendulum will swing much more dramatically after the next major attack.

The president must lead. No president can subordinate day-to-day decision making to others. In the days after 9/11, the president was confronted with an unprecedented danger. He has been criticized for justifying NSA's surveillance program on the basis of the power the Constitution provides him in a time of war. But the fear present in those early months and years has all but been forgotten.

Today we must all recognize that the campaign against terrorism will be of unlimited duration. It will require a different and enduring bipartisan legal foundation to carry us forward. The senior political leadership of our country should be asking together what we need to do now to increase our odds of deterring future attacks.

Beyond a discussion of what steps to take in the fight against terrorism, there must be honest and realistic expectations for the work of the intelligence community; there is no perfection in our business. Intelligence does not operate in a vacuum, but within a broader mandate of policies and governance. The men and women in the intelligence community are ready and willing to be held accountable for their work. But when policies are inadequate and warnings are not heeded, it is not "failure of imagination" on the part of intelligence professionals that harms American interests and the American people.

Terrorism is the stuff of everyday nightmares. But the added specter of a nuclear-capable terrorist group is something that, more than anything else, causes me sleepless nights. Marry the right few individuals with the necessary material, and you could have a single attack that could kill more people than all the previous terrorist attacks in history. Intelligence has established beyond any reasonable doubt that the intent of al-Qa'ida is to do precisely this. There is an abundance of nuclear material in the world, some of which may already be within reach of terrorist groups. It will require incredible alertness, foresight, and determination to prevent such groups from acquiring that material—a development that would have devastating consequences. Our nation ought to be moving heaven and earth to get a handle on all the deadly fissile material currently unaccounted for and possibly available to the highest bidder. If we do not quickly and completely snatch this material from our enemies' grasp, we will rue our lack of foresight and our misperception that "men in caves" lack the ability to acquire and employ such weapons.

Tactically, we can fight these extremists, and we will—for the next twenty-five years, person by person, cell by cell, bank account by bank account. One thing is certain: we have to continue the tactical elements of this campaign. And we cannot do it alone. There is no unilateral American solution to this problem. The relationships we nurtured with intelligence services around the globe and particularly in the Arab and Islamic world have been critical to the many successes we have enjoyed. The adversary we face will not negotiate, accommodate, or settle for peace. At the same time, we must recognize that you cannot kill or jail them all and hope to prevail.

The battle against terrorism must never be just about tactics. We will never get ahead of the problem unless we penetrate the terrorist breeding grounds and do something about promoting honest government, free trade, economic development, educational reform, political freedom, and religious moderation.

The first responsibility lies within Islam itself, to create and foster a religious dialogue that loudly repudiates the violence and radical thought that al-Qa'ida promotes and thrives on. No Westerner can shape this debate. It is the purview of governments and religious leaders and Islamic thinkers, who must no longer turn a blind eye to the extremist message. There must be a way to defeat the perversion of the Islamic faith that sends the message to its followers: "We have been humiliated because of the lack of opportunity and, as a result, our enemies—Christians, Jews, and apostate Muslims—need to die."

The second responsibility lies with the West and these same governments to facilitate educational and economic reforms that allow young men and women to have opportunities to live and flourish in a globalized world on terms in which they are respected and have a stake in their societies. Too often, such compacts have been broken.

Western governments, especially our own, must find ways to engage the mainstream Islamic world, focusing on common interests and objectives. And to do that effectively, we must have a multiyear, long-range commitment in resources, personnel, and deep expertise in Islamic cultures, societies, and languages. We must convince Muslims through their leaders and opinion makers that terrorism is their enemy as well.

Changes in the way we function operationally and diplomatically are urgently needed. But we must not fall prey to typical American impatience and rush into "solutions" that only make matters worse. To some extent, that is what happened with the 9/11 Commission. The commission did some very good work describing the nature of al-Qa'ida's plot. But it did not fully understand what actions were working against the terrorists prior to the attacks and did not fully analyze the actions taken in the months immediately after 9/11 that led to the successful takedown of two-thirds of Bin Ladin's top leadership.

The 9/11 Commission's mandate was not extended beyond the 2004 election as commissioners had requested. As a result, the politics of the moment demanded immediate action. John Kerry's campaign endorsed the commission's recommendations within twenty-four hours of the report being published. The Bush administration quickly followed suit, and thereby abdicated its obligation to lead and manage the executive branch in a responsible fashion.

A strong case can be made that the three roles in which I served—as head of the intelligence community, Director of CIA, and the president's principal intelligence advisor—were too much for any one person. Perhaps so. But to embrace a new structure without careful consideration of the implications was unwise. In the aftermath of 9/11, legislative changes were imposed before we had asked some fundamental questions. What was the world going to look like over the next twenty-five years? What threats and opportunities would we face? What capabilities would the country need to ensure its security? What kind of people would we need to recruit, train, and retain to accomplish the mission? These questions by themselves would have resulted in a vigorous debate and study. Then, and only after understanding the problem before us, should we have asked the question, "What architecture or structure should we put in place to maximize our potential to allow us to succeed?" Little of that was done. The legislation that was passed was based on structure, power relationships, and how they should be altered in Washington, rather than on what the country needed from intelligence to protect its future interests. The result was an over-centralized, multilayered structure that, at least where terrorism is concerned, lacked the speed and agility to meet the challenges we face.

From my perspective, the single biggest obstacle we needed to overcome was that there was no single place where foreign intelligence and domestic information could be put together and analyzed quickly to empower those who could do something about it—that is, CIA officers, FBI agents, foreign partners, or state and local police officers inside the United States.

In fact, prior to 9/11, there was precious little domestic data gathered. We had no systematic capability in place to collect, aggregate, and analyze domestic data in any meaningful way. Domestically, there were few if any analysts. There was no common communication architecture that allowed the effective synthesis of terrorist-related data in the homeland, much less the seamless flow of information from overseas to state and local officials inside the United States. At the beginning of the twenty-first century, U.S. intelligence officers in Islamabad could not talk to FBI agents in Phoenix.

While the 9/11 Commission stated that there were "fault lines within our government—between foreign and domestic intelligence, and between and within agencies," it focused almost exclusively on restructuring the American foreign intelligence community. Little if any attention was paid to the systemic deficiencies that existed on the domestic side.

The Department of Homeland Security was in place, and a new intelligence division was being created in the FBI, even before the 9/11 Commission had ended its inquiry. These changes were Washington-centric solutions that did not incorporate state and local officials, the men and women who could actually act on any data gathered—data they still don't receive.

What do I mean specifically? During the first Gulf war, our commanders complained about the disparate intelligence they received from separate civilian and military channels. In response to this, a major revolution in American intelligence occurred after that war, stimulated by a brilliant paper authored by President George H. W. Bush's Foreign Intelligence Advisory Board and by its chairman, Adm. Bobby Inman.

During the subsequent wars in Bosnia and Kosovo, we did not over-centralize the Washington power structure. Rather, we decentralized access to intelligence by pushing its analysis and exploitation as close as possible to the war fighter—whether in the foxhole or in the cockpit. Not only did we convey this data to the field in nanoseconds, but we also allowed our deployed forces to reach back into giant databases to pull the data they believed they needed to do their jobs. Military men and women far away from Washington actually know best what they need most, and today they have the ability to reach in and get it.

Today we are in possession of an enormous amount of data about how al-Qa'ida trains its members, operates, and thinks about the United States as a target. This is all rooted in what we have learned about them around the world through speedy and agile intelligence operations in concert with our foreign partners. Yet how much of this data is available on a daily basis, on one communications backbone, to the people who can do something about it? In reality, very little. It is simply not good enough to warn local police departments of imminent threats. We need to arm them with our knowledge of terrorists and their tactics. This can be done without compromising sensitive sources or methods. Technology today allows us to insert data with varying layers of access for those with a need to know. While some classified data will always be essential, the majority of the knowledge we impart should be unclassified. Without this information, the people most familiar with our cities and local communities have little, if any, basis to plan, allocate resources, and train and retain the right kind of people. The solution to the terrorist threat we face has little to do with structure. It is all about data.

My personal concern is that the head of the intelligence community, now known as the Director of National Intelligence (DNI), may be too distant from the people he is supposed to lead and may be divorced from the reality of risk taking and running operations. Still, the legislation that created the position of DNI is now law. For the good of the country, we must ensure that the DNI and American intelligence succeeds.

The current DNI, Mike McConnell, is an enormously capable former senior intelligence officer. His years in the private sector will give him useful experience to build the collaborative enterprise that American intelligence must become. His principal tasks will be to enable the constituencies that report to him to perform better without having a vast staff to micromanage every operation, collection decision, or piece of analysis.

He will understand that common policies with strong central management with regard to enabling the free flow of data, training and retaining intelligence professionals, enacting security protocols, setting priorities, and measuring achievement all can have a liberating effect on the intelligence community. As a former director of NSA, and chief intelligence officer to the chairman of the Joint Chiefs, he also understands that decentralization, the collection and processing of data, and analysis of that data as early as possible to get close to the beneficiaries of the intelligence, are essential elements of success.

Domestically, he will quickly understand that this decentralized model—in particular the linking of data and analysts, whether from the FBI or the intelligence community, to state and local police officers or to the private sector—is an essential element of deterring future terrorist attacks against the United States.

We as a country chose not to create a domestic intelligence service separate and apart from the FBI. In fact, little serious consideration was ever given to the proposal. At the time I left office, a time of crisis, it would have been difficult to create yet another wholly new entity on top of everything else we were establishing. Yet, the potential benefit of a domestic intelligence service should be debated. The answer to that debate should not be based on whether such a service detracts from the FBI but rather on whether the country would, over the long term, be safer and whether such a service, separate and apart from traditional investigatory and police work, could exist within our framework of laws and individual rights. At the very least, a dispassionate examination of the idea should be considered now, and not in the aftermath of another terrorist attack.

Whatever the challenges that face us, one of the most frequent questions I am asked is: "As a result of the steps that have been taken, are we safer today than we were on 9/11?"

The answer is yes, we are safer, but much danger still remains. We must not fool the American people into believing that reorganizing the structure of American intelligence has created an impenetrable shield. It has not. And much more work still needs to be done. My fear is that we have lost our sense of urgency.

One final lesson learned during my time as DCI over two administrations is that, despite what conspiracy theorists and political operatives would have you believe, people from both parties, with vastly different approaches, try to do what they think is right for our country. It is a great disservice when partisans on both sides of the aisle suggest that their opponents are willfully putting American lives at risk and playing into the hands of the enemy. As I said at the outset of this book, their methods can and should be debated, but not their motives.

My time as DCI ended with more than a Medal of Freedom hung around my neck. Not everything laid on my shoulders was welcomed or, I'd like to think, deserved. But certainly some of it was.

I rest easy knowing that I was in the arena and that I tried my damnedest to protect my country. Some people described me in the days prior to 9/11 as "running around with my hair on fire." If so, it was not because I was excitable but because we saw a threat and tried to do something about it. The work of American intelligence officers is a critical and largely thankless task. They share the dangers and uncertainties that are shouldered by our military. The country has many entirely appropriate and needed ways to thank our armed forces—but precious few to do the same for the men and women of the intelligence community. There are no parades to honor veteran spies or bands to welcome them home. Yet intelligence professionals willingly and enthusiastically embark on their important missions knowing that self-satisfaction for having fought the good fight will have to suffice in lieu of public thanks.

There is a tradition at CIA where fallen officers are memorialized with a marble star carved into the wall of our lobby. By the time I left, there were a total of eighty-three stars. Each May we hosted a memorial service, where we would read aloud the names of fallen officers and tell some of their stories, sometimes in the presence of family members who did not know of their exploits.

It was a way to bridge the past and the future, a way to teach our young officers about the meaning of service.

Eleven officers were taken from us during my time as director. I felt each of their deaths personally. Some, such as Mike Spann and Helge Boes, died in Afghanistan, on the front line of the war against terrorism. Others cannot be mentioned. Their lives, and deaths, must remain secret. But those of us who became part of the Agency family will always remember them.

Whenever the CIA's engraver, Tim Johnston, had to carve a new star, I would slip out of my office for a moment and watch him at work. The most valued gift I received during my tenure was a small marble star, presented to me by Tim. I still keep it on my desk.

When I was in office, I tried my best to represent the American people in thanking our intelligence professionals for what they do for all of us. Being the chief defender and spokesman for a secret organization was challenging—and I bear more than a few scars—yet I also experienced great moments of quiet joy with colleagues who took enormous risks on behalf of America and its allies. Considering it all, would I make that long journey again? Absolutely—in a heartbeat.

POSTSCRIPT

I would like to reflect on a few things that have happened—and not happened—in the time since *At the Center of the Storm* was first published in April 2007. First and foremost, I believe the threat of nuclear terrorism is greater today, nearly seven years after 9/11, than ever before. Yet, it appears to me the U.S. government is not giving this threat the highest priority it deserves. We cannot be willing to take our chances and trust that al-Qa'ida will never be successful in translating deadly intent into reality.

The intelligence is clear and compelling that al-Qa'ida has the intent to attack our homeland with a yield-producing nuclear bomb. The intelligence community also acknowledges that it cannot exclude the possibility that al-Qa'ida may already have the capability to construct a crude improvised nuclear device if it can obtain sufficient fissile material. This is a most disturbing judgment. No one should feel sanguine about the chances of preventing fissile material from falling into the hands of an adversary hell-bent on acquiring it. There have been well-documented incidents casting doubts on the security of fissile material worldwide. Moreover, there is much we do not know about the potential availability of expertise and material on the nuclear black market. Faced with such uncertainties, it would be foolhardy to gamble the nation's security on the odds that "men in caves" simply do not have the capacity to conduct a nuclear attack.

I regret that the threat posed by nuclear know-how in the hands of terrorists portrayed in this book—in chapter 14, "They Want to Change History," and chapter 15, "The Merchant of Death and the Colonel"—did not receive sufficient attention when it was first published. These chapters, in my view, are critically important not just because of what they say about past events, but because of what they tell us about the future peril presented by nuclear expertise, material, and devices acquired by those who wish us harm. Imagine if Pakistani unrest in late 2007 were to result in the transfer of nuclear materials to a terrorist group.

If a nuclear catastrophe were to befall a U.S. city, how loud would the outcry be? What price would we have been willing to pay to have prevented it? Why are we not doing everything possible *now* in order to forestall such an attack and to ensure that the unthinkable does not become reality?

The nuclear issue haunts me. As I reappraise the state of our war against al-Qa'ida, I see progress on many fronts. Yet, our adversary knows that a nuclear attack on U.S. soil would change everything. Indeed, a nuclear attack would change history. But I believe that we continue to fall far short in responding to this grave challenge.

What needs to be done?

The intelligence community, as well as the law enforcement community, the military, and other appropriate agencies of government, must move heaven and earth to thwart any potential weapon of mass destruction (WMD) attack before it materializes on U.S. soil. The best chance to avert a WMD attack is to penetrate and stop a plot in the earliest stages of the planning process. The odds of our success diminish precipitously over time, as terrorist planning draws closer to a specific date and place for an attack. For example, if terrorists were to procure sufficient fissile material, and if they were to acquire knowledge of how to use it, it would take a great deal of luck on our side—yes, luck—to avert a catastrophe. Rarely has the intelligence community faced such exacting standards for success. Each and every WMD plot must be uncovered and ruthlessly neutralized, using all means allowable under U.S. law. Failure is not an option. We should be seeking answers to key questions now:

- Have we made nuclear terrorism our highest priority, not merely on paper, but in terms of our response on the ground?
- Are we devoting sufficient resources to the problem?
- Have we put our best intelligence professionals on the problem?
- Are we conducting the sort of daring, creative, no-holds-barred operations that are necessary to penetrate and compromise nuclear plots in their early formation?
- In the absence of specific threats and leads, are we out "proving the negative" every day? That means verifying that our adversaries have not succeeded in a clandestine Manhattan project.

It would be irresponsible not to pose the right questions and insist on having reassuring answers to these and similar questions now. They will be of no use to us after an attack.

Currently, we face two obstacles to success in these efforts. First, nuclear terrorism falls between the seams of the counterproliferation and counterterrorism communities, not claimed by either and lying dangerously on the boundaries of both. Counterterrorism specialists tend to dismiss nuclear weapons as being a distant threat and remote possibility for terrorists, who have displayed a clear preference for "more practical" means of attack. Counterproliferation specialists, on the other hand, focus more on the sophisticated paths that states must take in order to achieve nuclear capability, rather than the improvisational shortcuts of the terrorist. Such institutional biases of the counterterrorism and counterproliferation communities may create gaps in our defenses that can be exploited by our adversaries.

Second, foreign and domestic intelligence efforts are managed by separate organizations with widely differing cultures that are insufficiently integrated in terms of action. In this context, information-sharing mechanisms are too slow and not sufficiently broad-based. This is particularly problematic vis-à-vis a potential nuclear plot, which would have a very small footprint; the smallest lead exposed to the light of day could help reveal ongoing planning for an attack. While there is certainly an ongoing effort by relevant organizations to improve cooperation within their existing authorities and constraints, in my view it represents a "muddle-through" approach that is inadequate for the unique challenges of this problem. We still have time to fix these problems, but to do so we must act with a sense of urgency.

As to the first obstacle, I believe we should create a small, tailored "skunk works" organization with all the capabilities and skills needed to handle all aspects of an aggressive, sustained intelligence hunt for WMD.

And as to the second obstacle, we need to put someone in charge of WMD terrorism who reports directly to the president. At present, it is unclear who bears ultimate responsibility for developing and implementing a national strategy to combat terrorist weapons of mass destruction. Appointing a senior level official with the responsibility and authority to address the issue of nuclear terrorism is essential if we are truly to focus on this urgent problem. The key role of this person would be to untie a Gordian knot of organizational challenges in order to unify governmental and nongovernmental efforts into a clear, coherent, common national cause.

Not since World War II has the United States been so in need of timely, accurate, and actionable intelligence. Although the intelligence community has undergone a major restructuring in the past three years, the jury is still out on whether the changes will lead to necessary improvement. Structure is only part of the problem. More important are recruitment of the best possible candidates; superior training, leadership, and support; consensus on what the nation expects from intelligence; effective oversight and support from Congress; and substantial investment in new technologies.

My hope is that this book will spark a discussion leading to strategic and tactical plans to address the threat of nuclear weapons in the hands of terrorists.

While I believe what I originally wrote about the threat of WMD in the hands of terrorists received too little focus, there were others parts of the book that suffered from no lack of media attention.

When *At the Center of the Storm* was first published, the intensity of emotions surrounding the ill-starred war in Iraq ignited a firestorm. Having the Director of Central Intelligence, one of the people who was at the table when the war originated, try to shine a light on the planning process—or lack of one—was like striking a match in a gas-filled room.

It is hard—if not impossible—to write a memoir and tell readers what you got right or wrong without appearing defensive. In the book, I tried to explain personal and institutional failures and successes with equal honesty and clarity. This is especially true regarding WMD in Iraq, a critical issue on which I believe the American intelligence community, which I led, let the country down.

Perhaps the biggest single criticism I heard after publication of the book could be summed up in the phrase "Now he tells us!" There was an all-too-common assumption that the events I wrote about in a book published in 2007 were crystal clear when they unfolded four or five years earlier. This was simply not the case. It was only after studying thousands of pages of classified documents, interviewing scores of my former colleagues, and reflecting on what I had seen, said, and done that I was able to come to some of the conclusions and many of the observations that are laid out in these pages.

Other critics said: "You told the Bush administration what

they wanted to hear on Iraq WMD, ignoring contrary evidence to please your political bosses." This too is untrue. In fact, we told the administration what we did about WMD because *that was what we and all of our allied intelligence services believed*.

Those who say we told the administration only what they wanted to hear are never able to square that allegation with the fact that, on the matter of alleged al-Qa'ida–Iraqi collaboration, we consistently told the White House and the Congress exactly what some in the administration did *not* want to hear—that we saw no operational connection. We were not willing to "cook the books" on that critical topic—so why would we have done so on the WMD question? The answer, of course, is that we did not.

There were a number of issues on which our analysis and reporting to the administration appeared to run counter to the course being pursued. This is particularly true with pre-combat assessments regarding the consequences of war in Iraq, which later turned out to be prophetic. More importantly, we provided the administration with critical intelligence reporting from our senior officers in Iraq, beginning in the summer of 2003 and at least through the time of my departure from CIA in July 2004. This intelligence was quite clear regarding the consequences of policies that would and, indeed, did lead to an ever-growing insurgency. An evaluation of this critical period would show that the available intelligence should have led to changes in U.S. policy in 2003 or 2004 rather than in early 2007.

Although Iraq dominated the coverage and reaction to the book, there was also a great deal of controversy surrounding what I wrote about the run-up to 9/11. The biggest misunderstanding centered around my July 10, 2001, briefing to Condoleezza Rice and the two months between that session and the attacks on the United States. A common critique was "If you felt strongly enough to ask for an urgent meeting with Condi Rice, why didn't you walk down the hall, grab the president by the lapels, and demand that he take immediate action?" Those making this argument suggest that the president was unaware of the underlying intelligence contained in my July 10 briefing. That was just not the case.

Throughout the spring and summer, the president received on a daily basis the latest information regarding our concerns about the al-Qa'ida threat. He understood our concerns—illustrated by the fact that he asked his briefers questions that in part led to the August 6, 2001, President's Daily Brief titled "Bin Ladin Determined to Strike in the US." After my July 10 briefing I was confident that the National Security Council staff would work to ensure that foreign and domestic agencies of the United States government were acting in concert. Indeed, the NSC started to coordinate such action, although, in hindsight, the process moved too slowly. In late July and August, the warnings and indicators from al-Qa'ida grew quiet. It was the lull before the storm.

Another development since publication of the book has been congressional passage of a law requiring CIA to release an unclassified version of a 2005 report by the Agency's Inspector General regarding accountability for pre-9/11 lapses. Both of my successors, Porter Goss and Mike Hayden, had previously resisted making the report public. My understanding was that their reluctance was not an attempt to avoid embarrassment but rather was based on legitimate concerns regarding the quality of the report itself.

Unfortunately, the IG report bought into the notion that CIA had no strategic plan to go after al-Qa'ida. As readers of this book know, that is not true. In fact, there was an extensive counterterrorism strategy that was called "The Plan." That plan was the basis for what the president authorized in the days shortly after the 9/11 attacks: an operation that led to the routing of the Taliban and the driving of al-Qa'ida from its Afghan sanctuary in a matter of weeks. The IG also criticized the Agency's leadership for not getting and using more financial resources for efforts against al-Qa'ida. But as I explained earlier in these pages, the intelligence community was so badly broken in the mid-nineties that we could not focus on just a single target. Although spending available for everything else at CIA went down or stayed flat, counterterrorism resources went up dramatically.

How can the IG's findings and my account be so dramatically at odds? It is hard to fathom. One thing I know: Despite the fact that the IG's report was years in the making, neither he nor his staff ever interviewed me for my insights. Nor did they interview others outside CIA or elsewhere in the administration (such as the FBI, the National Security Agency, the Defense Intelligence Agency, or the National Security Council staff) who could have provided the much-needed context that was absent from the report.

Moreover, if CIA's management of the terrorist threat was so badly flawed, as the IG alleged in his report, why did the Office of the Inspector General give me a report in August 2001, just three weeks before 9/11, telling me: "The DCI Counterterrorist Center (CTC) is a well-managed component that successfully carries out the Agency's responsibilities to collect and analyze intelligence on international terrorism and to undermine the capabilities of terrorist groups." The report went on to give CIA's counterterrorist efforts high marks across the board and made no recommendations to me for improvements.

I do not want my comments here to be misconstrued as an assertion that CIA's performance prior to 9/11 was beyond reproach. Certainly we and the entire government made many mistakes. The country had no systematic mechanism to translate foreign threats into a meaningful program of protection of the homeland.

Nor do I want to suggest that all criticisms of my book were without merit. I aided critics by getting a few dates and facts wrong in the earlier text. While these errors in no way changed the thrust of what I was saying, they gave naysayers a way to downplay the overall message of the book.

When *At the Center of the Storm* is read— and placed in proper context—I believe it helps illuminate CIA's actions during America's time of crisis and raises important questions about averting crises in the future.

George Tenet
February 2008

GLOSSARY

Aardwolf—Code name for a comprehensive written assessment of a situation submitted by a senior CIA field representative to Agency headquarters.

Alec Station—A "virtual station" set up within CIA's Counterterrorist Center in 1996 to focus specifically on al-Qa'ida. Originally designed TFL, for Terrorist Financial Links, it became known informally as the Bin Ladin Station.

al-Qa'ida—Arabic for "the base"; the umbrella name for Usama bin Ladin's Sunni Islamist group dedicated to driving Westerners out of the Gulf region and establishing a Muslim caliphate.

Ba'ath Party—The Arab nationalist secular political party that ruled Iraq from 1968 to 2003.

BND—The Bundesnachrichtendienst (Federal Intelligence Service); the external intelligence agency of the German government.

The Bubble—CIA's auditorium.

BW—Biological weapons; the use of bacteria, viruses, toxins, etc. as weapons.

Case officer—A member of CIA's clandestine service who recruits and directs foreign agents.

CBW—Chemical and biological weapons.

CENTCOM—The U.S. Central Command; the organization responsible for U.S. military operations in Northeast Africa and Southwest and Central Asia.

CIA—Central Intelligence Agency.

Covert action—An operation conducted in a way to conceal the role of the U.S. government behind those actions.

CPA—Coalition Provisional Authority; a government entity, under the direction of L. Paul "Jerry" Bremer, established in Iraq in April 2003 to provide transition to an Iraqi government. Disestablished on June 28, 2004.

CPD—Counterproliferation Division; the part of CIA's Directorate of Operations concerned with trying to stop the spread of weapons of mass destruction.

CTC—Counterterrorist Center; a portion of CIA made up of analysts and operators, which conducts operations against terrorists worldwide.

CW—Chemical weapons.

DC—Deputies Committee; NSC committee generally made up of the second ranking person from the NSC, State and Defense departments, CIA, etc.

DCI—Director of Central Intelligence; the head of the U.S. intelligence community and CIA. Position established in 1947 and disestablished in 2005 with the creation of the position of the director of national intelligence.

DDCI—Deputy Director of Central Intelligence.

DDI—Deputy Director (of Central Intelligence) for intelligence; head of the analytic arm of CIA.

DDO—Deputy Director (of Central Intelligence) for operations; head of the intelligence-collection arm of the CIA. Now called the National Clandestine Service.

DDS&T—Deputy Director (of Central Intelligence) for science and technology.

DIA—Defense Intelligence Agency; the Department of Defense's intelligence organization, providing foreign military intelligence to the warfighter.

EIJ—Egyptian Islamic Jihad; a terrorist group whose origins date back to the 1970s. Under the leadership of Ayman al-Zawahiri, EIJ essentially merged with al-Qa'ida in the late 1990s.

EXDIR—Executive Director of Central Intelligence; third ranking official in CIA.

Finding—A legal document, signed by the president, granting specific authorities to the CIA and the intelligence community to conduct covert actions.

FISA—Foreign Intelligence Surveillance Act; a 1978 law laying out specific authorities and procedures for the collection of physical and electronic intelligence regarding foreign intelligence.

FSB—The Federal Security Service of the Russian Federation; the domestic successor to the KGB in Russia.

GRC—Global Response Center; a watch center within CIA where overseas operations are monitored.

HPSCI—House Permanent Select Committee on Intelligence.

HVD—High-value detainee.

INA—Iraqi National Accord; an Iraqi political party founded in 1991 to provide opposition to the Saddam Hussein regime.

INC—Iraqi National Congress; an umbrella organization of Iraqi opposition groups set up in the early 1990s under the leadership of Ahmed Chalabi.

In-Q-Tel—A nonprofit organization, funded by CIA, to seek information technology solutions to the Agency's most critical needs.

INR—The Bureau of Intelligence and Research; a member of the intelligence community. A small organization within the Department of State providing intelligence analysis.

IRGC—Islamic Revolutionary Guard Corps. Formed in 1979 as a military force loyal to the Ayatollah Khomeini, the IRGC has become a large military organization focused on special operations.

ISG—Iraq Survey Group; a unit established in Iraq in 2003 to investigate whether Iraq had WMD stockpiles and programs.

JI—Jemaah Islamiya; a Southeast Asian Islamic militant group with close ties to al-Qa'ida.

JSOC—Joint Special Operations Command; a U.S. military organization charged with planning special operations missions.

KDP—Kurdistan Democratic Party of Iraq. One of the two major Iraqi Kurdish political parties, the KDP was founded in the 1940s and is led by Massoud Barzani, an influential Kurdish politician.

KGB—Soviet "Committee for State Security"; the Soviet Union's premier intelligence service and CIA's main rival during the cold war.

MOIS—Ministry of Information and Security; Iranian intelligence service.

Mossad—Israeli Institute for Intelligence and Special Operations; counterpart to CIA in Israel.

NALT—Northern Afghanistan Liaison Team; small CIA units deployed to Northern Afghanistan both before and after 9/11 to coordinate with members of the Afghan Northern Alliance.

NCTC—National Counterterrorist Center; established in August 2004 to serve as the primary organization in the U.S. government for integrating and analyzing all intelligence pertaining to terrorism and counterterrorism and to conduct strategic operational planning.

NGA—National Geospatial-Intelligence Agency; provides exploitation and analysis of imagery and geospatial information to describe, assess, and visually depict physical features and geographically referenced activities on Earth. (Formerly the National Imagery and Mapping Agency).

NIC—National Intelligence Council; the intelligence community's center for mid-term to long-term strategic thinking.

NIE—National Intelligence Estimate. Produced by the NIC, the NIE is the intelligence community's most authoritative written judgment concerning national security issues. It contain the coordinated judgments of the intelligence community regarding the likely course of future events.

NILE—Northern Iraq Liaison Element; small CIA teams that operated in Northern Iraq prior to the start of the 2003 war.

NIMA—National Imagery and Mapping Agency. (See NGA.)

Northern Alliance—Also known as the United Islamic Front for the Salvation of Afghanistan; an umbrella organization of Mujahideen who fought the communist and later Taliban governments in Afghanistan.

NSA—National Security Agency; U.S. cryptologic organization; coordinates, directs, and performs highly specialized activities to protect U.S. government information systems and produce foreign signals intelligence information.

NSC—National Security Council; the president's principal forum for considering national security and foreign policy matters with his senior national security advisors and Cabinet officials. The NSC also serves as the president's principal arm for coordinating these policies among various government agencies.

ORHA—Office of Reconstruction and Humanitarian Assistance in Iraq. Established just prior to the 2003 invasion of Iraq, ORHA was replaced by CPA.

OVP—Office of the Vice President.

PC—Principals Committee; NSC committee made up of the national security advisor, secretaries of State and Defense, DCI, etc. In the Bush administration, the vice president also attended.

PDB—President's Daily Brief; compilation of intelligence presented to the president each day.

Predator—Unmanned Aerial Vehicle (UAV) used for surveillance and, post-9/11, capable of delivering Hellfire missiles on targets.

PUK—Patriotic Union of Kurdistan. One of the two major Iraqi Kurdish political parties, the PUK was founded in the 1970s and is led by Jalal Talabani, an influential Kurdish politician and the current president of Iraq.

Red Cell—A group of CIA analysts established immediately following 9/11 to provide "out-of-the-box" and contrarian analysis.

Rendition—The practice of moving terrorists and other criminals from one foreign country to another, where they may be wanted by law enforcement officials, and interrogated.

Silberman-Robb Commission—Informal name of the Commission on the Intelligence Capabilities of the United States Regarding Weapons of Mass Destruction. Silberman-Robb issued its report in March 2005.

SSCI—Senate Select Committee on Intelligence.

Shin Bet (also known as Shabak)—The internal Israeli security service.

TTIC—Terrorist Threat Information Center; established May 1, 2003, the TTIC became NCTC.

UNSCOM—United Nations Special Commission; provided inspections of Iraq for possible WMD from 1991 until it was withdrawn in late 1998.

UTN—Umma Tameer-e-Nau; Pakistani nongovernmental organization ostensibly founded to provide humanitarian relief but which offered al-Qa'ida advice on nuclear weapons.

WINPAC—Weapons Intelligence Nonproliferation and Arms Control Center; CIA organization that provides intelligence support aimed at protecting the United States and its interests from all foreign weapons threats.

WMD—Weapons of mass destruction.

Yellowcake—An intermediate step in the processing of uranium ore. Yellowcake may be enriched to produce uranium suitable for use in weapons and reactors.

ACKNOWLEDGMENTS

Because of time, space, and security concerns, I cannot fully describe to readers of this book the terrific men and women with whom I was privileged to serve. Any successes we had during my time in office were the result of their fine work. Let me take this opportunity, however, to express my admiration and thanks to all of those who have served at the Central Intelligence Agency, and throughout the intelligence community.

While I cannot list by name all those who richly deserve my thanks and those of the American public, a few individuals and categories of people need to be singled out.

First, I was blessed to be surrounded by a superb senior management team at CIA, most notably my terrific deputy, John McLaughlin, and his predecessor, John Gordon. Dave Carey, Gina Genton, Buzzy Krongard, John Brennan, and Marty Petersen all carried on the day to day management of CIA, and the implementation of strategic plans, that not only restored agency morale but also laid a solid foundation for the future.

They were aided by superb leaders of our Directorates of Operations, Intelligence, and Science and Technology as well as our support officers whose contributions are always inadequately recognized, even in books written by their most fervent admirers like me.

I am also grateful to those people who provided especially close support to me, like my chiefs of staff and good friends John Moseman, John Brennan, and John Nelson; my tireless and terrific special assistants and PDB briefers; and my office administrative team led by Dottie Hanson and her unflappable sidekick, Mary Elfmann.

A special note of thanks to the men and women of the DCI security staff. They kept my family safe and made enormous personal sacrifices to ensure that my day was as smooth as possible. They were superbly led by Dan O'Connor, Mike Hohlfelder, and Tim Ward. I am grateful to speechwriters Lynn Davidson and Paul Gimigliano, who helped me communicate with clarity and honest emotion, not only to the public but also to the men and women of the CIA around the world. I also want to thank CIA's protocol staff, ably led by Sheila Siebert, which brought together Agency officers and visitors on countless occasions of both sadness and joy.

Don Cryer made CIA's vision for diversity a reality. Through his care and leadership, everyone at CIA knew that they would be valued and respected.

What makes CIA special is the sense of family that exists there. The Family Advisory Board did so much for our families.

I always had a special place in my heart for the men and women who served the Agency overseas.

My office was located at CIA's headquarters, but my responsibility and attention were spread across the broader intelligence community. My pride in the community and what they accomplished is deep. Few understand the strength of this community, or its unity of purpose. Standing together, American intelligence confers an enormous advantage to the United States. I thank Ken Minihan, Mike Hayden, Jack Dantone, Jim Clapper, Jim King, Pat Hughes, Tom Wilson, Jake Jacoby, Keith Hall, Peter Teets, Louis Freeh, and Bob Mueller for being great leaders and friends.

Joan Dempsey, Larry Kindsvater, Charlie Allen, John Gannon, Bob Hutchings, Mark Lowenthal, and Jim Simon were the workhorses who helped draw the community closer together. Their work was critical in maximizing the effectiveness of American intelligence.

As capable as the men and women of the U.S. intelligence community are, we would not have been able to achieve anything during my seven years as DCI without the help of some extraordinary friends and colleagues overseas. These friends are too numerous to name, and many, in fact, would prefer to remain anonymous—but they know they have my great thanks.

There are also countless people deserving specific thanks for their assistance in the production of this book. In writing *At the Center of the Storm* I interviewed scores of people who served with me at CIA. After all, this is their story as much as mine. I believed it was important to rely not just on my memory of events, but also on the views and observations of those who rode out the storm with me. The people who provided substantive insights to me (volunteers all) included many currently serving as well as former officials. Those still on the Agency payroll must go unnamed here, but they know they have my deep respect and thanks.

Other former officials, from inside and outside the intelligence community, who deserve my special thanks for their contributions to the book include (in alphabetical order) David Boren, Cofer Black, John Brennan, John "Soup" Campbell, Dave Carey, Hank Crumpton, Sir Richard Dearlove, Charles Duelfer, Louis Freeh, Tom Glakas, Bob Grenier, Dottie Hanson, Scott Hopkins, Martin Indyk, Buzzy Krongard, Anthony Lake, Jim L. (aka "Mad Dog"), Ken Levit, John McLaughlin, Regis Matlak, Jami Miscik, the late Stan Moskowitz, John Moseman, Rolf Mowatt-Larssen, Phil Mudd, Emile Nakhleh, Geoff O'Connell, Dan O'Connor, Marty Petersen, Rob Richer, Dennis Ross, Rudy Rousseau, Charlie Seidel, Winston Wiley, and Kristin Wood.

This book relies on more than just people's memories. Under Executive Order 13292, former presidential appointees are permitted to have access to classified documents from their period of service in order to conduct historical research. I relied on this privilege heavily and requested access to literally tens of thousands of pages of documents. These primary resources were of immense assistance to me in trying to make *At the Center of the Storm* as accurate as possible. My requests for the retrieval of these documents created considerable extra work for the already heavily burdened people in the CIA's Information Management

Office. I'd particularly like to thank Cindy Ferrari and her staff for their cheerful efforts to fulfill my many requests.

The CIA's Center for the Study of Intelligence was also quite helpful in identifying other documents and past research that helped inform my work. My thanks, particularly, go to David Robarge and Nick Dujmovic.

Performing the critical job of ensuring that those things that must remain secret stay so while allowing authors the freedom of expression, the CIA's Publication Review Board did its job well. I'd like to thank its director, Richard Puhl; his deputy, Jane Fraser; and their staff for their careful consideration of my submission.

Among the others who deserve special recognition are my students and graduate assistants at Georgetown University. Their questions during class have greatly helped me think through the issues I deal with in this book.

Arnold Punaro of SAIC graciously provided me with a secure workspace to review and work with classified material. I am grateful for his generous support.

I'd like to thank Jane Friedman and Jonathan Burnham of HarperCollins for believing in this project from the very beginning, and for assigning Kathy Huck and David Hirshey as editors—each displayed skill and patience in great measure while working with me on the book. Tina Andreadis, HarperCollins's director of publicity, has been a terrific advisor as we planned to bring *At the Center of the Storm* to the attention of the reading public. Bob Barnett of the Williams & Connolly law firm helped guide me through the legal shoals to bring this project to completion. There is no one more skillful than he to do so. I'd also like to thank Howard Means for his literary guidance and assistance in conceptualizing the structure on this project. And a special note of gratitude goes to the copy editor, Jenna Dolan, who dotted the *i*'s and crossed the *t*'s brilliantly; and to the cover photographer, Deborah Feingold.

Perhaps my best decision in pursuing this effort was in bringing

Drosten Fisher on board to help me. One of my original graduate assistants at Georgetown University, Drosten is a remarkable young man with boundless energy and a wealth of insights. Much more than a research assistant, he became a true partner in this project—not to mention a virtual hostage and member of my family.

I owe my collaborator on this project, Bill Harlow, an enormous debt of gratitude. Quite simply, I could not have written this book without him. Bill and I journeyed through the storm's center for seven years. He was by my side through the most difficult and trying times, so when I decided to write about my years at CIA he was the natural choice to help. More than anyone, Bill Harlow understands the secret world in which we operated. He combines a veteran intelligence professional's understanding of our business with a novelist's flair. Throughout this project, his patience and good humor kept us going and made sure we finished. I am proud to call him a friend for life.

The members of my family have always been the most important people in my life. I have spoken only briefly in the book about my parents, John and Evangelia Tenet, but they are the two greatest people I have ever known. While my dad has been gone for nearly twenty-four years, not a day goes by when I don't think of him. My brother Bill, who is more than my twin but also my alter ego; his wife, Alice; and their three wonderful daughters, Amy, Megan, and Joanna, lived vicariously through this tumultuous time with me.

I was lucky to marry into a wonderful family. Stephanie's beloved parents, John and Cleo Glakas, cared for me for nearly twenty-five years. Mom Glakas was thrilled to have someone else in the family who was a real Greek. John Glakas was the rock of Gibraltar. My two brothers-in-law, Nicky and Tommy, along with their wives, Katy and Maria Rosa, and their children, Gavin, Christian, Sara, Cristina, and Alexandra, were a source of love and support throughout it all.

My son, John Michael, is the best son a dad can have. He will always be my pride and joy. My wife, Stephanie, is simply remarkable. Her devotion to the men and women of CIA and to their families made me a better Director. Her love for more than twenty-five years sustained me. She is my greatest treasure.

INDEX